Architecture of the Universe

Necia H. Apfel
Associate Director, Astro-Science Workshop
The Adler Planetarium and Northwestern University

J. Allen Hynek
Professor Emeritus of Astronomy
Northwestern University

The Benjamin/Cummings Publishing Company, Inc.
Menlo Park, California · Reading, Massachusetts
London · Amsterdam · Don Mills, Ontario · Sydney

Cover: A photomicrograph of a thin section of Apollo 17 lunar sample number 70017. This sample was among those brought back from the Taurus-Littrow landing site on the Moon. Sample 70017 was described as a loose rock-vesicular, porphyritic, coarse-grained basalt.

Architecture of the Universe is a revised edition of *Astronomy One* by J. Allen Hynek and Necia H. Apfel, published by W. A. Benjamin, Inc., 1972.

Library of Congress Cataloging in Publication Data

Apfel, Necia H.
Architecture of the universe.

Edition for 1972 by J. A. Hynek and N. H. Apfel published under title: Astronomy one.
Includes index.
Bibliography: p.
1. Astronomy. I. Hynek, Joseph Allen, 1910– Astronomy one. II. Title.
QB43.2.H96 1979 520 78-57260
ISBN 0-8053-4747-X

Library of Congress Catalog Card No. 78-57260

The Benjamin/Cummings Publishing Company, Inc.
2727 Sand Hill Road
Menlo Park, California 94025

ABCDEFGHIJK-MA-79

preface

Our rapidly increasing knowledge of the universe is very exciting and its pursuit is high intellectual adventure. Although its subtle beauty and elegance cannot be fully grasped without mathematics and an insight into physical theory, much of the knowledge and the excitement it generates can be appreciated and enjoyed by those whose major fields of interest are nonscientific. We address this introductory text to these people—our future teachers, journalists, lawyers, political leaders, entertainers, and to all those to whom we look as informed citizens.

We feel that it is not enough merely to present the raw facts of modern astronomy. It is the spirit in which these facts have been gained and the manner in which astronomy has influenced the way man thinks about himself and his relation to the world about him that is important. Therefore, what we do not yet know or understand is emphasized equally with what is regarded as verified. We have stressed that today's established facts and theories may someday be shown to be in error or in need of revision. We encourage the desirability of maintaining an open mind, ready to accept new ideas and discoveries. We have tried to convey the spirit of science as a spirit of inquiry and attack upon the unknown.

For those students who will work in fields not associated with science and for whom an introductory course in astronomy may be their only formal contact with science, such a method of presentation is important. These students deserve to know something about the nature and development of science and to recognize it as "organized curiosity" about the world rather than as a series of formalistic statements unrelated to the affairs of human life.

Since students in the humanities do not usually study advanced math or science, we have kept mathematical concepts simple, and have employed only basic arithmetic, and elementary algebra and geometry. The use of formulas has been minimized, and wherever possible a verbal description of the mathematical relationship has been given before the formula is stated symbolically. There are some who feel that no mathematical formulas are necessary in an introductory text and with that we disagree. Astronomy is a science that depends quite heavily upon mathematical computations, and students should be aware that such formulas do exist and are used, even if the students never have need of them.

We have emphasized the metric system of measurement throughout the text because it is used in science and is becoming more and more familiar to everyone. However, to make such units more meaningful to those who have not yet learned to "think metric," the English equivalent is included, usually in parentheses.

Throughout the book we have included suggestions for activities, projects, or experiments which the student can do, either alone or as part of a group. For the most part no special equipment is required, although a small telescope or a good pair of binoculars are indicated for several exercises. How these activities are to be incorporated into the curriculum is, of course, up to the instructor.

We start our scientific adventure with an overall picture of the modern view of the universe—a form of "road map" to guide the student on the journey. Galaxies, stars, planets, the sun, and moon are ordered according to scale, distance, and astronomical significance. The earth is introduced both as man's home and "spaceship," and as his seat in the "cosmic theatre." The basic questions of cosmology, relativity, and the architecture of the universe are discussed early in the book. In most texts these exciting and challenging aspects of astronomy are relegated to the very last chapters. This robs the student of a vista of the broad sweep of modern astronomy. Furthermore, in a one-semester course these exciting chapters are sometimes never reached—and therefore never taught.

After the architecture of the universe and a short study of relativity, we progress to a study of the stars, starting with their appearance as "points of light." From constellations we proceed to an analysis of the Hertzsprung-Russell diagram and finally to stellar evolution, at which point the student has been introduced to much of the physical nature of the universe.

A chapter on galaxies covers a subject that was not included in the earlier chapter on the architecture of the universe. Radio galaxies and quasars are discussed as well as some of the recent discoveries of gamma ray and x-ray sources.

Next, the tools of the astronomer are introduced. This topic has been put far back in the text (and perhaps might well be further back if we were to arrange all the topics by their popularity alone). It enables the reader to come in from the vast expanse of the universe and prepare for a much closer look at the sun and the solar system.

So much has happened in the field of planetary astronomy in recent years that the section in this book that deals with it is quite large. An elementary school teacher said some 35 years ago, "This is a well-researched report on Mars, but really, no one is terribly interested in the subject." How times have changed! Planetary astronomy may now excite the student as much as cosmology and relativity, and the authors sincerely hope that these chapters will not be relegated to the "read them on your own" brush-off.

The final chapter completes the course with a discussion of mankind's place in the universe. A section on UFOs has been included because it is a topic of great interest to most students regardless of their field of study. Although the subject is not strictly one of astronomy, it must be studied because of the close association made by the general public between UFOs and extraterrestrial life.

We hope that the student's exposure to astronomy as an arena of scientific exploration will be stimulating, exciting, informative, and potentially useful to them no matter where their careers may lead them.

Necia H. Apfel

J. Allen Hynek

July 1978

contents

Appendices

1 *The March of Astronomy*

From the earliest times, human beings have been aware of the heavens above them and have attempted to explain what they saw in terms of their own cultural concepts. The universe has changed extremely little since that time, but our understanding of it has undergone tremendous changes.

Primitive cultures were well aware of the celestial world above them. They left pictorial evidence of their observations of stars, eclipses, comets, and other events in the sky. Their folklore was filled with references to the constellations, brighter stars, and the sun, moon, and planets. Most of them associated deities or folk heroes with these celestial objects, and they all had some explanation of the creation of the universe. These legends

1.1 *An ancient concept of the universe.*

were attempts to describe and explain the changes in the world around them, including changes in the sky. Because of their belief that such changes had an influence on human affairs, the orientation of their places of worship was usually aligned on the risings of specific stars or the direction of the rays of the sun on specific dates.

These ancient structures were therefore both astronomical and religious in purpose. Superstitious and ritualistic beliefs were intermingled with very good astronomy. The alignment of stones with respect to specific celestial events for the purpose of indicating seasonal changes—and perhaps predicting eclipses—required accurate observations and record keeping. Stonehenge on the plains of Salisbury in England is perhaps the most famous of these monuments, but there are many others throughout

1.2 Stonehenge on the Salisbury Plain in England. On the morning of the longest day of the year (the Summer Solstice) the sun can be seen rising over the "heelstone," the dome-shaped stone in the middle arch (British Crown Copyright: reproduced with permission of the Controller of Her Britannic Majesty's Stationery Office).

the world. Scientists of "archeoastronomy" are currently investigating these early observatories in hopes of learning more about ancient astronomy.

Somehow, what was observed in the heavens came to be explained in terms of its influence here on earth. Celestial events were thought to be related to human affairs, to the religious beliefs or cultural values of the community. For example, the planets were named after the early gods and were credited with having the same powers and influences as their namesakes. It was therefore important that the position of each planet be well known and that its future position be predictable. Our word "disaster" comes from this concept. *Astrum* (Latin) meant a star or planet, and a

1.3 A clay brick from the Sumerian city of Larsa in Mesopotamia, which dates from about 2000 B.C. It depicts the Zodiac figure of Leo, the lion. Note the star Regulus in the lion's shoulder (courtesy Adler Planetarium, Chicago).

"dis-aster" configuration of stars and planets was one unfavorable for events on earth. All of this required intensive observations and lengthy calculations, and eventually an acceptable theory of the cosmos.

The ancient philosophies of the cosmos nearly always placed the earth at the center of the universe. The rare exceptions did not gain wide acceptance and were eventually discarded. Early people were well aware of their dependence upon the universe and the forces of nature and therefore their role in their philosophical theories was that of a puppet at the mercy of capricious gods. It was very important for them to know how these gods would behave, and this was possible only through the interpretation of the planetary positions and motions. The roots of astronomy and astrology are thus intertwined. In fact, it was due to the needs of astrology that such good astronomical observations and records were made.

This philosophical attitude persisted throughout the Middle Ages and was codified by the Church. It taught that the architecture of the universe was sacred because the architect was sacred, and was not meant for man to know. Science and reason could only go as far as the revelation of God

1.4 Frontispiece from Poeticon Astronomicon, *the first printed book on astronomy (1482). The text was originally written by the Roman author, Caius Julius Hyginus in 40 B.C. The woodcut depicts Ptolemy exhibiting an astrolabe to his handmaidens (courtesy Adler Planetarium, Chicago).*

as interpreted by His deputies here on earth. To step beyond certain bounds and challenge theological doctrine was heresy and therefore punishable.

It was not until the sixteenth and seventeenth centuries that the restrictions on scientific progress were broken down. The decline of the medieval political structure, the schisms within the Church, and the early stirrings of the industrial revolution all had an effect. Men such as Nicolaus Copernicus, Galileo Galilei, and Johannes Kepler found evidence of a new fact that could not be ignored: the sun was the center of the solar system. The earth was merely one of the planets revolving about it. Humankind therefore was no longer at the center of the universe; humanity was small and insignificant. Yet through reason, through science, humans could know the cosmos. There was no limit to our ability to understand. The universe was indeed vast, but it functioned by laws—natural laws—and these could be discovered through scientific processes.

1.5 *An astrolabe made by Jean Fusoris in Paris (c. 1390). (The face is shown on the left and the reverse on the right.) Astrolabes were the most versatile of the early instruments for navigation and surveying. They were even more important, however, as analog computers for astronomers. A star's position could be found for any moment in the year, or the time could be determined if the position of the sun or a star were known (courtesy Adler Planetarium, Chicago).*

Thus, although human beings lost their geocentric (earth-centered) universe, they believed they had gained the mastery of their environment. They gained the dignity of human-centered rather than god-centered existence, and the concept of control over their own fate. Planetary motion was no longer the capricious whim of the gods or an astrological predictor. Instead, it was part of a huge machine or clock whose every tick was predestined. This concept began the Age of Reason and the philosophy of determinism, an idea that took about 200 years to become fully conceptualized. During this mechanistic period the construction of "orreries" became very popular. These were devices which demonstrated planetary motion by the use of precise clock-driven gears. They were machines to demonstrate the celestial machine.

By the end of the eighteenth century, professional astronomy had completely broken away from its astrological and religious background and was beginning to branch out into three major areas: observational, gravitational, and descriptive astronomy.

Observational or positional astronomy, which was later to be called "astrometry," dealt with mapping the exact positions of celestial objects. This information was applied to timekeeping, surveying, and navigation—all very practical matters necessary for the development of business and commerce both at home and in faraway lands.

Gravitational astronomy, which was to become "celestial mechanics," applied Sir Isaac Newton's laws of gravitation to the motions of celestial objects. Orbits and the gravitational effects of planets on each other and on other objects in the solar system were calculated and predicted. The existence of the planets Neptune and Pluto was indicated by their gravitational pull on the planet Uranus long before either was sighted through a telescope. Using these laws, Edmund Halley was able to predict the return of the comet named after him many years before it was actually seen again.

Physical astronomy studied the nature and physical make-up of the celestial objects. With improving telescopes, planetary features, nebulae, and faint stars were explored in ever-increasing detail. However, until the advent of photography in the mid-1800's, physical astronomy (or "astrophysics," as it was later called) played a minor part in astronomy. Astrometry and celestial mechanics constituted practically all of the field.

Modern astronomy began about 150 years ago when the foundations of our current view of the universe were laid. Photography opened the door to exploration which previously had been impossible. Large telescopes, long photographic exposures, and other technological advances made the depths of space more accessible. However, relatively few astronomers were concerned with the new information that was being obtained. At first, the new discoveries were largely of an impractical nature and, with the exception of observational astronomy, no immediate use could be seen for them. Only a few foresighted astronomers paid much attention to the stars as physical bodies. For the others, stars remained merely "points of light" on the celestial sphere.

In the last 50 years, astronomy has undergone yet another radical change. More knowledge of our universe has been acquired in that short time than in the whole previous history of astronomy. The very founda-

1.6 A large silver polyhedral dial made in Germany (c. 1720) from a drawing by Oronce Fine in his De Solaribus Horologis, *published in Paris, 1565. This instrument demonstrates the Renaissance mathematician's interest in solving complex astronomical problems rather than the practical application of knowledge (courtesy Adler Planetarium, Chicago).*

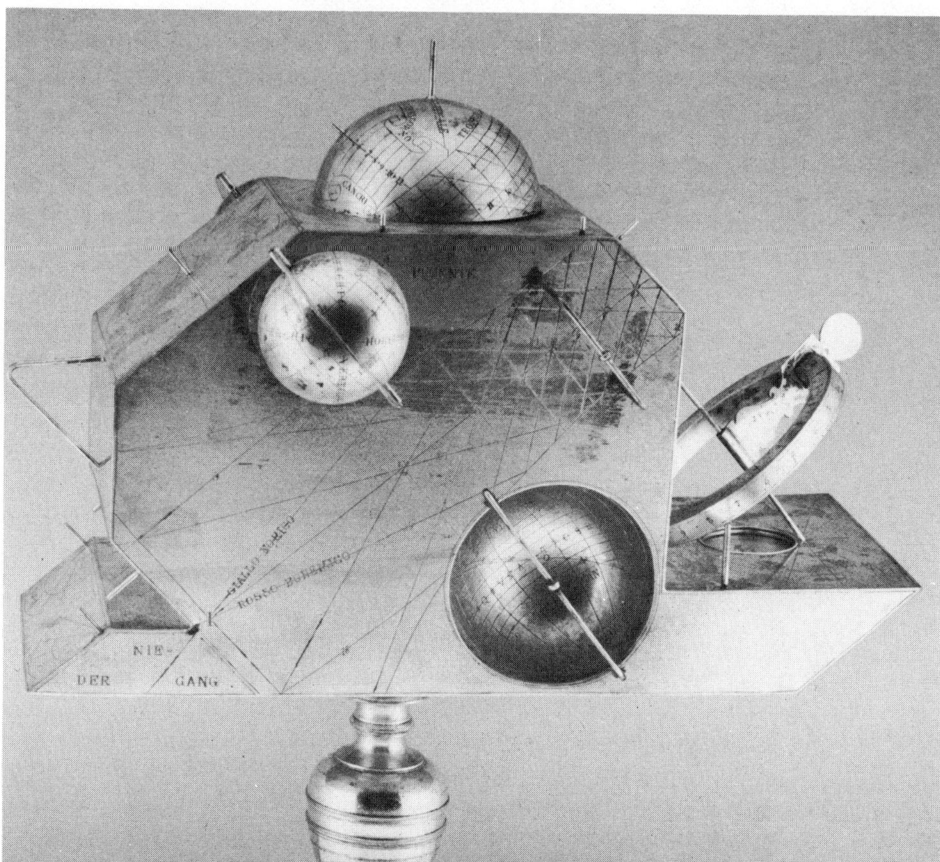

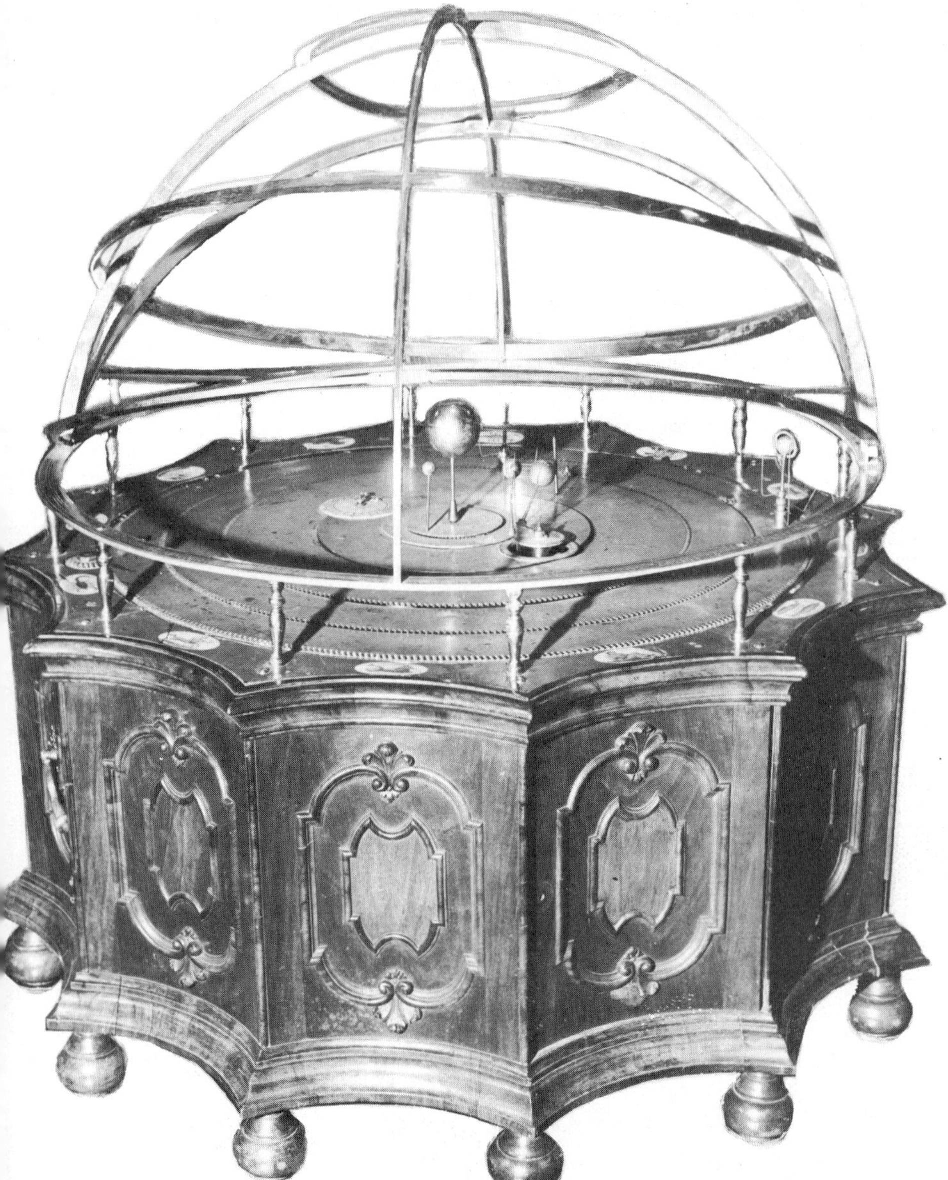

1.7 An early planetarium made by Pieter Isenbrock of Haarlem in 1737. It depicts a sun-centered solar system and includes the earth and the five other planets known at that time (courtesy Adler Planetarium, Chicago).

tions of the science have been rocked with exciting discoveries, planetary explorations, and challenging new theoretical models.

The mechanistic "clockwork" view of the universe has disappeared. Instead a picture of an extremely complex and immensely vast universe emerges—a universe which once again raises doubts of humankind's ability to know it all. And it is science itself which has led to this realization. The cosmos is not a machine. Indeterminancy, a theory which results from modern physics, states that by merely studying or exploring some facet of our universe, we wittingly or unwittingly influence it in some manner. This was first realized in the study of subatomic particles, when the Heisenberg principle of uncertainty was discovered: our very observations disturb the objects being studied. We have sent space probes to explore beyond our own Spaceship Earth. From our probes we have learned much of the moon and nearby planets. But we also realize that our probes, by landing on the moon and planets, have altered them from their original state.

In addition to the uncertainty or indeterminancy of science, we must also consider the implications of Einstein's theory of relativity. How fast an

1.8 *The Apollo 15 lunar rover, the first manned vehicle to travel on the surface of the moon (NASA photograph).*

object moves depends upon its velocity in relation to something else. Nothing is "at rest" in the universe; in one form or another everything is in a state of flux. Time, size, and velocity are relative quantities, not absolutes as was previously thought. Today, physical scientists and mathematicians accept these new concepts and work with them. Slowly, very slowly, these concepts are being incorporated into other areas—sociology, psychology, and philosophy.

The study of astronomy has witnessed the decentralization of the earth and the elevation of humanity's image of itself. With the possibility of life existing elsewhere in the universe, the last vestige of our cosmic provincialism should disappear. Our first steps into space have opened new vistas to explore. Our realization that we are an integral part of all that we

1.9 Half-earth as seen from the moon (NASA photograph taken by the Apollo 11 astronauts).

study will lead to a new perception of the nature and extent of the universe and our place in it.

The practical value of astronomy in shaping our thought and life throughout the ages cannot be underestimated. Its influence has been intangible, yet incalculable. Astronomy has been and continues to be a crucial intellectual adventure for mankind. If one should ask what good is astronomy, one might as well ask what good is a symphony orchestra or an art museum or a flower garden, or why, indeed, man undertakes anything. They all help us enjoy and enrich our lives, and beyond, give our lives meaning. There are some who finance and build centers for the performing arts, and others who build observatories in space. Mankind is the better for both, and in any healthy society both must be nurtured.

Discussion Questions

1. Discuss the ways in which astronomical discoveries and concepts have influenced humanism.

2. How do you think the view of our earth as seen from the moon (Fig. 1.9) changes our concept of ourself and the universe we live in? If this photograph had been available 500 years ago, how might history and philosophical thought been different?

3. What was the general attitude toward scientific research among astronomers and other scientists during the last 2000 years or more? When did this attitude start to change to the modern approach? Are there still vestiges of the old attitude in our thinking today?

4. What is your concept of astronomy?

2 The Architecture of the Universe

The architecture of the universe is a description of the arrangement in space of celestial objects: the different kinds, their distances from us and from each other, and their sizes. In any society, what this architecture has been thought to be has depended not only upon the information available at the time but also upon whether this description was acceptable to the society into which the description was introduced. Obviously, both factors have caused the description of the architecture of the universe to undergo tremendous changes throughout the ages.

2.1 *A woodcut depicting an ancient astronomer peering behind the celestial sphere in an attempt to discover the working of the universe.*

Early mankind placed the earth at the center of the cosmos. The celestial sphere was considered a reality, and some even believed that it consisted of a hard, crystalline material. Some thought that the stars were embedded in this sphere, while others declared that the stars were holes through which the celestial fires shone. In either case, the sphere completely surrounded the fixed earth and had a finite radius. It was believed to rotate around the earth, thus making the stars rise and set each night. As late as the seventeenth century, the stars on this sphere were thought to be only some 65 million miles away. (Today we know the sun to be 93 million miles away!)

Early astronomers regarded all objects in the sky as "stars," including the seven "wandering stars" or *planeta.* We know these seven today as the sun, moon, and the five visible planets, Mercury, Venus, Mars, Jupiter and Saturn. Each was believed to revolve around the earth on separate rings or spheres, which were arranged concentrically within the outer "fixed stars" sphere. Some thought that the turning of these spheres produced tones, from which we get the term "music of the spheres."

Even when the sun instead of the earth came to be placed at the center of the solar system, there was little doubt that the sun was then the center of the universe. When astronomers realized that the solar system was merely part of a huge galaxy of stars, that galaxy was presumed to be the center of the universe. It was difficult to give up the concept of an

2.2 *Armillary sphere made by Alex Ravillius in 1542. Originally used as observing instruments, these spheres later became teaching devices, mainly for the Ptolemaic system. The earth is shown as the central body (courtesy Adler Planetarium, Chicago).*

ego-centered cosmos. It was not until the twentieth century that the knowledge of our place in the galaxy and the nature and distribution of the galaxies in the universe was gained—and accepted.

Using a roll of adding machine tape, plot the distances of the planets from the sun. Use the scale 1 AU = 1 meter (39.37 inches). The sun can be represented at the beginning of the tape by a circle about one centimeter in diameter. The planets will be no more than dots on the paper roll. How long a roll of paper would be necessary to put Alpha Centauri, the next nearest star to the sun, on this map? The Andromeda galaxy?

Cosmic Distances

With the decentralization of the earth came the discovery that the universe was very much larger than had ever been imagined. Visualizing the great cosmic distances is virtually impossible, but they can be understood by using a scale model. Let one inch represent the distance between the sun and the earth. In reality, this distance is about 93 million miles (1.5×10^8 kms) and is called an Astronomical Unit (AU). According to our scale, Pluto, the farthest planet from the sun, is 39 inches away. The nearest star, however, would be 4⅓ miles away (Fig. 2.3). Using this scale model, the Andromeda galaxy[1] would be two million miles away, or more than eight times the actual distance to the moon. Astronomers now study objects which even on our model would have to be placed several billion miles away—and remember that these distances are on a scale in which the earth-sun distance is represented by just one inch.

The choice of this scale is a very convenient one because of an interesting coincidence: on it one inch equals one Astronomical Unit and one mile is almost equivalent to one light year, the commonly used yardstick in astronomy. The nearest star is in actuality 4⅓ light years away, and the Andromeda galaxy is two million light years away. A "light year" is defined as the distance that light travels in one year. The velocity of light, which is considered the greatest speed possible in the physical universe, is approximately 186,300 miles per second or 300,000 kilometers per second. If we multiply this amount by the number of seconds in a year, which is about 31 million, we will have the distance that light travels in one year—

---

[1]The Andromeda galaxy is a large nearby galaxy very similar to our own Milky Way galaxy.

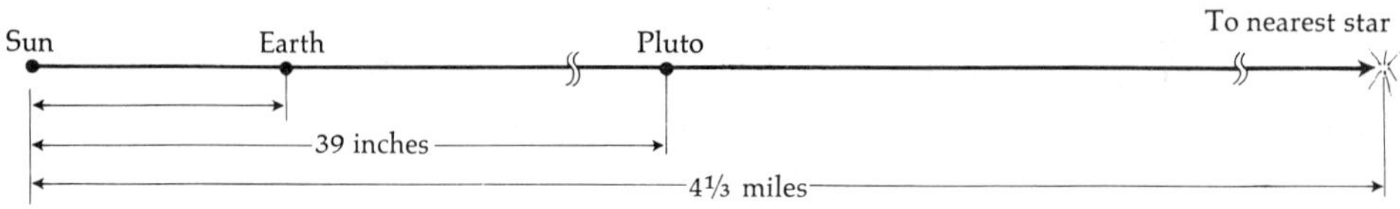

2.3 On the scale 1 inch = 93 million miles (1 AU), the star nearest to our sun would be 4⅓ miles away. On this same scale, the Andromeda galaxy would be 2 million miles away.

about six trillion miles. The length of a string wrapped 236 million times around the equator of the earth would equal the length of one light year (Fig. 2.4).

We use these large yardsticks in astronomy because of the great distance involved. To use miles or kilometers would be like using inches or millimeters to measure the distance between New York and San Francisco. For the distances of objects in our own solar system we can use the Astronomical Unit, but when we consider the stars, we must have a much larger unit.

Therefore, when we say that the nearest star is 4⅓ light years away, we mean that the light we receive today from that star left there 4⅓ years ago, and the star that we see now is the way it looked when that light left it. We never see anything in the universe as it actually is "now." For example, if the Andromeda galaxy had exploded a million years ago, we would not know about that event for still another million years. As the astronomer looks deeper and deeper into space, he is also looking farther and farther back into time.

2.4 A string wrapped around the earth's equator 236,000,000 times would be almost one light-year long. The nearest star is 4⅓ light years away.

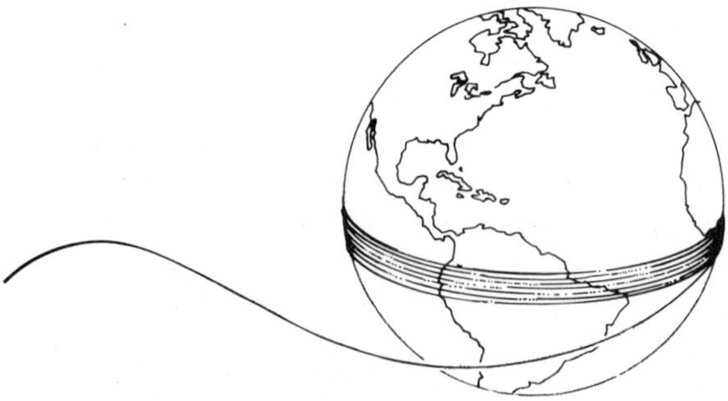

In one sense then, the astronomer can be said to be looking at a four-dimensional universe, the fourth dimension being that of time. We never see the cosmos as it is today, right now. Even the sun's light is 8 minutes old. As we look into space, we look back in time.

Similarly, the geologist, as he probes deeper and deeper into strata of the earth, also "looks" farther back into time. In fact, the time scales of geology and astronomy are very closely related (Fig. 2.5). For every fossil that the geologist can unearth, the astronomer can point his telescope toward a galaxy and say that the light left there when that ancient animal or

2.5 *Time scale for geology and astronomy.*

Corresponding geologic era or period	*Years ago*	*Events on earth*	*Light seen today, left stellar object, light-years ago*
Archeozoic Precambrian	4.5–2.6 billion	Before life existed on earth	Quasar 3 C 9
Devonian	400–350 million	Fishes	Cluster of galaxies in Hercules, 340 million
Jurassic	180–135 million	Dinosaurs	Cluster of galaxies in Perseus, 173 million
Tertiary	70–10 million	Ancient mammals	Cluster of galaxies in Virgo, 75 million
Pleistocene	2–1 million	Earliest man	Andromeda galaxy over 2 million
Recent epoch	*Date* A.D.	*Historical event*	
	300–400	Fall of the Roman Empire	Orion nebula, 1600
	1609	Galileo's first telescope	Pleiades, 400
	1875	First telephone	Canopus, 98
	1939	World War II began	Arcturus, 36
	1957	Sputnik, beginning of the Space Age	Altair, 17
	4⅓ years ago	College freshman was still in Jr. High School	Alpha Centauri, 4⅓ 11a

plant was alive. Some galaxies are so distant that the light we receive from them today left them before any life had evolved here on earth.

The Milky Way

Look at the picture of the Andromeda galaxy in Fig. 2.6. The bright foreground stars in the picture are in our own galaxy, the Milky Way, and we are looking "through" them to the much more distant galaxy beyond. The effect is much like sitting in a screened-in porch and looking through the screen (with all the bugs and leaves on it) to a distant garden beyond. The Andromeda galaxy, like our own, contains on the order of 100 billion stars, and it is estimated that there are perhaps one billion galaxies throughout the visible universe.[2]

Can you locate the Milky Way in the night sky? Unless it is very dark and you are far from city haze and lights, this will probably not be possible. However, if it is visible, can you determine which parts appear brighter than others? In which constellations is the Milky Way the brightest? (See the Star Charts—Appendix VII.) Is it continuous or irregular and ragged in appearance? Draw a sketch of the portion that is visible. Now view the Milky Way through binoculars or a small telescope. What differences do you now observe?

We cannot see our own galaxy as a whole because we are inside it, but after viewing other galaxies plus learning more about our own, we can tell that the Milky Way galaxy is similar to many others. Figure 2.7 indicates its lens shape, with the sun (which is just one of its billions of stars) located in one of its spiral arms quite far from the center. We are not sure why many galaxies have this spiral shape, although some theories indicate that this is part of the evolutionary process in galaxies.

[2] In order that we keep such numbers in perspective, it is helpful to remember that one billion seconds represents about 33 years!

2.6 *The Andromeda galaxy (M31) (photograph from the Lick Observatory).*

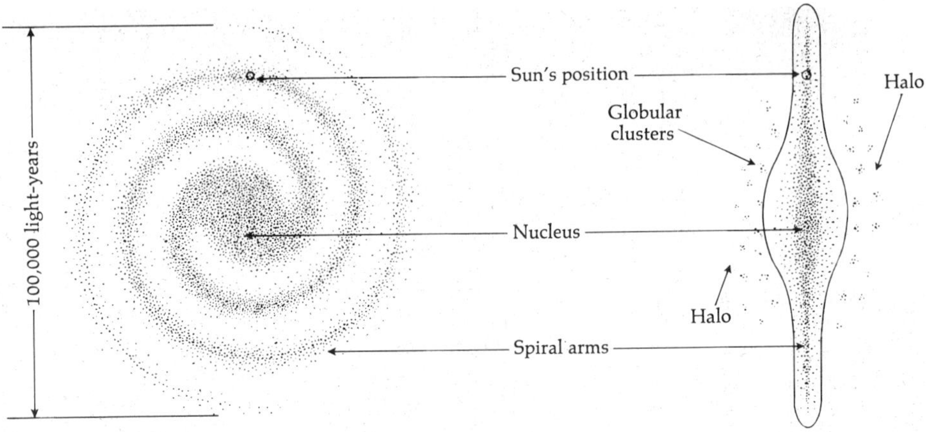

2.7 Schematic representation of the Milky Way galaxy as viewed from above (left) and from edge-on (right).

Because we are located within this lens-shaped galaxy, when we look "through" the lens surface (see Fig. 2.8), perpendicular to the plane of the galaxy, we see few stars. But when we look "along" the lens, toward the edge, the number of distant stars becomes so great that all we see with the naked eye is a faint, whitish, rather ragged band cutting across the entire sky. This is what the ancients called the "pathway of the gods" or the Milky Way, thus naming our entire galaxy. Only with a telescope or good binoculars can we see separate stars in this area of the sky.

Turn your binoculars to different parts of the sky. Do you see the same number of stars wherever you look? If not, where do you see the most? Try counting the stars visible through your binoculars in one area and then in another.

When we look at the photograph of the Andromeda galaxy, the most obvious objects are the stars just as they are in our own galaxy. But we can also detect bright clouds of dust and gas, mainly in the spiral arms. These *nebulae* (Latin: "mists" or "clouds") are also present in our own galaxy, and, as we shall see in the next section, once caused a great controversy in astronomical theory.

We do not know of any other planetary system either in our own galaxy or in any other, but it is widely believed that others exist and con-

2.8 *Our lens-shaped galaxy. Far fewer stars are seen when we look in the direction of the "lens surface" than when we look along its plane. When we look along this plane we see the faint band in the sky which the ancients called the Milky Way.*

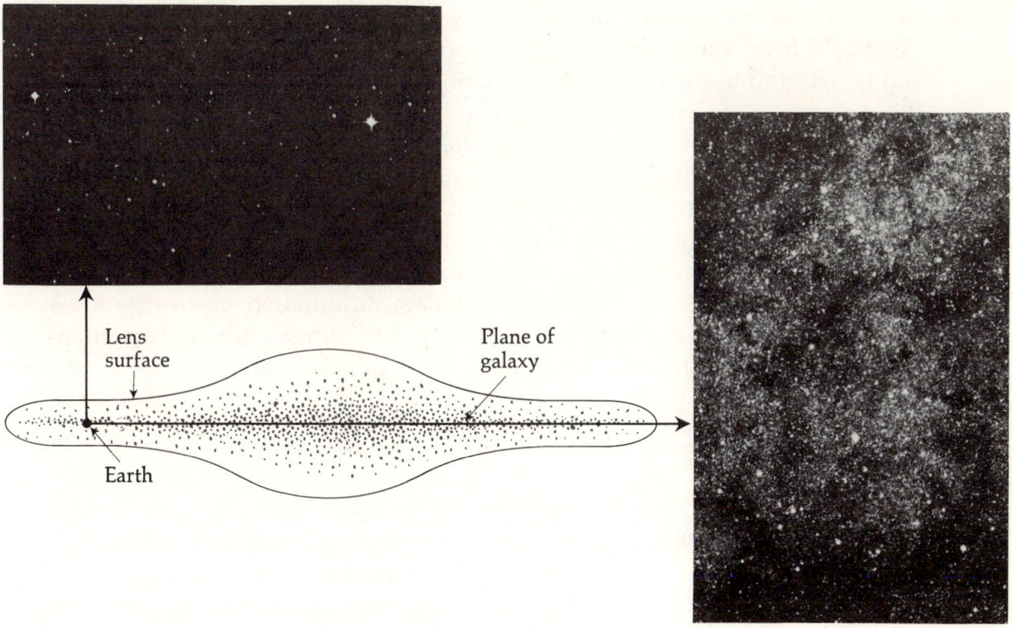

sist, like ours, of planets, moons, and many minor members surrounding a central star. All the stars and other bodies in the galaxy revolve around the galactic center. In our galaxy, the sun is moving at about 240 km/sec (150 miles/sec) relative to the center, and takes approximately 200 million years to make one complete revolution. Of course, the rest of the solar system, including the earth, moves around the galactic center along with the sun.

Locate the Andromeda galaxy, which is best seen in the autumn sky (see Star Chart—Appendix VII). View it with a telescope or a good pair of binoculars and draw a sketch of your observation. When did the light that you see start out from that galaxy?

The Nebula-Galaxy Controversy

It was not until the seventeenth century, when the telescope became the basic tool in astronomy, that these thousands of previously unseen stars were revealed. At the same time, with the use of the telescope, many fuzzy, nebulous patches in the sky were discovered. Although distinguishable from the stars through these early telescopes, they all looked similar enough to each other to be categorized as the same type of object—the nebula. Most astronomers believed these nebulae to lie within the confines of our own star system or galaxy. Most denied the existence of other galaxies, although some philosophers, notably Thomas Wright and Immanuel Kant, argued that there could indeed be vast, independent systems of stars. Today we know this to be the case, and that some of the patches are distant galaxies very far removed from our own. Other patches are in fact large gaseous clouds or nebulae, illuminated by nearby stars within our own galaxy (Fig. 2.9). Until the late nineteenth century there was no known method to distinguish one from the other.

Locate the Orion nebula, which is best seen in the winter sky (see Star Chart—Appendix VII). View it with a telescope or good pair of binoculars and draw a sketch of your observation. If you have already viewed the Andromeda galaxy, what differences did you observe between the two objects? (Of course, knowing what they are and what they should look like will bias your observations, but try to put yourself in the position of the late nineteenth and early twentieth century astronomers whose telescopes did not allow them to easily distinguish these differences.)

By then telescopes had improved such that some of the nebulae were found to contain individual stars. Their spiral shapes were also noticed and they were referred to as "spiral nebulae." But there was still no evidence that either the spiral or the gaseous nebulae lay beyond the confines of our own galaxy; they all continued to be classified as nebulae. At the turn of the century, Agnes Clerke wrote:

> *The question whether nebulae are external galaxies hardly any longer needs discussion. It has been answered by the progress of discovery. No competent thinker, with the whole of available evidence before him, can now, it is safe to say, maintain any single nebula to be a star system of coordinate rank with the*

2.9 The Orion Nebula (M42) photographed with the 100-inch telescope at Mount Wilson (photograph from the Hale Observatories).

> *Milky Way. A practical certainty has been attained that the entire contents, stellar and nebular, of the sphere belong to one mighty aggregation.*[3]

The debate and controversy continued, and even as late as 1920, the question of the true nature of the nebulae and the size of the galaxy was being argued. At a famous debate in Washington, D.C., two noted astronomers, Harlow Shapley and Heber Curtis, presented opposing arguments. But the best thing to result from this event was the realization that it

---

[3]Agnes Clerke, *The System of the Stars,* London: Longman, Green & Co., 1890, p. 368.

is futile to debate anything in science. What is really needed in such cases is observations, not arguments. Such observations were presented just three years later. Edwin Hubble at Mount Wilson obtained proof of the real distance of the galaxies. He did this by observing a special kind of star, a pulsating star called a Cepheid variable, whose regularity is so important that it has sometimes been termed a "celestial lighthouse." In order to understand what Hubble learned from the Cepheids he was studying, it is necessary to go back to 1912 when the importance of these special stars was just being realized.

The Cepheids

In 1912, Henrietta Leavitt, an astronomer at Harvard University, made a most important discovery during her examination of photographic plates of what today are known to be two small satellite galaxies of our own Milky Way, called the Magellanic clouds (Fig. 2.10). These were named after the navigator Magellan, who first described them on his historic round-the-world journey. Visible only from the Southern Hemisphere, they were known by the early twentieth century astronomers to be composed of stars, many of which could be individually studied with the aid of a telescope. In addition, because these "clouds" were known to be sufficiently far away, their component stars could be treated essentially as though they were all at the same distance from us.

We use this same approximation when we consider the distance between, say, New York and San Francisco. We are not concerned with the fact that one street in New York may be closer to San Francisco than another. The cities are far enough apart to be treated as single units.

Leavitt, in her study of the plates of the Magellanic clouds, found a group of variable stars of the Cepheid type. Variable stars are those whose light intensity or brightness increases and decreases, usually in regular intervals which are called the star's *period*. The Cepheids, many of which had already been identified in our own galaxy, have their own particular characteristic way of varying in brightness and therefore are easily recognizable. The star "delta" in the constellation Cepheus (δ Cephei) was the first star discovered of this type, hence the name "Cepheid."

When Leavitt plotted the apparent brightness of the group of Cepheids in the Magellanic clouds against the length of their periods, she discovered that the longer the period, the brighter the star (see Fig. 2.11). Such a relationship would have been extremely difficult to determine with the Cepheids in our own galaxy because they are at different distances from us and therefore their apparent brightness depends in part on their

2.10 *The Large Magellanic Cloud (photograph from the Lick Observatory).*

distance. The Magellanic Cepheids, on the other hand, could all be considered as at the same distance, and the relationship between the Cepheid's period of variation and its brightness (the period-luminosity relationship) could be estabished. At that time, however, the distance to the Magellanic clouds had not been established, so Leavitt could only use the apparent brightness of these stars in plotting this relationship.

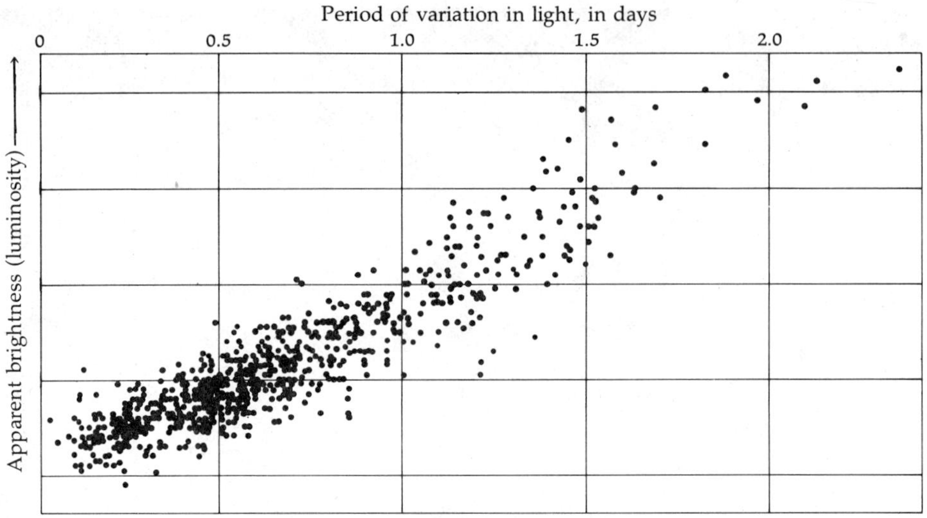

2.11 The correlation between period and luminosity (or brightness) for Cepheids in the Magellanic Clouds. Note that the brighter the star appears to be, the longer the period in which its brightness varies.

When she made this period-luminosity relationship of the Cepheids known to the Harvard astronomer Harlow Shapley, he realized that the full impact of her discovery was that it could be used as a distance indicator. He listed Cepheids in our own galaxy whose distances and apparent brightnesses had already been determined. He could then easily calculate their true brightness. (Obviously, if we know how bright any light source really is, we can gauge its distance by noting how faint it appears to us. And conversely, given the apparent brightness and the distance of a star, its intrinsic or true brightness can be determined.)

Now Shapley was able to use the true brightness of the Cepheids in the period-luminosity relationship instead of their apparent brightnesses. He then went back to the Cepheids in the Magellanic clouds, and using this new knowledge, he calculated the distances to these stars and thus to the clouds themselves.

A powerful method of determining celestial distances beyond our own galaxy was thus discovered. Once a Cepheid variable star was located in any galaxy, the star's distance and hence the distance of the galaxy could be found. All that was needed was: (1) the time period in which the Cepheid goes through one complete cycle of light variation, and (2) the Cepheid's apparent brightness. Both of these are obtained from observation. Using the period-luminosity relationship, the true brightness of the

Cepheid can be determined. Then by comparing the star's true brightness with its apparent one, its distance can be calculated.

With this background, let us return to the nebula-galaxy controversy. In 1924, just four years after the debate between Curtis and Shapley, the American astronomer Edwin Hubble discovered many Cepheids in the outer regions of the Andromeda galaxy (it was still called a nebula, of course). This was made possible by the use of the then-new 100-inch telescope at Mount Wilson in California and by the fact that Cepheids are among the brightest of the stars. At that time the Mount Wilson telescope was the most powerful in the world (Fig. 2.12).

2.12 The 100-inch Hooker reflector telescope of the Mount Wilson Observatory. It was this "workhorse" that finally provided the answer to the nebula-galaxy controversy (photograph from the Hale Observatories).

Hubble used the period-luminosity relationship to determine the distances of the newly discovered Cepheids. He was thus able to establish that the Andromeda galaxy was definitely outside our own Milky Way galaxy. We know now that it is more than two million light years away, whereas our own galaxy is less than 100,000 light years in diameter.

This method of using Cepheids as yardsticks was later used to find the distances of other relatively nearby galaxies, but in galaxies farther away individual stars could not be distinguished even by the most powerful telescopes. Still, there was no longer any doubt that our galaxy was only one of many other galaxies throughout the universe. Other methods of determining distances in space were discovered and it was found that the universe was immensely more vast than had ever been imagined even a half century before.

The Universe of Galaxies

The fundamental building blocks of the universe were found to be the galaxies themselves. Each galaxy has the same basic components that are present in the Milky Way, yet no two are exactly alike. As astronomers reached further and further out into space, they found that instead of a uniform distribution of galaxies, galaxies were in clusters.

The Milky Way galaxy is a member of a small cluster of galaxies known as the Local Group (Fig. 2.13). Spread over a region of about three million light years in diameter, the Local Group consists of about two dozen galaxies. Two of the largest members are the Milky Way and the Andromeda galaxy. Until the late 1960s, the Local Group had only 17 known members. Then two more were discovered by the Italian astronomer Paolo Maffei, although there is now some question as to whether these small members are truly part of our group. In 1971 the Canadian astronomer Sidney van den Bergh discovered three more small galaxies near the Andromeda galaxy. Then in 1975 a very small galaxy was discovered very close to our own, and in deference to the candy bar industry was dubbed "Snickers." (It is another galaxy but "peanuts" compared to the Milky Way.)

The Local Group is a relatively small cluster. Others, such as the Virgo cluster have thousands of members. The clusters are named after the constellation in which they are found. Remember that, as with the picture of the Andromeda galaxy, we are looking through our own galaxy of stars out beyond the nearest galaxies to groups of galaxies millions and billions of light years away.

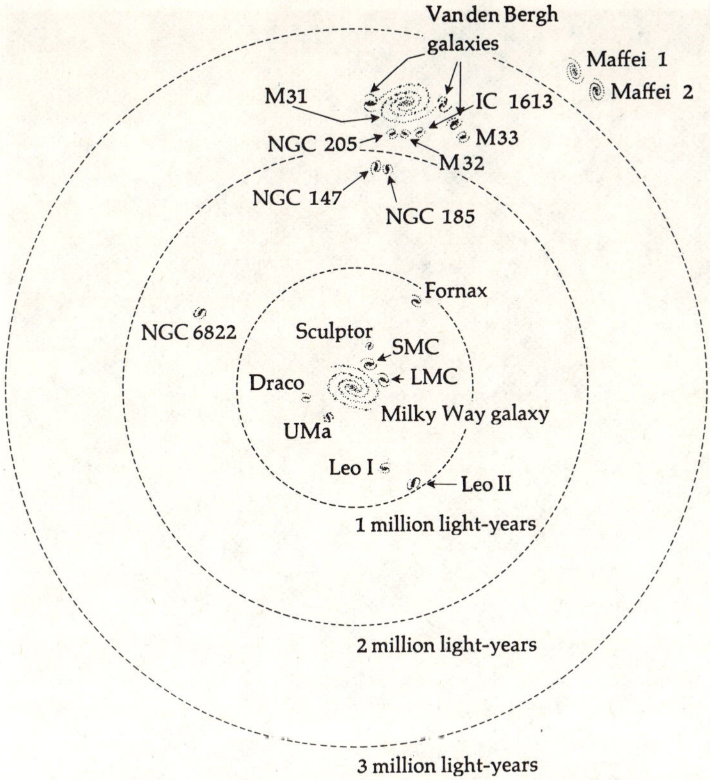

2.13 The Local Group consists of nearly two dozen known galaxies located within 3.6 million light years (one megaparsec) of the Milky Way. This diagram is plotted with the Milky Way at the center to show the relative positions of other galaxies. The actual center is about midway between the Milky Way and the Andromeda galaxy.

There are several clusters of galaxies within 250 million light years and they are therefore close enough for us to distinguish the features of individual members. (Fig. 2.14). But beyond these relatively nearby clusters lie many rich ones which, due to their great distances, appear much more compact (Fig. 2.15 and 2.16). Their individual members look almost starlike. Undoubtedly, many smaller distant clusters are not yet recognized as such.

As in our own Local Group, the galaxies in a cluster are separated by distances many times greater than their individual diameters. Within each

2.14 Part of the cluster of galaxies in the constellation Coma Berenices. These galaxies are about 440 million light years distant (photograph from the Hale Observatories). Note: *Figures 2.14, 2.15, and 2.16 were all taken with the 200-inch telescope. The photographs were all enlarged by the same amount to show how the apparent size of galaxies decreases with increasing distance.*

2.16 Cluster of galaxies in the constellation Hydra. These galaxies are about four billion light years away from us (photograph from the Hale Observatories).

2.15 Cluster of galaxies in the constellation Corona Borealis. This cluster is about 1400 million light years away (photograph from the Hale Observatories).

galaxy, however, the stars are relatively a great deal farther apart. In our neighborhood of the Milky Way galaxy, for example, the distance of the nearest star to our sun is about 26 million times the diameter of the sun!

Between the stars and between the galaxies there are atoms and molecules of dust and gas, but these accumulations of matter are so rarified that the space they occupy is a better vacuum than scientists can create here on earth. One is thus led to believe that the greatest component of the universe is empty space. This is basically true, but it is the material in the universe and not its empty space that determines its architecture.

There has been some speculation that the nearby clusters of galaxies may form a *supergalaxy*. There are some 65 such clusters within 50 million light years of our Local Group and most of them lie on or near the same plane, similar to the way most of the stars in the Milky Way galaxy lie on or near the galactic plane. Most of the galaxies in this huge, seemingly flattened supergalaxy are also flattened like our own galaxy and the flattening appears to be generally in the same direction as that of the supergalaxy. This supergalaxy is estimated to have a diameter of some 100 to 200 million light years (as compared with our Local Group which is only about 3 million light years in diameter) and its center appears to lie in the direction of the constellation Virgo. There is also some evidence that clusters of galaxies beyond this local supergalaxy are also grouped into supergalaxies.

It is well to remember at this point that on our original scale of 1″ = 1 AU, a cluster of galaxies four billion light years away would have to be placed four billion miles away on our map. This is a distance about 40 times greater than the actual distance from the earth to the sun.

The pictures we obtain today of these distant objects show us what they looked like millions or billions of years ago. Some recently discovered objects, named *quasars,* appear to be even farther away than the most distant galaxies. (These mysterious objects and their discovery are discussed in Chapter 11). Some astronomers think that they may represent an early stage in the evolution of some galaxies, but since we are seeing them as they were billions of years ago, we have no way of knowing what their form is today. It is possible that in the early stages of the universe, there may well have been quasars in our neighborhood as well. It is even possible that the very early stages of the Milky Way were quasar-like. If so, is it also possible that astronomers within the most distant galaxies are now seeing us as a quasar? Do they perhaps see us as some fuzzy spot much like any one you might select in the center of Figure 2.17?

2.17 *At the center of this photograph is a faint cluster of galaxies in the constellation Pisces. This cluster is at a distance of well over four billion light years and is near the limit of the observable universe (photograph from the Hale Observatories).*

Discussion Questions

1. Why do we say that when we look up at the stars we are looking back in history? Why is the light from galaxies sometimes referred to as "fossil light?"

2. If our sun were closer to the center of the galaxy, would we still see the white band in the sky which we call the Milky Way? Would we still see the Big Dipper constellation?

3. Why did the term "nebula" prove to be a source of debate in the early 1900's? What does this term refer to today? Why is the Andromeda galaxy referred to as the great Andromeda nebula in older texts?

4. List the steps by which the galaxies were shown to be extragalactic.

5. What is the speed of light in miles per hour? In kilometers per hour?

6. Two Cepheid variables, *A* and *B*, are observed to have the same period of pulsation. The average brightness of *A* appears much less than that of *B*. What is the relation of the true brightness of *A* to *B*? What can be said about their respective distances from us?

7. Is it possible that the Milky Way might appear as a quasar to someone looking at it from a galaxy billions of light years away? Explain.

8. On a tape measure, label the distances of the planets from the sun, using 1 inch = 1 AU as your scale. On this scale, where would Alpha Centauri (the star nearest to the sun) be located? What about the Andromeda galaxy? (Use the planet chart on page xxxx)

3 *Radiation from the Universe*

We have just taken a cursory look at the domain of astronomy. It stretches from the surface of our own planet to the most remote celestial objects. Until very recently the study of this vast cosmos was limited to earth-bound observations. Unlike other scientists, astronomers could only use their sense of sight to gather data. The telescope extended their observational abilities and the camera recorded what their eyes could not detect. Other instruments attached to the telescope analyzed the light being received in ways impossible to the human eye. But astronomers were still limited to visual radiation.

Today, through technological advances in such fields as electronics engineering, computer design, and space aeronautics, the field of astronomy has been tremendously enlarged. Astronomers now study all types of radiation from outer space. The designers of new astronomical instruments, the astronauts who walked on the moon, the computer experts who programmed the space missions, the teachers who inspired new generations of scientists—these and many others are all actively involved in astronomy. In fact, it would be hard to define the modern field of astronomy since it encompasses and intertwines with so many other disciplines. Perhaps we can say that anyone who uses his or her talents and scientific training to study cosmic problems could justifiably be called an astronomer.

Before we can more fully understand the domain of astronomers, we must look more closely at the basic prinicples with which they work.

The Electromagnetic Spectrum

Until about 50 years ago, light was the only form of radiation studied by astronomers. But light constitutes only a very small portion of what is called the *electromagnetic spectrum*. As you can see in Figure 3.1, this spectrum extends from the very short gamma ray and x-ray waves through the familiar region of light to the ultra-long waves of radio. Although the

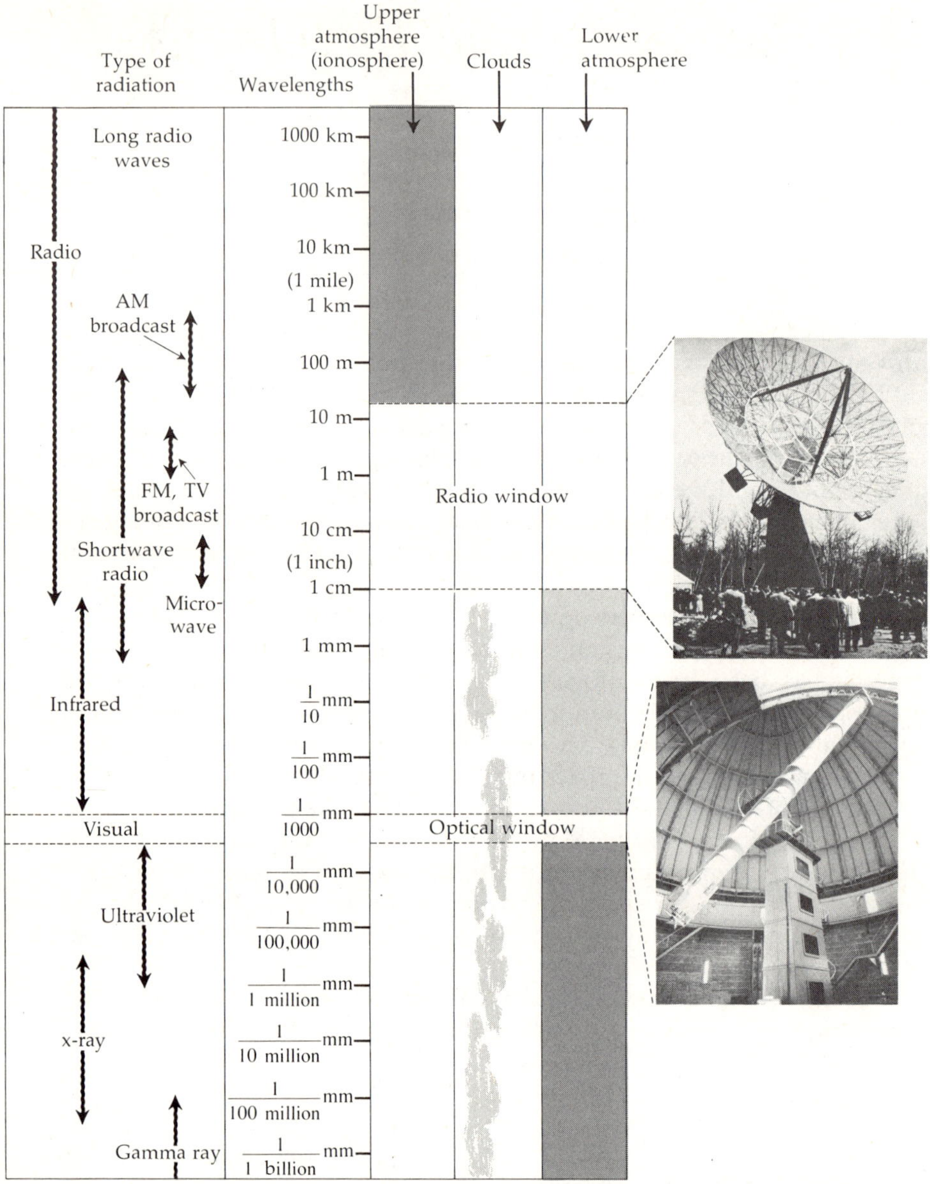

3.1 The electromagnetic spectrum. Only two "windows" in the atomsphere allow radiation to reach earth.

specific wavelength determines how the particular radiation will manifest itself to us, all of these types of radiation conform to the same basic laws of light. Whether it be radio, microwave, or ultraviolet radiation, it will travel at the speed of light.

This speed is considered to be the greatest possible in the physical universe, and is also believed to be a universal constant. That is, regardless of who is measuring it, from whatever position in space or time, it will not vary. It is also independent of the velocity of its source or of that of its observer. A stewardess moving down the aisle of an airplane may be going only one or two miles per hour from the point of view of the passengers but to someone on the ground she is moving some 600 mph. However, if it were possible for the passengers to measure the velocity of the light coming from either inside or outside the cabin, they would find agreement with such measurements made from the ground. Light and other forms of electromagnetic radiation travel at 186,282 miles per second or approximately 300,000 kilometers per second. That is about 670 million miles per hour.

As the radiation travels from its source, its speed remains constant, but it does diminish in intensity in a very predictable manner. Its intensity or brightness diminishes by the square of its distance from the source. The measurement of a source of electromagnetic radiation at two meters will be ¼th as intense as it was at one meter. At three meters it will be 1/9th as intense as at one meter. As is illustrated in Figure 3.2, the radiation energy spreads out over a larger and larger area, so that for any unit area, the

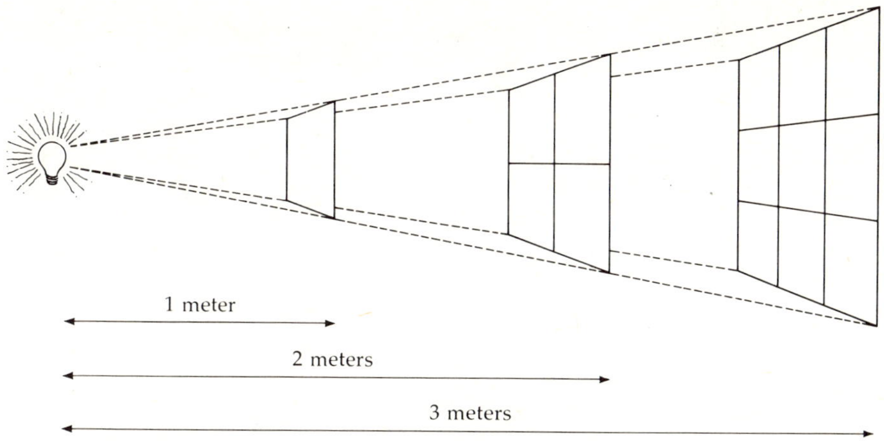

3.2 *The inverse-square law of light. At 2 meters each square area receives ¼th the radiation received at 1 meter. At 3 meters each square receives 1/9th as much radiation as at 1 meter.*

Until the seventeenth century, many people believed that light traveled instantaneously. Attempts to measure its velocity had been unsuccessful. Then in 1675, the Danish astronomer Olaus Roemer, while studying the eclipses of the larger moons of Jupiter, found that these eclipses occurred about 16 minutes later when Jupiter was farthest from the earth than when it was at the nearest point (see figure below).

This could only mean that light did not travel instantaneously but did indeed take a finite time to cross a specific distance. Earlier experiments had failed because the distances involved were not great enough to measure such a velocity.

According to Roemer's measurements, it took the light 16 minutes longer to travel to point A *than it took to travel to point* B *in the figure. If the earth is 1.5* $\times$ *10^8 kilometers from the sun (1 AU), what is the speed of light according to Roemer's measurements? How does this figure compare with velocity of light measurements today?*

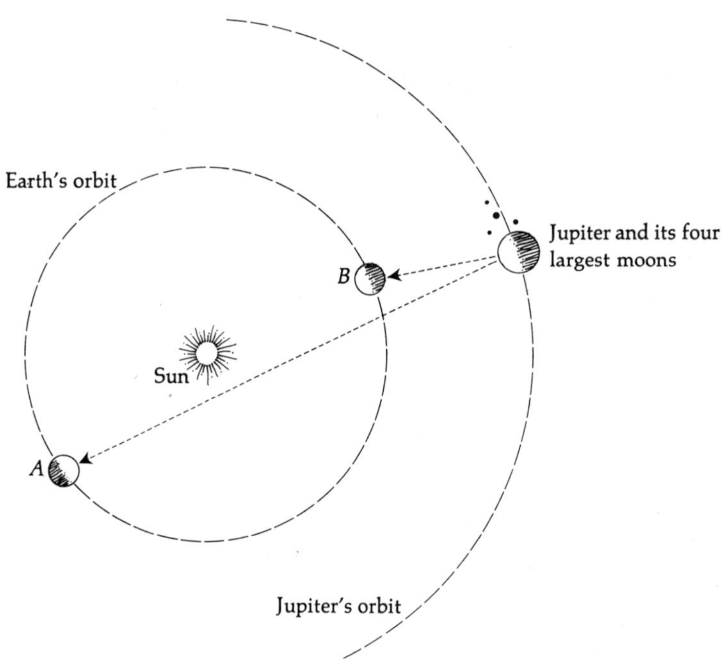

amount of radiation received becomes progressively weaker and weaker with distance. This is called the *inverse-square law.*

Using an ordinary photographic exposure meter, record the light received from a flashlight at several distances (this is best done in a dark room). Make a graph showing the relationship between the distances involved and the meter reading.

The small portion of the electromagnetic spectrum that encompasses visual light covers the entire color spectrum that we usually associate with the rainbow. Color, as we perceive it, is determined by the wavelength of the light rays entering our eyes. The longest waves in this optical region of the spectrum produce the sensation of red, and the shortest ones violet. Any wavelength in between will normally be perceived as some basic color of the rainbow as our brain interprets what our eye has received.

This visible radiation is one of the very few portions of the electromagnetic spectrum that can penetrate the atmosphere. We live at the bottom of an ocean of air which is almost entirely opaque to most forms of radiation. Only over very short distances can we employ radiation such as x-rays, microwaves, or ultraviolet. At distances greater than perhaps a few feet, these radiations are effectively blocked. However, the atmosphere is almost perfectly transparent in the wavelength region we perceive as light and color, and in part of the radio region. The yellow region of the visual spectrum is the one in which our sun radiates the most energy, and is also the one to which our eyes are the most sensitive. This phenomenon is undoubtedly due to the process of biological evolution. If our eyes were sensitive instead to a totally different region of the electromagnetic spectrum, we would not be capable of seeing through our atmosphere.

It is fortunate that we are protected by this opaque blanket of air because much of the radiation from the sun and from outer space that reaches the top layers of the atmosphere is deadly. However, for astronomers, the atmosphere presents a serious handicap. Except for the optical "window," and a somewhat larger radio "window" which was discovered about 50 years ago, astronomers in observatories here on earth had no way of recording the much wider spectrum of radiation from outer space.

Then, with the advent of high-altitude balloons and rockets, more "windows" through which radiation could be recorded were opened up (Fig. 3.3). With the coming of spacecraft, the entire electromagnetic spectrum became available for study. Above the blanket of our atmosphere, manned and unmanned satellites and space probes gathered and are continuing to gather data with specially designed equipment (Fig. 3.4). X-rays, gamma rays, ultraviolet and infrared are all getting their share of attention as new and sometimes mysterious sources of radiation are detected.

3.3 Project Stargazer being made ready for launching (photograph from Dearborn Observatory, Northwestern University).

3.4A A close-up view of the Skylab space station (top right) photographed from the Skylab command module during the "fly-around" inspection prior to docking (NASA photograph).

3.4B Astronaut Owen K. Garriott, science pilot of the Skylab 3 mission, stationed at the Apollo Telescope Mount console (opposite) from which he can control the solar physics telescopes while in earth orbit (NASA photograph).

Not only has the equipment aboard these spacecraft given us previously unavailable observations, but it has also made possible radio and visual observations without the growing interference of pollution. This pollution takes many forms. City lights and industrial smog have already greatly reduced the efficiency of observatories in populated areas. The increased use of the radio spectrum for commercial broadcasts, CBs and the like have left very few wavelengths free for astronomical use. Therefore, the future of astronomy lies in the development and use of spacecraft and perhaps in a moon-based observatory.

Wavelength and Frequency

Electromagnetic radiation may be thought of schematically as a wavy line (Fig. 3.5). The length of the wave is measured from crest to crest. The *amplitude* or height of the wave is a measure of its intensity. Radio waves are much longer than those of light, yet both forms of radiation travel at the same speed—the speed of light. Therefore, if we could measure the number of waves passing some specific point, we would find that more

3.5 *The length of the electromagnetic wave determines the type of radiation that reaches us. The wavelength multiplied by the frequency equals the speed of light. In each case in the diagram, the time interval is one second.*

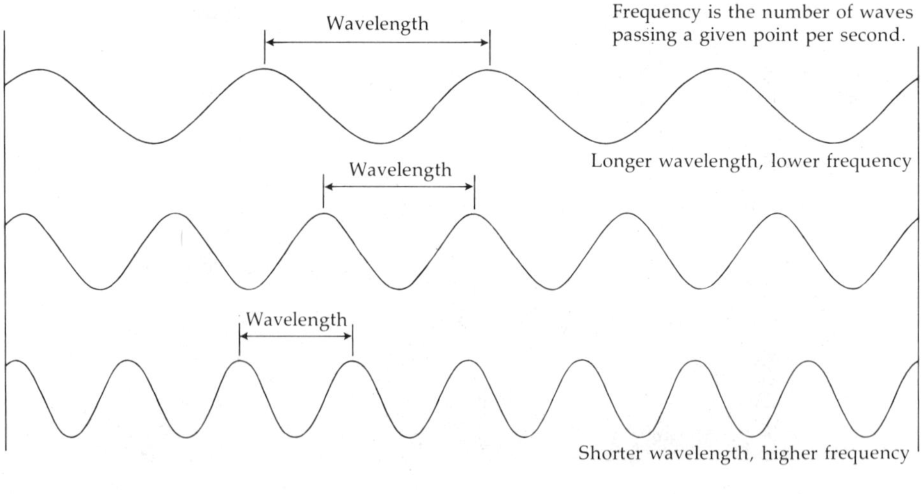

light waves passed in a given unit of time than radio waves. The number of waves passing per second is called their *frequency*.

If we multiply the frequency by the wave length, regardless of the type of radiation we are dealing with, we will obtain the speed of light. This means that the shorter the wavelength, the higher the frequency and the longer the wavelength, the lower the frequency. For example, the wavelength of yellow light is approximately 5×10^{-5}cm. Since the speed of light is 3×10^{10}cm., the frequency of yellow light must be 6×10^{14}hertz[1]. A radio station broadcasting on 1000 kilocycles or kilohertz is using a wavelength of 3×10^{4}cm. "Shortwave radio" is just that, with much higher frequencies than normal broadcasting.

The Doppler Effect

We are all aware of the change in pitch that occurs as a train passes us with its whistle blowing. When the train approaches, the whistle's sound is of a noticeably higher pitch than when the train has passed and the source of the sound is receding (Fig.3.6). The sound of the whistle travels continuously at the speed of sound, yet the pitch changes. Pitch is determined by the *length* of the sound wave. Similar in this respect to light waves, the longer the sound wave, the lower the frequency, and vice versa.

If the train is standing still and the whistle blowing, a specific number of sound waves of a certain length reaches your ears each second. Your brain interprets this as a specific pitch. However, when the train is approaching, although the velocity of the sound itself does not change, the waves arriving at the ear become more frequent and each is consequently shorter in length (Fig. 3.6*b*). More of these shortened wavelengths reach your ear per second, causing you to hear a higher pitch. As the train recedes (Fig. 3.6*c*), the waves arrive less frequently and are therefore of greater length than they would be when the train is at rest. You hear these waves of sound as a lower pitch. The faster the train moves, the greater the change in pitch.

This is called the *Doppler effect*, named after Christian Doppler, an Austrian physicist who first explained the principle in 1842. Light waves and other electromagnetic radiation behave in a similar manner to sound waves, but instead of perceiving a change in pitch, we observe a shift in the spectrum.

---

[1] The hertz is the unit of measure of frequency. It is the number of cycles or wavelengths passing a specific point per second.

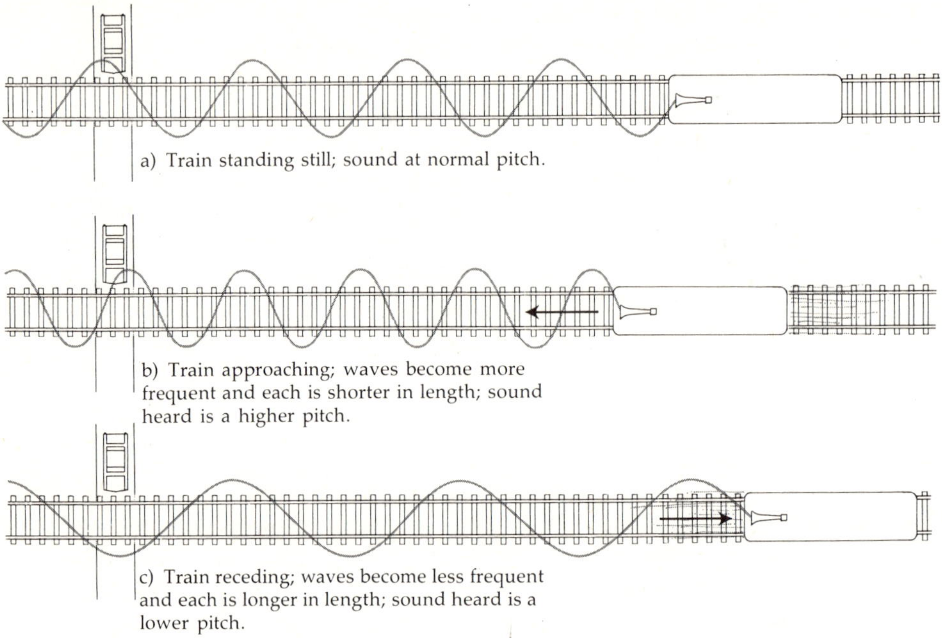

3.6 *The Doppler effect. The velocity of sound through the air remains a constant and is the same in all three cases. It is the frequency (waves per second) and wavelengths that change.*
(a) *Train standing still; sound at normal pitch.*
(b) *Train approaching; waves become more frequent and each is shorter in length; sound heard is a higher pitch.*
(c) *Train receding; waves become less frequent and each is longer in length; sound heard is a lower pitch.*

It should be remembered that although we used sound to illustrate the Doppler effect since sound is a wave motion, sound is *not* a part of the electromagnetic spectrum. Sound is a wave motion in air (or in some other medium), whereas light and other radiation travel in empty space. Sound and light waves differ also in that light travels almost one million times faster than sound. That is why we see lightning and hear radio static long before we hear the rumbling of thunder.

For the moment, let us consider only the visual part of the electromagnetic spectrum, although the same principles apply to the entire range of radiation.

Using a tape recorder, record the sound of a model airplane. Record it first while standing at the center of its circular path. Then record its noise while standing outside of its path. Listen to the tapes. What differences do you note in the two? Draw a diagram showing why such a difference occurs.*

**If a model airplane is not available, its noise can be simulated by whirling a small object rapidly around your head.*

Because light travels so much faster than sound, the speed of any terrestrial object is too insignificant compared with the speed of light for us to notice any effect in our daily lives. But when we are dealing with the much greater velocities often encountered in astronomy, this change in wavelength and frequency can be measured. However, very rarely can we detect a shift in wavelength (color) with our eyes alone. Even with what may seem to be extremely fast velocities to us here on earth, the minute shifts that occur in wavelengths require much more sensitive measuring devices.

First we must photograph the spectrum of a star or other celestial object by attaching a spectrograph to a telescope. The spectrograph is an instrument which "breaks up" the starlight into its component colors or spectrum (see Fig. 3.7). This is done with the use of a prism or very fine grating. On the photograph of the spectrum, called a *spectrogram,* we find dark or black lines, or gaps, which represent places in the spectrum of the star where light has been absorbed by specific chemical elements in the star's atmosphere. These are called *absorption lines*.

Since each of these lines represents a specific wavelength, if the spectrum is shifted by the Doppler effect, the lines must also be shifted by the same amount. By measuring this shift with a microscope, as compared with stationary laboratory lines, we can obtain the amount of wavelength change (see Fig. 3.8). Then by using a ratio formula[2] we can determine the velocity of the object toward or away from us. The change in wavelength is in the same proportion to the stationary wavelength as the object's velocity

---

[2]The radial velocity ratio formula is:

$$\frac{\Delta\lambda}{\lambda} = \frac{v}{c}$$

λ is the Greek letter "lambda" used to designate wavelength.
Δ is the Greek letter "delta" used to mean "the change in."
v is the radial velocity of the object.
c is the internationally accepted symbol for the velocity of light.

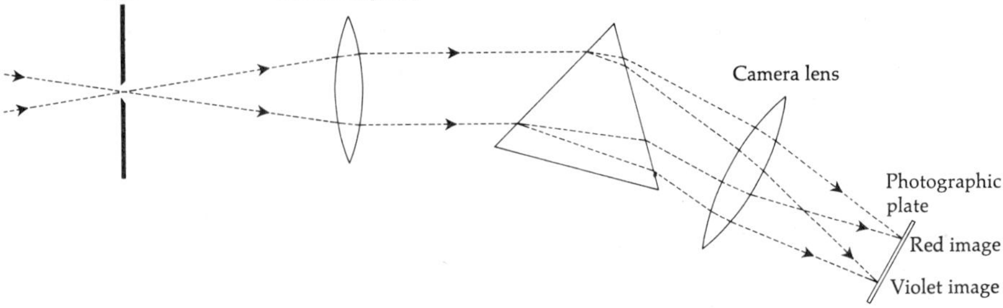

3.7 A simple prism spectrograph.
Slit: the objective lens of the telescope focuses light at the slit opening.
Collimating lens: makes light a beam of parallel rays.
Prism: breaks up each of the parallel rays of light by bending different wavelengths or colors by different amounts (violet is bent the most, red the least).
Camera lens: forms the image on the photographic plate.

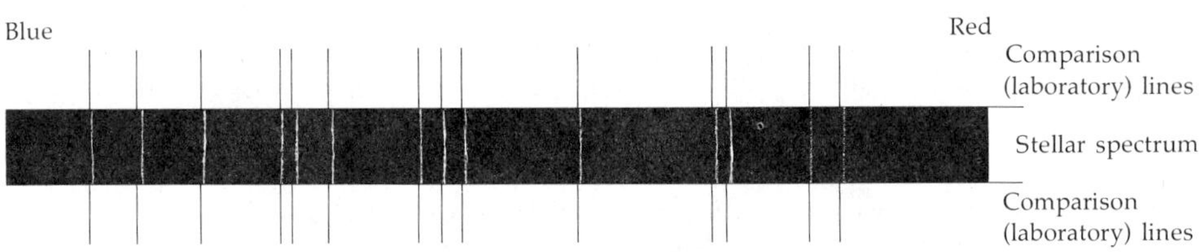

3.8 Schematic drawing illustrating spectral line shift (Doppler effect). The stellar spectrum shows the absorption lines shifted toward the red end (red shift). The amount of shift has been greatly exaggerated for the purpose of illustration. (The stellar spectrum represents a photographic negative.)

is to the speed of light. Thus, if a wavelength of 5000 angstroms[3] is measured at 5100 angstroms (i.e., found to have shifted by 100 angstroms), the velocity of the object involved is 100/5000 or 1/50th the speed of light.

---

[3]*Angstrom*: a unit of length commonly used in measuring electromagnetic wavelengths. It is equal to 10^{-8}cm (one ten-millionth of a millimeter).

Note that we can measure only approaching and receding (radial) velocities, not sideways (tangential) motion.

When the approaching or receding velocity of a celestial object is measured, this is called its *radial velocity*. A sideway motion across our line of sight (which cannot be measured by spectral line shift) is called *tangential velocity*.

Radial Velocities of the Galaxies

By the beginning of the twentieth century, the radial velocities of some of the stars in our own galaxy were being obtained. Then in 1913, V. M. Slipher at Lowell Observatory in Arizona started to obtain the spectra and subsequently the radial velocities of what were then still unidentified "fuzzy patches" or nebulae. For the brighter ones he got both positive and negative values. (By convention, a positive radial velocity is obtained from a receding object, whereas a negative one is obtained from an approaching object.) As Slipher observed fainter and fainter nebulae, he found that their spectra showed greater and greater "red shifts" corresponding to greater and greater positive radial velocities. Presumably this meant that the fainter ones were receding from us faster than the brighter ones.

By 1924, when it was finally determined that many of the nebulae were in fact outside of our own galaxy, there seemed to be some evidence that there was a correlation between radial velocity and distance. By the early 1930's, Edwin Hubble, with the help of M. L. Humason, who was also at Mount Wilson, established the truth of this relationship. The more distant a galaxy is, the greater its red shift, and hence its radial velocity.

In Fig. 3.9 the photographs of several individual galaxies are shown on the left. Each is identified by the name of the cluster of galaxies in which it is a member. These names come from the constellations in which each cluster is located. On the right are the spectra of these galaxies, with an arrow to indicate how much two specific spectral lines have been shifted toward the red end of the spectrum. The radial velocities have been calculated using the ratio formula and the distances have been estimated from the faintness of the galaxies. Note that the fainter the galaxy appears and, therefore, presumably the farther away it is, the greater its red shift.

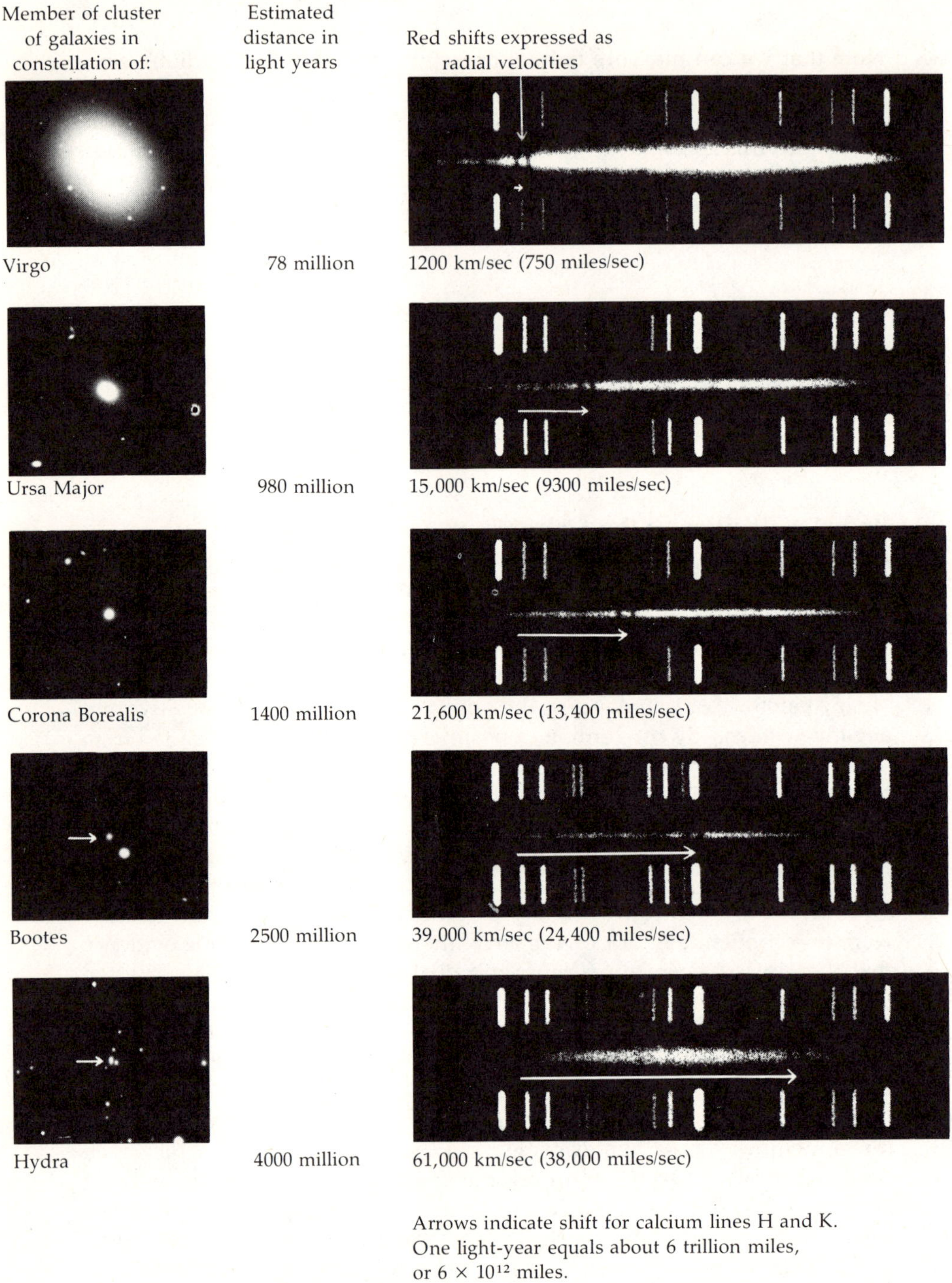

3.9 Velocity-distance relationship. Relation between the red shift and the distance for distant galaxies (photograph from the Hale Observatories).

The Expanding Universe

When Hubble plotted this relationship between radial velocity and distance, he obtained a straight-line graph (Fig. 3.10). The inclination or slope of this line is called the "Hubble constant." It represents the amount by which the speed of the galaxies increases with increasing distances of the galaxies. Note that the speed of the individual galaxy does not increase, but that the galaxies farther away from us recede at a faster pace than those nearer.

In 1929, Hubble made his first calculation of this constant. Because the estimate of galactic distances on which he based his graph was much too small, he arrived at a figure of 164 kilometers per second (or 100 miles per second) per million light years. Over the years, as more and better observations have been made, the distances to the galaxies have been drastically revised. Our universe was found to be tremendously larger than previously imagined. The Hubble constant has therefore also been changed to reflect this revision. The currently accepted value is 16 kilometers per second (or 10 miles per second) per million light years. This means that for every million light years we go out into space, the objects we observe there will be receding from us 16 kilometers per second faster. Clusters of galaxies that are billions of light years away are receding thousands of times faster than those in our immediate neighborhood of the universe.

3.10 *The radial velocity-distance relationship based on clusters of galaxies.*

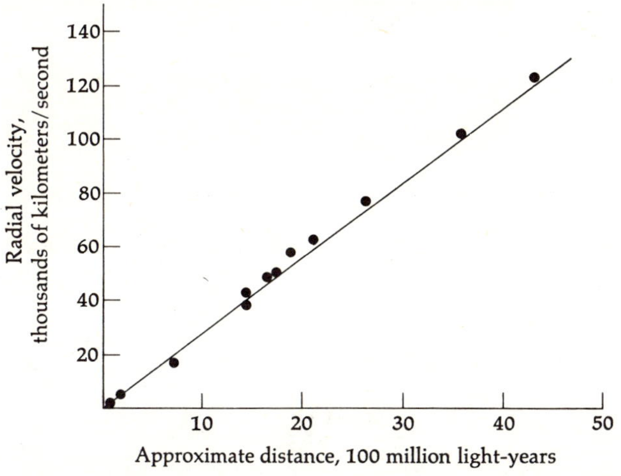

If the more distant galaxies and cluster of galaxies are receding from us faster than those closer to us, we must conclude that our universe is expanding. The first impression is that we are at the center of the universe and that everything is rushing away from us. But this is not the case—it is merely an illusion. Actually, the universe is expanding uniformly in all places and no matter where we are standing in the universe, we will experience the same illusion.

To better understand this, the following experiment will help. Using Figure 3.11 as a guide, imagine yourself in a vèry large classroom which has no walls. You are seated in Desk *A*. The rest of the students are seated about two feet apart from one another.

During every 10-second interval, the space between the students doubles (but the students remain the same size). After the first 10 seconds, your immediate neighbor (*B*) would be four feet away instead of two feet, and the person on the other side of your neighbor (C) would be eight feet away from you instead of the original four feet. In other words, although *C*'s distance from you has also doubled, his actual increase in distance is

3.11 *The expanding classroom.*

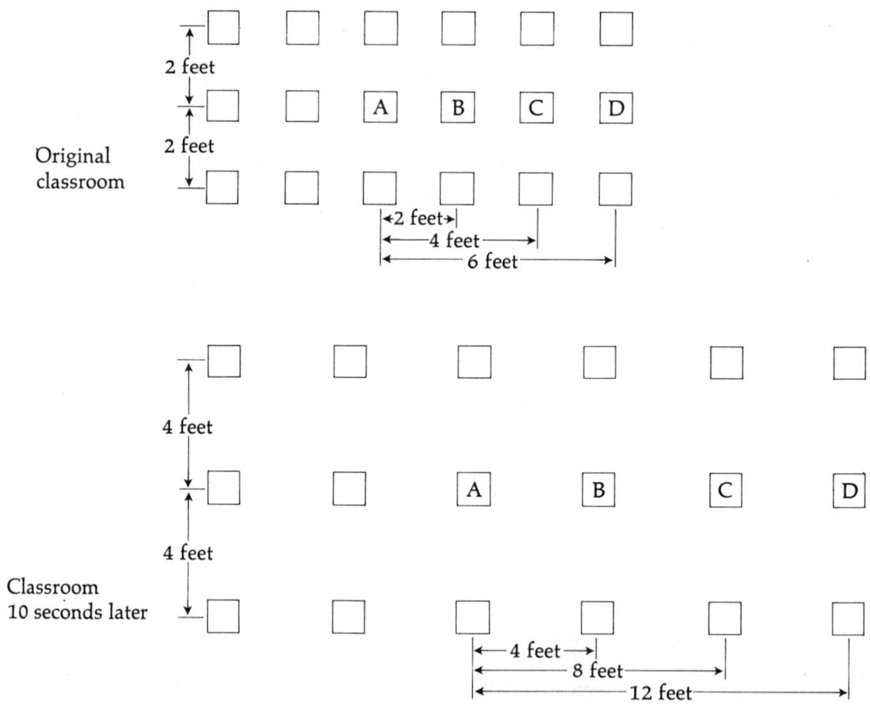

twice that of *B*'s, which means that C must have receded *twice* as fast from you in the 10-second interval as *B* did in order to cover twice the distance. And similarly, the third person from you (*D*) would have receded three times as far as *B* did, traveling, therefore, three times as fast.

Thus, the farther away a student is from you, the faster he would appear to be moving away from you. This expansion would be experienced by everyone in the room. Each would feel that he was at the center and that everyone else was receding from him, with the more distant people receding faster. Note again, however, that although the space between the students continues to increase, the students and their chairs remain the same size.

Take a partially blown up balloon (use one that is round in shape) and glue many small bits of cotton on it to represent galaxies. Choose three widely separated bits, designating them A, B, *and* C. *Measure the distances from each one to several other bits of cotton. It might be helpful to use ink marks on the balloon to keep track of which distances have been measured.*

Now finish blowing up the balloon. Repeat your measurements from cotton bits A, B, *and* C. *Be sure to measure along the surface of the balloon in all cases. Chart the changes in distances of various cotton bits from* A, B, *and* C *respectively. How do these differences relate to the original distances? Can you confirm that the farther away a cotton bit (galaxy) is, the greater is its change of distance from cotton bit* A, B, *or* C? *Make a graph plotting the distance moved on the* Y*-axis (vertical) and the original distance on the* X*-axis (horizontal) as below.*

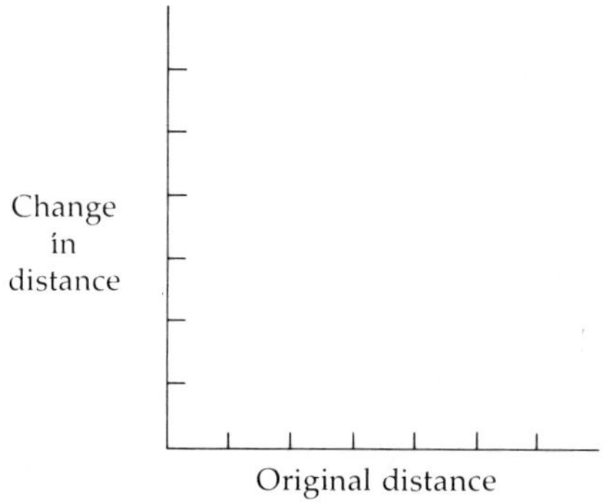

Olbers' Paradox

Although the radial velocities of the galaxies provided the major evidence that the universe is expanding, there were other lines of reasoning that pointed in the same direction. For example, the obvious fact that the night sky is dark leads to this conclusion. If we assume for a moment that the universe is static, unchanging, and infinite, and that the galaxies are distributed uniformly throughout all space, then wherever we look in the sky, we should see, however faintly, the light of a galaxy. Of course, from the inverse-square law of light we would expect that the farther away the galaxy was, the dimmer it would appear, but with uniform distribution, the number of galaxies should increase with distance by the same proportion, namely, the square of the distance. This would compensate for the dimming effect and the sky would therefore be bright. Why then is it dark?

Heinrich Olbers, a German astronomer, pointed out this paradox in 1826; it has ever since borne his name—Olbers' paradox. Olbers' argument dealt with the stars since galaxies were unknown in his day, but the principle is the same. It was not until the expansion of the universe was discovered that the answer to the paradox could be given.

In an expanding universe the radiation received from the galaxies is not only shifted toward the red end of the spectrum, but it is weakened as well. Radiation energy at the red or longer wavelength region of the electromagnetic spectrum is weaker than at the blue or shorter wavelength region. As we have seen, the further a galaxy is from us, the larger its red shift, and therefore, the greater the weakening of the radiation that finally reaches us. In other words, the apparent brightness of a galaxy falls off more quickly when it is receding from us than it would if it were standing still. The brightness of a galaxy about two billion light years away is only two-thirds what it would be if the universe were not expanding. Indeed, if a galaxy were receding at the speed of light, it could not be seen at all because its brightness would be zero. Thus we do not receive all the radiation from the distant galaxies and our night sky is dark.

All the evidence therefore points to the fact that the universe is expanding. This is a basic assumption in any explanation of the architecture and dynamics of the universe. It is part of the continual flux to which every part of the cosmos is subjected. And it is fully accepted by the scientific community. If it has not already subtly done so, it will eventually be felt in our philosophical thought, in our value systems, and even in our daily lives.

Discussion Questions

1. Make a chart of the various kinds of electromagnetic radiation. List the ways in which each type is used in our daily lives. What type of matter (or material) will block each type?

2. In addition to the whistle of the passing train, what other examples of the Doppler effect might we encounter in everyday affairs?

3. What are the differences between sound and light?

4. What is meant by "radial direction?"

5. What do we mean by "red shift?"

6. Define the term "optical window."

7. Light takes about eight minutes to travel from the sun to the earth. How long does sunlight take to get to Pluto? To the next nearest star (Alpha Centauri)?

8. Light of laboratory wavelength 5000 angstroms is observed to have a wavelength of 5200 angstroms coming from a celestial object. What speed of recession or approach does this represent?

9. If two lightbulbs with the same wattage are seen and one appears 10 times as bright as the other, what can be said about their distances from the observer?

10. If a galaxy is found to have a radial velocity of about 30,000 km/sec, what is its distance from us according to the velocity-distance relationship? What is 30,000 km/sec in miles per hour?

11. A spaceship with a very powerful light is traveling towards the earth at $^{9}/_{10}$ths the speed of light. If an astronomer on earth measures the velocity of this light, what will his results be? How can he determine the velocity of the spaceship?

12. Two clusters of galaxies are observed; *A* is receding at 40,000 km/sec; *B* at 80,000 km/sec. How much farther away from us is cluster *B* than cluster *A*?

13. If the red shifts exhibited by quasars can be taken as distance determiners, how far away would a quasar have had to be if its light which we observe today left there when the earth was newly formed? (Estimated age of the earth is about 4½ billion years.)

14. The sun is about 160 billion (16×10^{10}) times brighter than the average bright star seen in the night sky. Using the inverse square law, how many times farther away would the sun have to be to appear as an average star?

4 The Cosmology of an Expanding Universe

A cosmological theory is an attempt to explain in a coherent manner how our cosmos is structured, how it originated, and how it evolved. The theory must incorporate all of the evidence available at the time and show why such evidence supports its own explanation rather than other explanations. Throughout the history of mankind, cosmological theories have been propounded, and although today we look upon most of them as naive and simplistic, in their time and in the society into which they were introduced, they "worked." New theories which revolutionized the accepted concept of the universe were usually resisted because they threatened to change not only the concept of the universe, but also the concept humans had of themselves—their self-image. Cosmological theories face a similar challenge today.

The Age of the Universe

If all of the clusters of galaxies are speeding away from each other, and if this speed was maintained throughout the past, how long ago was everything back at the starting point? And what was that starting point?

The distance that an object has traveled can be calculated by multiplying its rate of speed by the duration of its journey. This is the familiar "distance equals rate times time" formula. In this case, however, we know the distance and the rate if we use Hubble's constant. We are looking for the time involved. To find it, we must divide the distance by the rate.

Let us take a cluster of galaxies one billion light years away. According to Hubble's constant, this cluster is moving away from us at 16,000 km/sec. (A billion is a thousand million, so we multiply Hubble's constant by 1000). To get the time involved, we must divide the distance of one billion light years by the rate of 16,000 km/sec. However, we must use the same units of measure if our answer is to be meaningful. One light year is approximately 9.5×10^{12} km. One billion light years is thus $9.5 \times 10^{12} \times 10^{9}$ or 9.5×10^{21} km.

$$\frac{9.5 \times 10^{21}\text{ km}}{16{,}000\text{ km/sec}} = 5.94 \times 10^{17}\text{ seconds}$$

There are 3.156×10^7 seconds in a year.

$$\frac{5.94 \times 10^{17}}{3.156 \times 10^7} = 1.88 \times 10^{10}\text{ or }18{,}800{,}000{,}000\text{ years}$$

Assuming that the expansion of the universe has been constant, this is approximately the currently accepted age of the universe, the time when the expansion began.

The age of the universe is determined here by the current value of the Hubble constant. When this amount was thought to be much larger, the universe was correspondingly calculated to be much younger.[1] Obviously, if the Hubble constant is found to have a different value, then the age of the universe derived from it must also change. At present, this figure reasonably agrees with estimates of the age of the universe derived by other methods.

Cosmological Theories

If the radial velocities of the galaxies are being interpreted correctly, then approximately 19 or 20 billion years ago all the matter and energy in the universe must have been clumped together in a single primordial mass. The proponents of the *Big Bang* cosmological theory believe that this was the case and that the violent explosion of this mass started the expansion outward.

An enormous amount of energy was released in this explosion. It was originally in the form of very short, high-energy gamma rays and x-rays. Physicists George Gamow and R. H. Dicke predicted that if this radiation from the original explosion were still coursing through the universe, it would be detectable today as very weak, long microwave radio waves, having been elongated and weakened by the expansion. In 1965, engineers at the Bell Telephone Laboratories detected just such radiation. It reaches the

[1] If we use Hubble's original calculation of 164 km/sec (see page 49), the age of the universe, according to the above method of determination, is only 1.8 billion years.

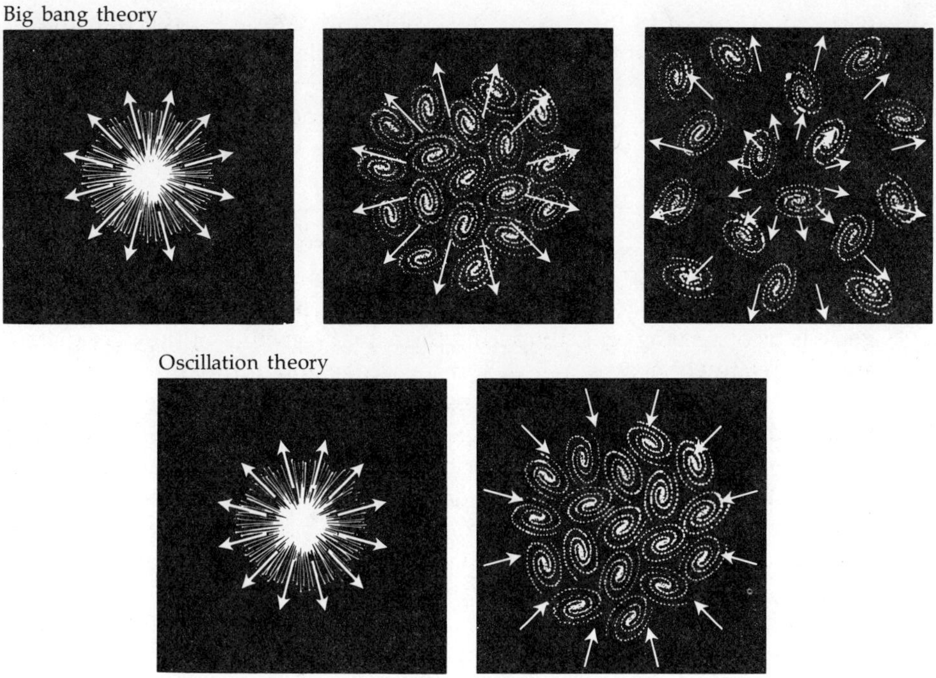

4.1 Schematic representation of the two major cosmological theories generally accepted today. Thus far, most evidence tends to point towards the Big Bang theory.

earth from all directions in equal intensity. This was one of the main pieces of evidence that gave the Big Bang theory more credence than its competitors.

Although some cosmologists believe that the expansion will continue endlessly, others think that it will eventually slow down and stop and that the universe will then start to collapse back in upon itself (Fig. 4.1). Once fully collapsed, it will again explode outward and repeat the cycle over and over again. The collapsing stage will begin when the mutual gravitational attraction of its parts slows down the expansion and eventually reverses the process. One of the major problems of the *Oscillating Universe* theory is that it requires a certain amount of gravitational attraction to cause such a reversal and this depends upon the actual amount of matter in the universe. Thus far, cosmologists have not been able to account for enough mass to "close" the universe. The visible universe, i.e., the one that is detectable through optical telescopes, is insufficiently massive for the task. However, recent research gives evidence of more mass in the universe than that which is visible. For example, x-ray telescopes aboard recent satellites have collected data which indicates that some clusters of galaxies have large masses of very hot, invisible gases. But these gases alone are still not enough mass. About 100 times as much mass as is currently known to exist would be necessary to insure a "closed" universe. Otherwise, the universe must be presumed to be "open," i.e., expanding endlessly.

Another approach to solving the problem of the open or closed universe is by observing and comparing the actual velocities of the very distant and the relatively nearby galaxies. If the expansion of the universe is actually slowing down sufficiently to eventually stop it completely, then the Hubble diagram would not be an exactly straight line. The farther out into space we look, the farther back in time we go. Hence we should see the most distant galaxies less slowed down than the nearer ones. The result would be a slightly upward curve on the upper portion of the Hubble diagram. Although an extremely slight curve upward has been noted in recent research, it is not enough to indicate a closed universe.[2]

Another cosmological theory, the *Steady State* theory, should be mentioned since it was a serious contender until a few years ago. Today, most of the evidence tends to go against it. The proponents of this theory agreed that the universe was expanding, but they claimed that as the galaxies spread out, new galaxies were constantly being formed to fill in the subsequent spaces. New matter was constantly being created to form these new galaxies. This is not so phenomenal as it may seem. If every fifty years or so only one atom of hydrogen were created in each volume of space equal to the volume of the Empire State Building, sufficient galactic matter would be created to maintain the proper density.

Thus according to the Steady State theory, no matter where or when an observer was in the universe, he would observe it much as we see it today. New stars and galaxies might be created as older ones spread out, but the distribution and density of these celestial objects would remain steady. However, as we noted at the end of Chapter 2, some very unusual objects called *quasars* have been discovered in the last two decades that seem to refute this assumption. We will study them more thoroughly in

---

[2]An absolutely straight line for the Hubble diagram would indicate an open and completely empty universe. The small amount of matter that is known to exist causes the diagram to assume its very slightly curved shape at the upper portion.

later chapters (see page 216), but their importance in this discussion is the fact that they all have extremely large redshifts. This places them at the furthest limits of the visible universe, if we assume that the radial velocity-distance relationship is correct (and most evidence seems to uphold this assumption).

None of these quasars are located near us. There are none in or near the nearby galaxies or the nearby clusters of galaxies. Therefore, we must conclude that our part of the universe is unlike some other part, if not in space, then in time. In studying these very distant objects, we are actually looking back in time and seeing cosmic conditions as they were billions of years ago. These conditions are not like what we see nearer to us and, therefore, in a different time period. The existence and location of these remote quasars is evidence contrary to the basic assumption of the Steady State theory.

The Big Bang cosmological theory (whether with an opened or closed universe) is currently the most accepted one. The original fireball of matter and radiation exploded and in a sense is still exploding. It cannot be thought of as having specific boundaries or a location. It was everywhere—and still is. Although it has expanded tremendously, the original matter and energy (in perhaps different form) are still in existence. We are all part of this original material. It is all around us.

It should also be remembered that these cosmological theories are just that. They have not been proven and all that can be said is that the evidence points toward their acceptance today. Like those who lived before us, we try to explain our cosmos by the information available to us. One difference we have from many of our predecessors is that we understand that our theories are not necessarily the final answers, the ultimate truth. And we continue to search.

Concepts of Relativity

To further understand the structure or the architecture of the universe, it is necessary to study some of the aspects of *relativity*. When we examine everyday objects or events, we do not concern ourselves with relativistic matters. We know that the stewardess in an airplane and a person on the ground have different measurements of the stewardess's forward motion. We are also aware that although we think of the airplane as flying in a straight line, it is actually making a huge curve as it follows the earth's curvature (see Fig. 4.2). When we say that a car is traveling at 55 miles per

4.2 An airplane actually moves along a curve as it flies the shortest route between Chicago and Paris.

hour, it is understood that this speed is in relation to the road, and we need not specify so. We have adjusted to these concepts and to the context in which they are perceived.

But when we consider the vastness of the universe, the effects of matter and energy upon this universe, and the enormous velocities of distant galaxies and quasars, we must employ relativistic concepts. Nothing in the universe is static. Everything moves at a velocity relative to some other object. The car moves at 55 mph relative to the road, but the car as part of the earth moves at 18 miles per second relative to the sun, which, along with the entire solar system, in turn moves at 12 miles per second relative to the nearby stars, and at 150 miles per second relative to the center of the Milky Way galaxy. Can we say therefore that the absolute or true velocity of the car is 55 miles per hour?

The only absolute value, according to the theory of relativity, is the speed of light. Regardless of when or where it is measured, the value remains the same: 2.998×10^{10} cm/sec or 186,262 miles/sec.

As another experiment, let us imagine two physicists, each at the bottom of a very high and narrow tower. At the top of each tower there is a mirror facing downward. Each man has a flashlight which he points upward, turns on for an instant, and then measures the time the light takes

Imagine yourself in a smoothly flying 747 jet airplane. (If you are planning such a trip, you can actually try some of these experiments). The shades have been drawn for the in-flight movie. Other than the fact that you know the seats face the front of the plane, can you tell in which direction the plane is flying? Do you experience any feeling of motion (the vibration or noise of the engines does not prove that the plane is moving)? When you walk down the aisle, is it any easier to walk toward the front or the rear of the cabin? If a fly were flying around the cabin, would it be flying faster when it was going toward or away from the front of the plane? Would it have any greater difficulty flying in either direction? If you could have a game of catch the length of the cabin, would it be easier to throw the ball one way rather than the other?

4.3 *All objects in the universe move at velocities relative to some other object. There is no absolute or true velocity except for the speed of light.*

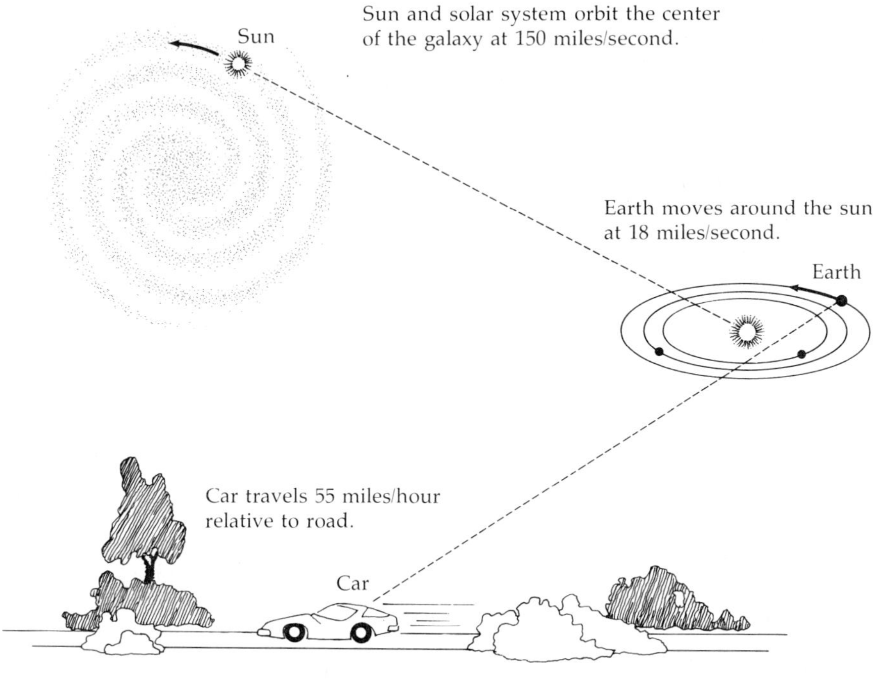

to go up to the mirror and be reflected down again (see Fig. 4.4). (Actually, because light travels so very fast, such an experiment would be possible only if the towers were impossibly tall.)

While this is going on, the towers are moving away from each other at a very high speed. Since neither tower has any windows and their velocities are constant, the physicists will have no sensation of motion. Each will claim that his light went straight up and came straight down again. And each will record exactly the same amount of time the light took to make the round trip.

Let us now suppose that each tower has a peephole through which each physicist can observe the other tower receding from him. Each will claim that the other physicist's light did not go straight up and down but rather followed a longer zig-zag course. Distance equals rate times time, and the rate, being the velocity of light in this case, is constant. Therefore, Scientist *A* says that if the distance that *B*'s light traveled is greater than the distance his own light traveled, the time it took must also be greater. *B*'s

4.4 *The Moving Towers. From Physicist* A*'s viewpoint, Physicist* B *is rapidly moving away from him. To* A, B*'s light seems to follow a zig-zag path, although to* B, *it is going straight up and down (grey dashed line).*

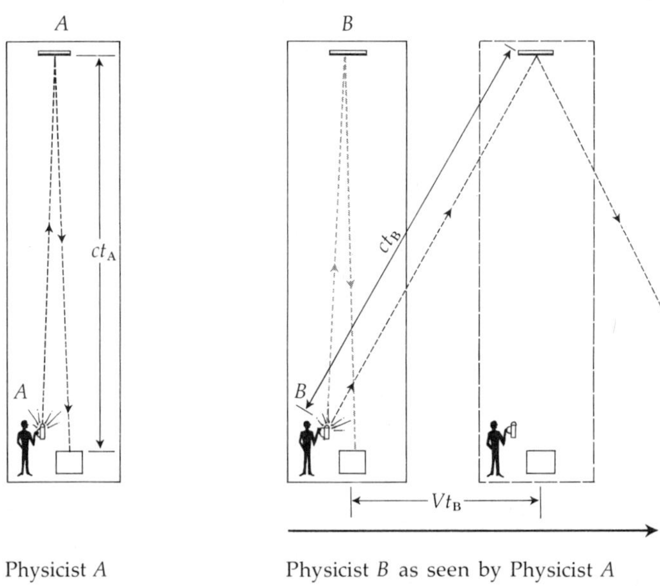

timekeeping must be wrong. Physicist *B* will come to the same conclusion about *A*. Each will conclude that the other's clock ran slower than his own. Neither will be aware of the slowing of his own clock relative to the other's. This is because the motion and time of both are relative values, not absolute quantities.

Imagine a large boxcar moving very rapidly down the track. Its motion is smooth and uniform. A light is turned on in the exact center of the car. For an observer inside the car, which wall will be illuminated first, the front or the rear? Would an observer outside the car agree with this? What would the outside observer say about the distance the light had to travel in each direction to reach the wall? Draw a diagram to illustrate this situation.

In this example the motion was constant and uniform. Therefore, each party's timekeeping apparatus slowed down relative to the other's. But when one object is accelerated away from another, then it is the only one affected by this relative slowing process. In many science fiction stories, space travelers are able to reach distant stars in what seems to them very short periods of time. Not only do the actual clocks slow down as the travelers are accelerated out into space at almost the speed of light, but their aging processes are also slowed down. They are not aware of the time passage back on earth, but only of what they consider normal time passing aboard their craft. Back on earth 1000 years may go by while the travelers age and actually live only 20 or 30 years.

Fantastic as this may seem, it has actually been shown to happen to μ (mu) mesons. These are subatomic particles that are naturally produced when cosmic rays strike molecules in the earth's upper atmosphere (see page 196). When artificially produced by high-energy "atom-smashers," these mesons are found to be extremely short-lived. So short, in fact, that by our clocks the naturally-produced mesons should not live long enough to travel through the atmosphere and reach ground. But they do, and the reason is that their velocity is so close to that of light their "clocks" have been slowed down relative to ours—and by our timekeeping they actually live longer!

Let us suppose that two galaxies are rapidly receding from us in opposite directions, each traveling at ¾ths the speed of light (Fig. 1 below). Would an astronomer on Galaxy A *see Galaxy* B *receding at 1½ times the speed of light? No, we know from the well-tested concepts of relativity that no material objects can travel faster than the speed of light. When velocities are small compared to the speed of light, we can use ordinary formulas since the relativity effect is minute. But when velocities approach that of the speed of light, as with our physicists in their towers, relativistic formulas, although more complicated, must be employed.*

The nonrelativistic formula is simply: $V_A + V_B = V_{AB}$

The relativistic formula is: $\dfrac{V_A + V_B}{1 + \dfrac{V_A V_B}{c^2}} = V_{AB}$

If two cars are receding from you (see below), each at 100 km/hr., we can quickly calculate that they are receding from each other at 200 km/hr. If we use the second formula above, our answer will not only be more difficult to obtain but will differ by only a very small fraction. Try it.

On the other hand, for the two galaxies we use the relativistic formula:

$$V_{AB} = \frac{V_A + V_B}{1 + \dfrac{V_A V_B}{c^2}} = \frac{\frac{3}{4}c + \frac{3}{4}c}{1 + \dfrac{(\frac{3}{4}c)(\frac{3}{4}c)}{c^2}}$$

$$= \frac{1\frac{1}{2}c}{1 + \dfrac{\frac{9}{16}c^2}{c^2}} = \frac{\frac{3}{2}c}{\frac{25}{16}} = \frac{24}{25}c$$

Thus, Galaxy A *will see Galaxy* B *receding from him at 24/25th the speed of light and vice versa.*

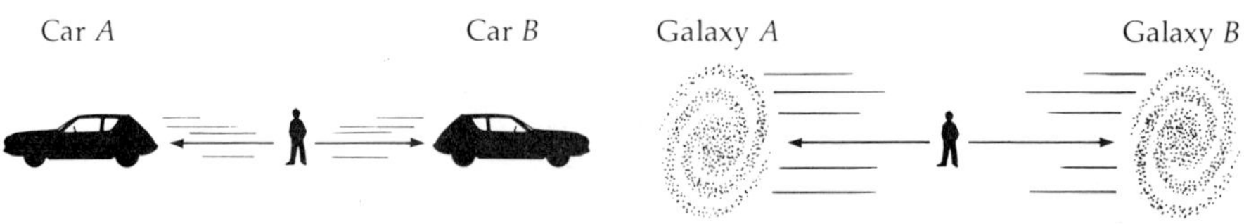

The Curvature of Space

We spoke earlier of the curved path that an airplane takes as it flies, say, from Chicago to Paris. This route is a *geodesic*: the shortest distance between two points in any given geometry. In *Euclidean* or *plane geometry*, which only deals with flat surfaces, a geodesic is a straight line. But the earth is not flat and therefore the airplane's path, as determined by the shape of the earth, is curved. In the same way, the path of light (and other electromagnetic radiation) is determined by the geometry of space. Light's path is "bent" by the presence of matter in any given location. The overall curvature of space—the large-scale geometry of the universe—depends upon the overall distribution of matter, that is, the mean density of matter in the universe.

The fact that light curves in the presence of matter has been supported by observations taken during total eclipses of the sun, when the brilliant solar rays are blocked by the moon for a few minutes. Starlight, normally invisible during daylight hours, can then be observed. Very precise measurements taken during total eclipses have shown that light from stars passing very near the surface of the sun is slightly deflected. The presence of matter, in this case the mass of the sun, has caused the path of the light to bend (see Fig. 4.5). Since the path of light is determined by the geometry (or shape) of space, this implies that space itself is curved by the presence of matter.

To better visualize the concept of curved space, consider the earth as a sphere. It has a finite (not infinite) area and yet is unbounded, i.e., its surface has no center or end. It is also known that for large triangles on the surface of the earth, the angles do not add up to 180°, as they do in Euclidean (or plane) geometry. In a similar way, curved space has no center or end but may have a finite volume.

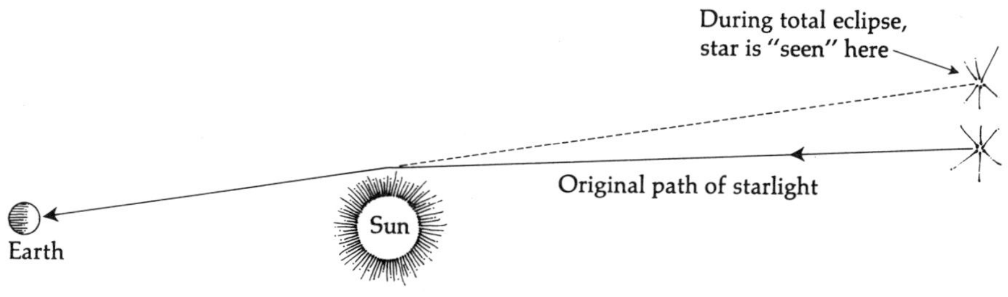

4.5 Observation of starlight during a total solar eclipse gives proof that light is bent by the presence of matter.

We could prove that the earth is round by measuring the sum of the angles in a large triangle on the earth's surface. Similarly, if there were some way in which we could station three astronauts on three remote galaxies, we could have them calculate the sum of the angles of the triangle that they would form. A flat, uncurved Euclidean universe would result in 180°. If the sum were more than 180°, the shape of the universe would be spherical. If the sum were less than 180°, the universe would be saddle-shaped or hyperbolic (see Fig. 4.6). A spherical universe would mean that a space traveler, if he traveled long and far enough, would eventually come around to his original starting point, much like the round-the-world tourist does. This would not be true in the flat or the hyperbolic universe.

Once again we are confronted with very strange concepts. We can appreciate how the populace must have reacted when word reached them that the earth really did move. Their senses told them otherwise. So do ours. They also tell us that the desk we write on is a solid object, and that

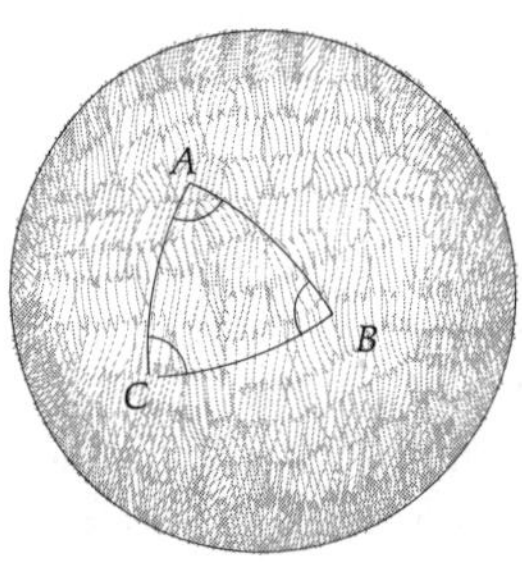

$\measuredangle A + \measuredangle B + \measuredangle C > 180^\circ$
Spherical

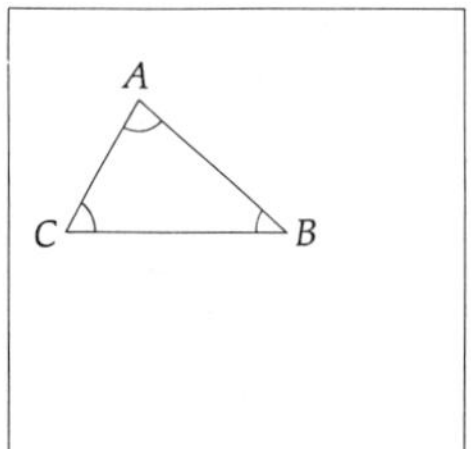

$\measuredangle A + \measuredangle B + \measuredangle C = 180^\circ$
Euclidean (flat)

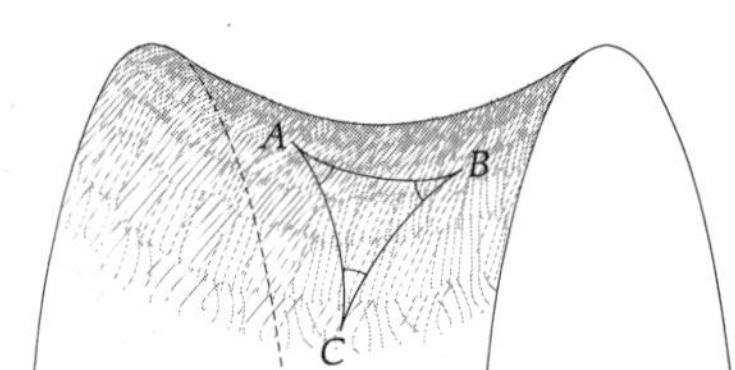

$\measuredangle A + \measuredangle B + \measuredangle C < 180^\circ$
Saddle-shaped (hyperbolic)

4.6 *What is the shape of the universe?*

those twinkling objects we see in the night sky are stationary and everlasting. For everyday existence on this planet such perceptions are adequate and probably necessary. But the discoveries of science have shown us that the information we receive through our senses is not the entire picture. Our ability to accept this puts us above the most intelligent primate and is one of the main characteristics of human beings.

Discussion Questions

1. Compare the current cosmological theories with others which have been considered in the past. What theories were there before scientists knew about the expanding universe?

2. In the famous "twins paradox," one twin travels at nearly the speed of light and then comes back to earth to find himself much younger than the twin who stayed home. One might think (as in the case of the towers) that either twin could say that the other was moving. Why then would one twin age more rapidly? What is the essential difference between this example and the moving towers? (Hint: has the speed of the traveling twin always been uniform, as was assumed in the traveling towers?)

3. What would the geography of space have to be for a space traveler to leave the earth in one direction and come back from the opposite direction?

4. What would it mean if some distant galaxies were found to have a blue shift instead of a red shift? Discuss the consequences of such a discovery.

5. The earth's velocity around the sun is about 18½ miles/sec. What is this in mph? In km/hr?

5 *Points of Light*

On a clear night with an unobstructed view of the horizon, the sky appears as a large inverted bowl, liberally dotted with tiny shining lights. Some of these lights are brighter than others; some even have a faint color. But against the dark sky there is no depth perception. They all appear to be on the inside surface of the bowl. This is the celestial sphere.

Because the earth rotates from west to east, the celestial sphere appears to turn in the opposite direction, from east to west, causing the stars and other celestial objects to appear to rise in the east and set in the west. The axis of the earth around which the earth rotates extends from the north pole to the south pole, directly through the globe. This imaginary line can be extended out to the celestial sphere. The points where it intersects the celestial sphere are the north and south celestial poles (see Fig. 5.1). Because the earth turns on its axis, the celestial sphere appears to turn on this larger axis and all the stars appear to circle these imaginary points in the sky. If a camera is directed toward the north celestial pole, a time exposure will produce a photograph like Figure 5.2.

Using a camera on a tripod, take time exposures between one and two hours long of various parts of the sky. Point the camera at the north celestial pole, the zenith point (directly overhead) and at the celestial equator (90° from the north celestial pole). What differences are there in the star trails thus produced? What would such photographs look like if you were at the equator? At the north pole?*

**The camera must be completely free of any vibrations and away from any extraneous light.*

The earth's equator can similarly be extended outward to the celestial sphere. The place where it intersects the celestial sphere is, not surprisingly, called the *celestial equator.* If we were standing at the north pole of the earth, the celestial equator would be at the horizon (Fig. 5.3*a*). During the night, all the stars would appear to make large circles parallel with the horizon as they revolved around the north celestial pole directly overhead.

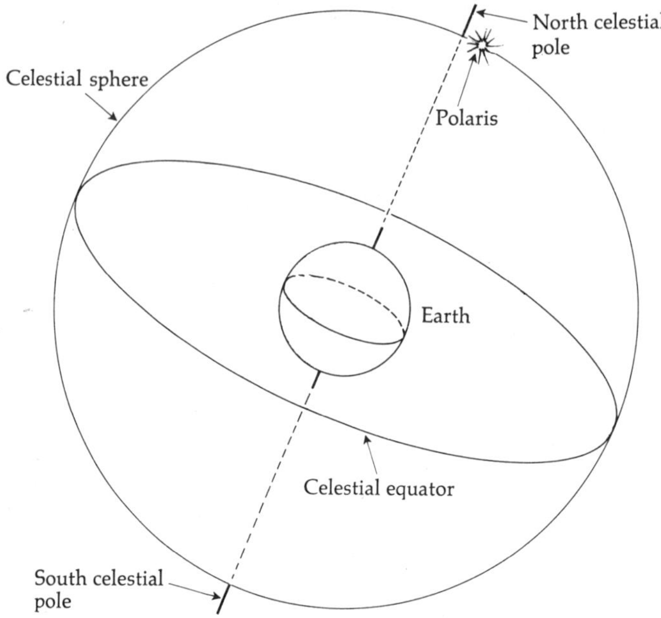

5.1 *The celestial equator and celestial poles.*

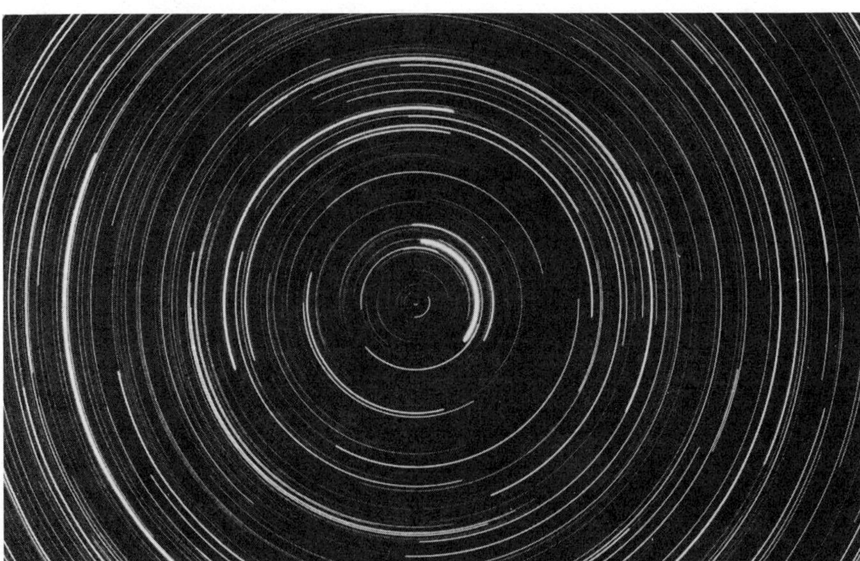

5.2 *A long-exposure photograph of the north celestial pole produces star trails as the earth rotates on its axis (photograph from the Lick Observatory).*

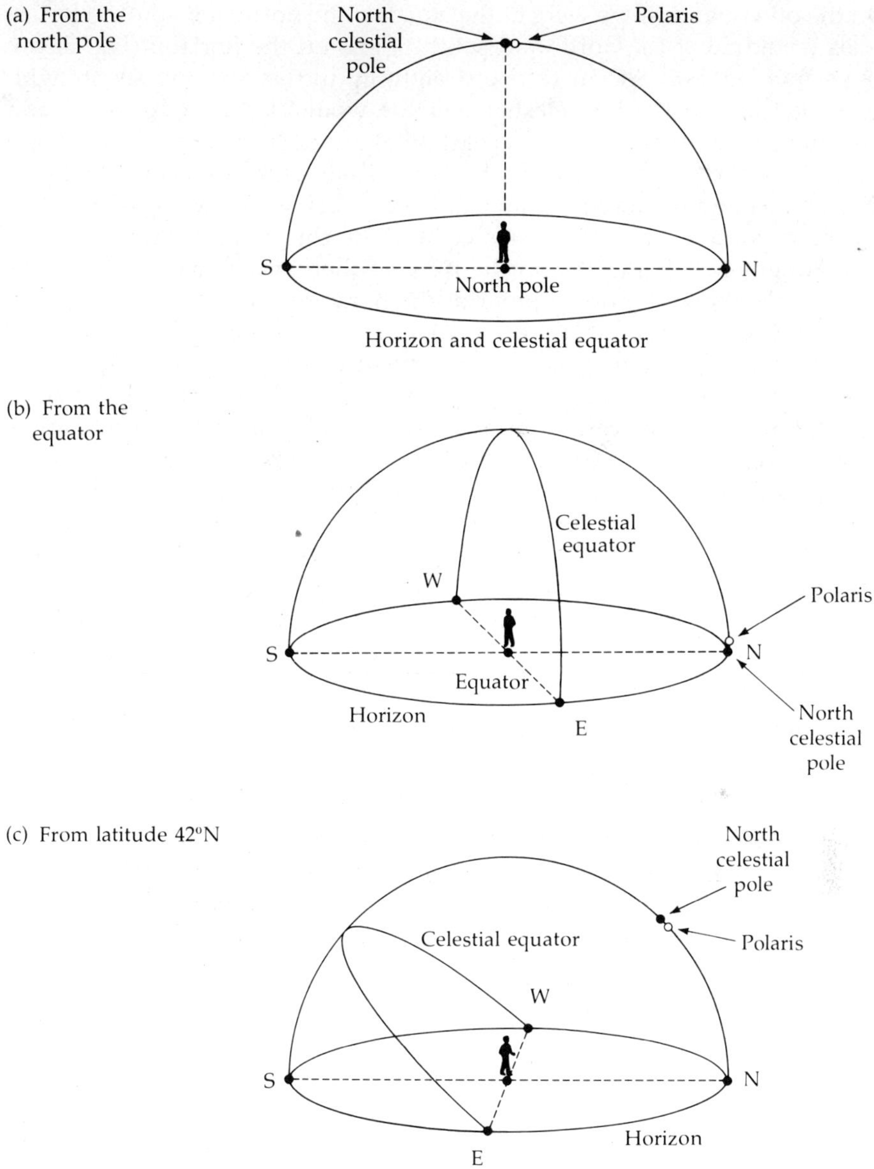

5.3 The relationship of the north celestial pole and the celestial equator from different observing points on earth.
a. From the north pole.
b. From the equator.
c. From Latitude 42° N.

On the other hand, if we were at the equator, the north and south celestial poles would lie at the north and south points on the horizon (Fig. 5.3*b*). The stars would all appear to rise straight up in the east and set straight down in the west. The celestial equator would lie directly overhead, stretching across the sky from the east point on the horizon to the west. At the equator the sun also rises and sets vertically with respect to the horizon, thus spending an equal time above and below the horizon. For this reason, night and day are always equal there throughout the year.

For any location in the northern hemisphere, the north celestial pole lies directly above the north point of the horizon. Its distance in degrees above the horizon is equal to the observer's latitude (Fig. 5.3*c*). Therefore, the nearer our location is to the north pole, the higher in the sky is the north celestial pole. In fact, if we know how many degrees above the horizon this point is, we can determine our latitude on earth. Very conveniently, the star Polaris or North Star lies within one degree of the north celestial pole on the celestial sphere. It is an easy star to find and has therefore been a major navigational aide to explorers, sailors, pilots, and scouts throughout the history of mankind (Fig. 5.4). Unfortunately, there is no such star at or near the south celestial pole.

Using the star charts in Appendix VII, locate several constellations in the night sky. It would be helpful to first locate the Big Dipper and the North Star (see Fig. 5.4). Then follow the curve suggested by the three stars in the handle of the Big Dipper away from the bowl to the next bright star on the celestial sphere. This is Arcturus in the constellation Boötes. If you follow the curve still further, you will find Spica in the constellation Virgo. In a similar manner, locate other stars and constellations. By devising your own methods for locating them, you will remember them more easily.

To any observer in the northern hemisphere, the stars all appear to circle the north celestial pole during the night, just as they do at the north pole, but those that are farther (in degrees) from the pole than the latitude of the observer must dip below the horizon and come up again in their 24-hour apparent circular journey. Because their paths are curved, they rise and set at an angle to the horizon, curving across the sky towards the south. In the southern hemisphere, the same phenomenon takes place around the south celestial pole.

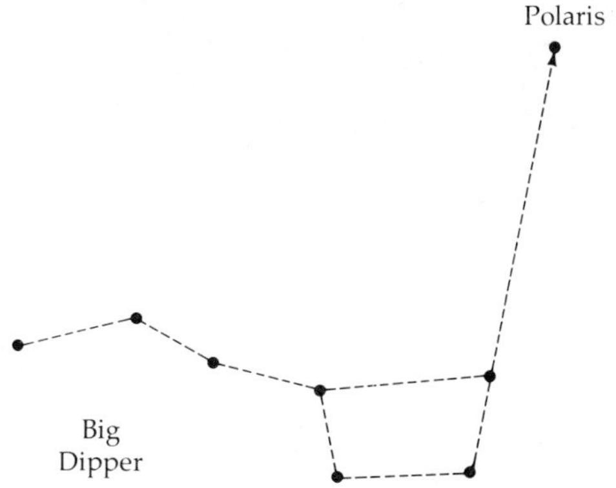

5.4 *How to find the North Star (Polaris).*

In the night sky, point to the North Star. This is very close to the north celestial pole. With your other hand, trace the celestial equator. To do this, your arms should describe a right angle (90°). Identify some of the constellations and stars that lie along the celestial equator. With your arms try to divide the 90° on the celestial sphere into halves, thirds, or sixths. An aspirin tablet held at arm's length will cover about ½° in the sky (the apparent diameter of the moon or the sun). How many degrees on the celestial sphere does your fist span when your arm is held straight out in front? How many "fists" are there from horizon to zenith? How many degrees across is the bowl of the Big Dipper?

Coordinate Systems

Just as with the North Star, any star's distance above the horizon can be measured in degrees. This is called the star's *altitude*. (Do not confuse this with the altitude of an airplane above sea level.) But since there may be many stars at a given altitude on the celestial sphere, another directional position is needed. In astronomy, we start at the north point on the horizon and measure eastward in degrees until we reach the point directly beneath the given star. This is the star's *azimuth* (see Fig. 5.5).

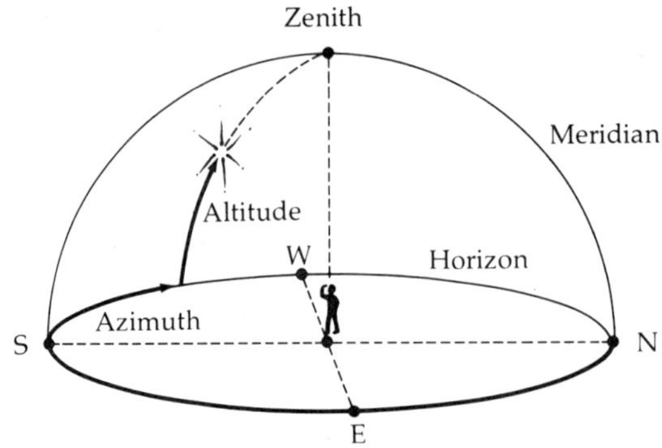

5.5 *The altitude-azimuth system*

At about 9:00 on a clear night, locate the Big Dipper and draw a diagram of its position above the north point on the horizon. Locate three other bright stars and draw their positions on the chart. Try to use one which is above the Big Dipper and one on each side of it (as seen on the celestial sphere). Estimate each star's azimuth and altitude. At 10:00 on the same night repeat your observations. Do the same at 11:00. One month later, repeat all three observations and drawings. In which direction did each star move nightly? Make a sketch of this motion. What were the monthly changes? Can you estimate how many degrees they moved each hour? Each month? What causes the nightly motion? The monthly motion?

In Figure 5.5, two other terms are introduced which are also important in astronomy. As you can see on the diagram, the *zenith* is the point directly over your head on the celestial sphere (see Fig. 5.6). Everyone has a slightly different zenith point and obviously, the farther from you someone is on earth, the farther away his zenith is on the celestial sphere. Your *meridian* is the imaginary line from the north to the south point on the horizon that passes through your zenith. If someone is directly north or south of you, he will have the same meridian. The meridian is especially important to us in timekeeping. The ante-meridian (A.M.) hours refer to those before the sun has passed the meridian; the post-meridian (P.M.) hours are those after this time.

The obvious disadvantage of the altitude-azimuth system of coordinates, which is based upon the observer's zenith and meridian, is that as

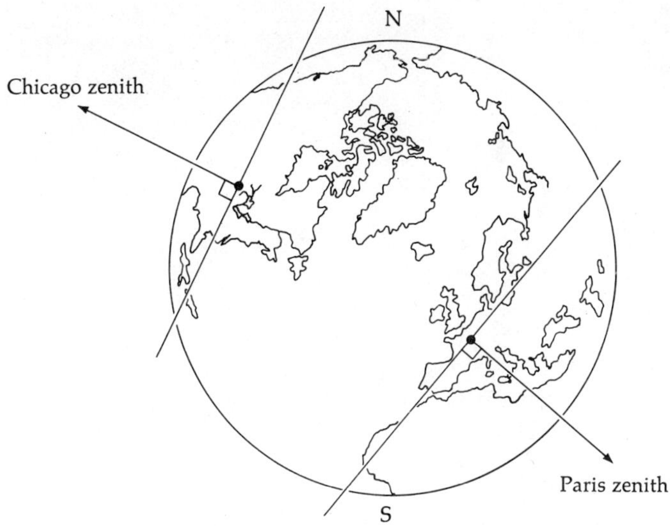

5.6 The zenith point (directly overhead) is unique to the observer. The meridian is shared only by those on exactly the same north-south line.

Record the time that several stars cross your meridian on a given night. Time the same stars two weeks and one month later. What changes do you observe? How could such observations be used in timekeeping?

---

the earth rotates, the coordinates of any object change continuously. Also, the coordinates you measure for a particular celestial object at a specific time will not be the same as those measured by someone else some distance away. Your horizons will not be the same and therefore any measurements from the horizon will not be the same either (see Fig. 5.7).

A separate coordinate system using reference points on the celestial sphere rather than on the earth is much more useful. The most logical such points would be the north and south celestial poles and the celestial equator. These can be used much like we use their terrestrial counterparts in latitude and longitude.

The position of a star or other object on the celestial sphere can be measured in degrees above (north) or below (south) the celestial equator. This is called the star's *declination* and corresponds to latitude on the sur-

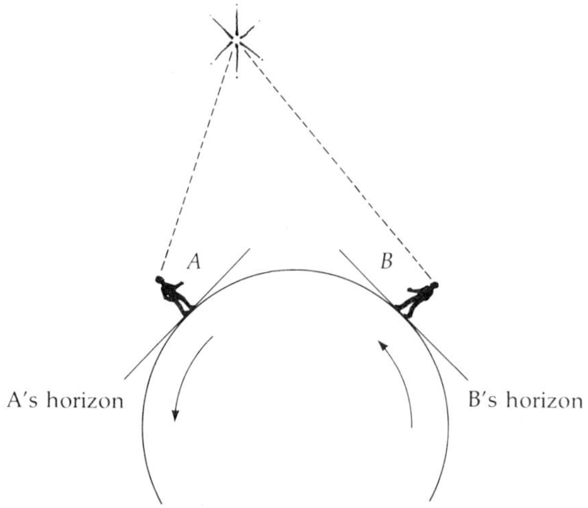

5.7 Observer A's measurement of the star's altitude will be greater than B's at this particular moment. A sees the star already past the meridian but still far above the western horizon. B sees the star just rising in the east.

face of the earth. We use a positive declination for stars north of the celestial equator and a negative declination for those below it (Fig. 5.8).

For the other directional measurement—corresponding to longitude—we must have a beginning point from which to measure, much like Greenwich, England. We use a point on the celestial sphere called the *vernal equinox*. This is the location of the sun on the celestial equator at the beginning of spring in the northern hemisphere.[1] The angular distance along the celestial equator measured eastward from the vernal equinox is called the *right ascension*. (This term can be traced to ancient astrology.) To find a star's right ascension, one draws a line from the north celestial pole through the star down to the celestial equator. The angular distance on the

---

Using a protractor or other such measuring device, try to measure the declination of several bright stars. Knowledge of the location of the celestial equator is essential for this. Check your answers with the list of brightest stars (Appendix IV).

---

[1]The vernal equinox is more specifically defined on page 84.

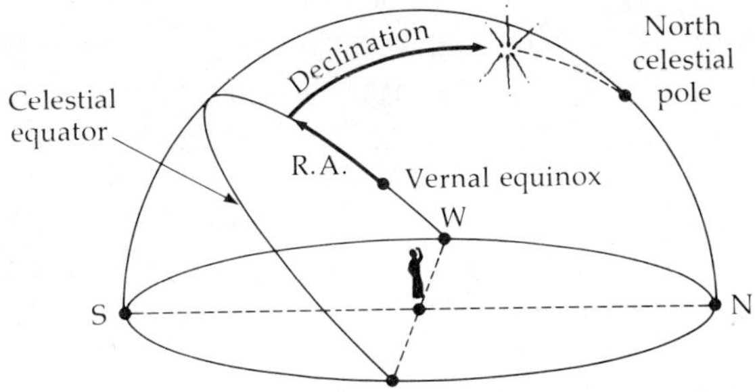

5.8 *The right ascension-declination system of coordinates.*

celestial equator from the vernal equinox to that intersection point is the star's right ascension (Fig.5.8).

Right ascension is measured not in degrees, but in hours, minutes, and seconds like a clock. Thus the sky is divided into 24 hours measured along the celestial equator. Since the sky rotates in approximately 24 hours, every hour each celestial object moves 1/24th of the way around the sky—or 15° of arc (Fig. 5.9). Thus, Arcturus in the constellation Boötes has a right ascension of 14 hours, 13.4 minutes ($14^h\ 13^m.4$) and a declination of +19 degrees 27 minutes (+19° 27′) (1950 coordinates).

Learn the right ascension of several bright stars that you can easily recognize in the night sky. This is easier than using the vernal equinox as a reference since it is not marked by a specific star and is not always above the horizon in the night sky. Try estimating the right ascension of other stars. Check your answers with the list of brightest stars (Appendix IV).

5.9 *Units of time and of arc.*

24 hours (24^h)	= 360 degrees (360°)
1 hour (1^h)	= 15 degrees (15°)
4 minutes (4^m)	= 1 degree (1°)
1 minute (1^m)	= 15 minutes (of arc) (15′)
4 seconds (4^s)	= 1 minute (of arc) (1′)
1 second (1^s)	= 15 seconds (of arc) (15″)

The Constellations

Let us return to the observation of the night sky. With a little imagination, it is easy to start grouping the stars into shapes, forms, and pictures. This is what practically every early civilization has done. Over the generations these configurations became more elaborate and were associated with folk heroes, gods, and wild animals. Many of these *constellations* are still used today. Their names are the Latin translations of Greek names, representing characters from Greek mythology. It is perhaps difficult to see any resemblance between the pictures of the mythological figure and the star grouping, which leads to the suggestion that the constellations were not supposed to depict characters accurately. The Greeks may have merely named areas in the sky to honor mythological characters, many of whom were their gods and goddesses. If so, the seemingly futile attempt to fit the pictures to the stellar arrangements came much later (see Fig. 5.10).

It is also possible that the naming of sections of the sky by the early Mediterranean sailors had some navigational purpose. The peoples of the

5.10 The constellations Cygnus, Orion, and Scorpius as depicted by the German astronomer Johann Bayer in his Uranometria, *published in 1603.*

ORION

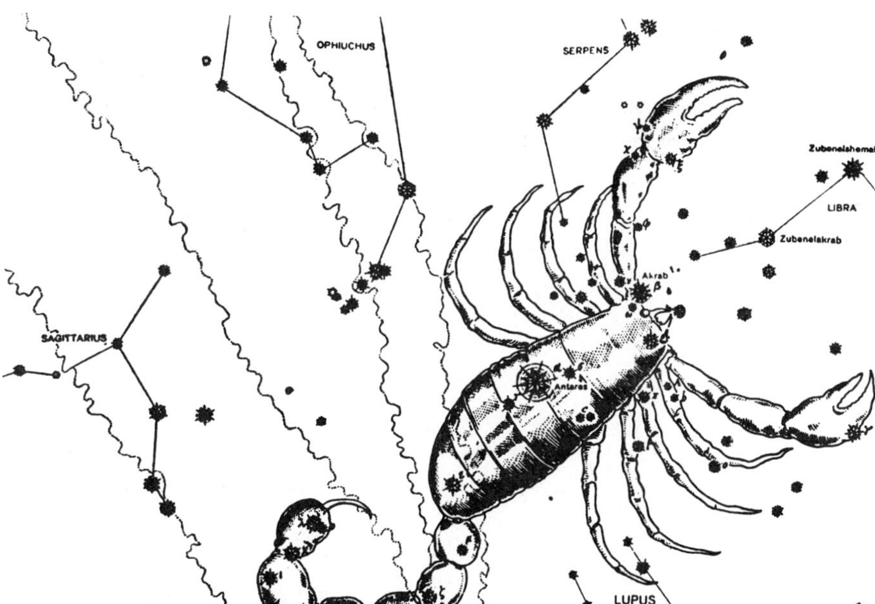

SCORPIUS

Pacific islands used the stars and specific star groups to guide them over the wide expanses of water where no landmarks were visible. Runaway slaves in pre-Civil War days in this country were told to "follow the Drinking Gourd" to the north and freedom. This was the constellation The Big Dipper, two of whose stars point directly to the North Star (see Fig. 5.4).

Become familiar with the mythological stories of several of the constellations. Locate these constellations in the night sky. Can you distinguish the character that the group of stars is supposed to represent?

We saw how the earth's rotation caused most of the stars to rise and set each night. The earth's revolution around the sun each year causes other apparent changes on the celestial sphere. Because the earth is on different sides of the sun in, say, winter, than it is in summer, its night sky view is different. Of course, the daytime sky changes also but the brilliance of the sun keeps us from seeing anything but its light (see Fig. 5.11). Thus in the winter, the constellation Orion is most conspicuous, whereas in summer it is easy to pick out the three bright stars of the "summer triangle"—Vega in the constellation Lyra, Deneb in Cygnus, and Altair in Aquila. Similarly, the spring and fall night skies have their own charac-

5.11 Throughout the year, different constellations are visible at night as the earth revolves around the sun.

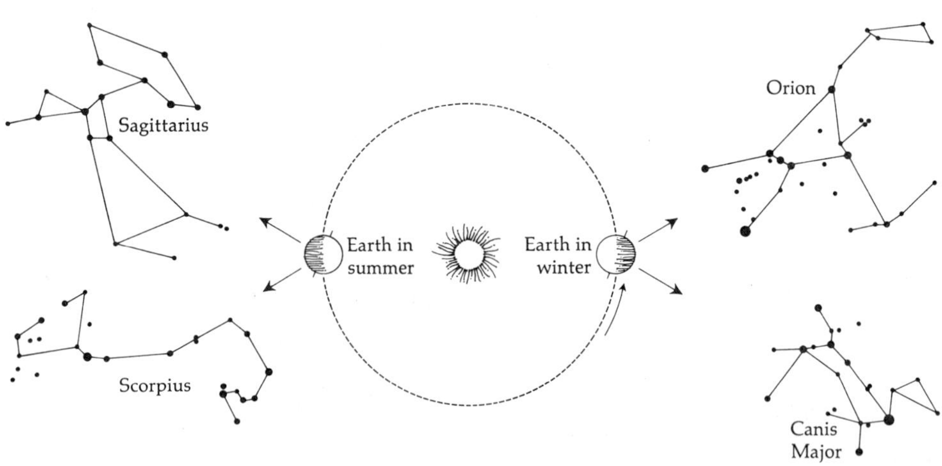

teristic constellations and stars. In Appendix VII, the star charts show the major constellations according to the seasons of the year. Note that although a constellation such as Orion is associated mainly with winter skies, it is visible at some time during the night in the early spring and late fall. During the rest of the year it lies in the same direction as the sun and is therefore not visible.

As is shown in Figure 5.12, the time of the stellar risings and settings depends upon the position of the earth on specific dates. A star that is on the meridian at midnight in the summertime will set at midnight three months later in autumn. It will not be seen in the winter and then will be seen rising at midnight in the spring.

Early civilizations were well aware of the changing sky throughout the year, and in fact used these changes to establish calendars and to predict annual events. For example, the early Egyptians knew that when the bright star Sirius rose and was visible just before dawn, the Nile was ready to overflow its banks. This was a very important event in their agricultural communities.

5.12 The shift in the time of stellar risings and settings is due to the earth's revolution around the sun.

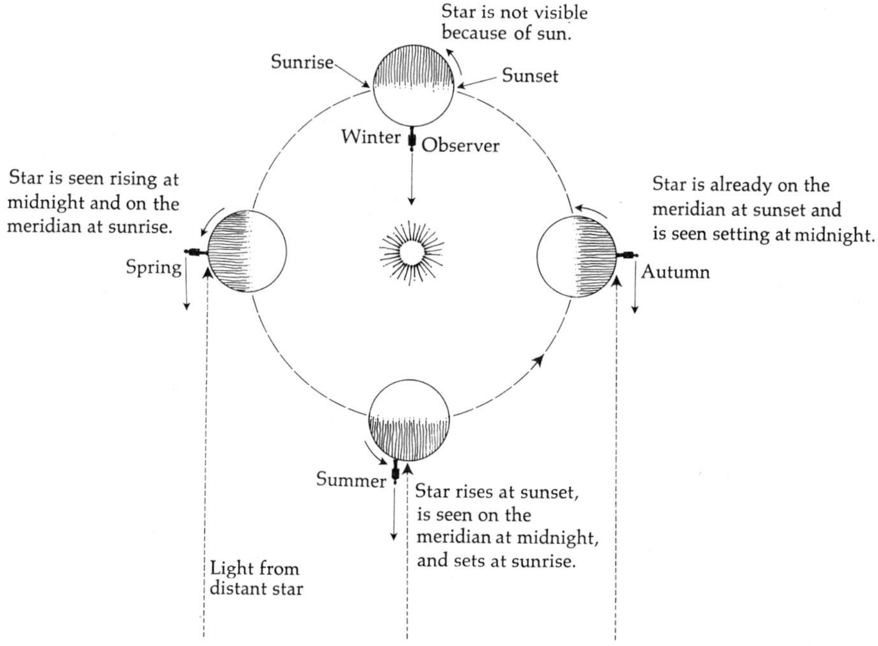

The Path of the Planets

All of these constellations were made up of *fixed stars,* stars that did not change their positions relative to the other stars on the celestial sphere. The ancient peoples noticed, however, that there were other "stars" which were not fixed but instead moved on definite paths through specific constellations. Each night their positions relative to the fixed stars slowly changed, some moving a noticeable distance on the celestial sphere and others barely creeping along. These were the "*planeta*" or "wandering stars." The moon and sun were part of this group; they were believed to be closer than the rest of the *planeta* and therefore appeared larger. The others were Mercury, Venus, Mars, Jupiter, and Saturn, each deriving its name from a god or goddess whose characteristics it was thought to possess.

The paths these "wandering stars" seemed to follow on the celestial sphere took them through twelve of the constellations. Because of this, these twelve constellations became much more important to man than any others. These were the constellations of the *Zodiac* (Fig. 5.13). Today we know that the reason the planets appear to move through the zodiac constellations is that their orbits all lie on about the same plane in space.

Locate several constellations of the zodiac. If possible, locate the vernal equinox. Can you trace the ecliptic and the celestial equator on the celestial sphere?

Early mankind did not understand why the *planeta* followed these particular paths in the sky, but since they considered themselves the central figure in the cosmos, it was naturally assumed that motions on the celestial sphere had some significance to the lives of men and nations. The heavens were watched closely and graphs and charts were drawn up so that predictions of where each "wandering star" would be could be made. Although the sun blots out the stars in the daytime, these early astronomer-astrologists were even able to tell in which constellation it lay at any time. For example, in those days, the sun was in the constellation of Aries between March 21st and April 21st and in Leo between July 21st and August 21st. The sun no longer is in these locations on those dates, but astrology, which has persisted into modern times, still uses these ancient positions. There is no scientific evidence to support astrological beliefs.

Place a lamp in the center of a darkened room. Walk around the light, keeping your face turned toward the walls. You are viewing the changing "night sky." Each turn around the lamp represents one revolution of the earth, i.e., one

5.13 *The Zodiac is the twelve constellations that lie in the same plane as the earth's orbit around the sun. As viewed from the earth, the sun and all of the planets appear to move through this band of constellations. In the above diagram, the sun is "seen" in Leo, whereas Mars is "seen" in Sagittarius. Six months later, when the earth is on the other side of its orbit, the sun will be "seen" in Aquarius.*

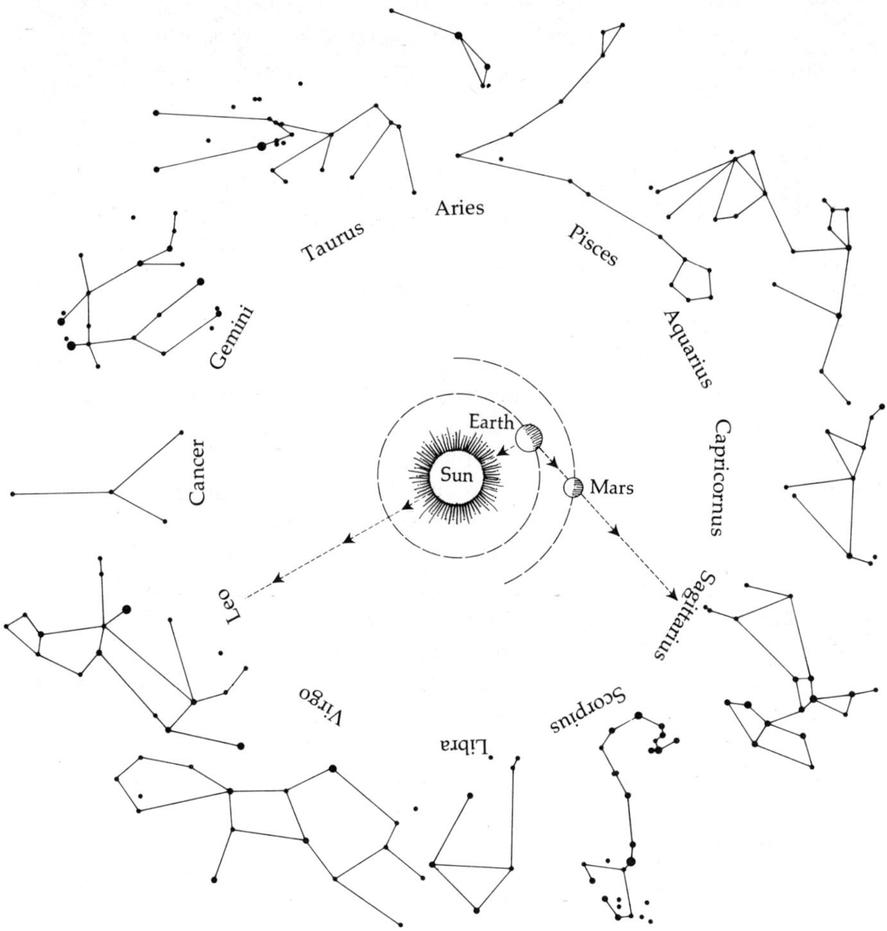

year. Certain objects will appear as you walk around, much as different constellations are seen at different times of the year. At the level of your eye will be a specific group of objects—a pane of a window, a part of a poster, one week of a calendar, a shelf in a bookcase. These are all on the same plane or level in your room—eye-level or the "plane" of your eye. If you had hung various size balls at this eye level, they would all appear in front of the specific objects on the walls. Similarly, as the earth revolves about the sun, different constellations are seen throughout the year, and because the planets all have orbits on about the same plane, they appear to move through the same constellations—the zodiac.

The Path of the Sun

Of all these apparent motions, the sun's is the most important. Actually it is the earth orbiting around the sun that caues the sun to appear to move along the celestial sphere. If we could observe the Earth from Venus, we would see it move through all the twelve Zodiac constellations just like the other planets. But because we live on the earth, the specific path that the earth takes has greater importance and use than the paths of the other planets. The revolution of the earth around the sun causes the sun to move in an apparent path along the celestial sphere, called the *ecliptic* (Fig. 5.14).

In Figure 5.15 the relation between the ecliptic and the celestial equator is shown. Both are imaginary circles on the celestial sphere but the ecliptic is tilted 23½° to the celestial equator. This is why classroom globes are always positioned on pedestals with 23½° of tilt. (Remember that by definition the celestial equator and the earth's equator are on the same plane.)

Because these two circles are at an angle to each other, they must meet at two points, as shown in Figure 5.15. These points are called the *vernal equinox* and the *autumnal equinox*. We have already seen how the vernal equinox is used in determining the right ascension of a celestial object. It is the point where the sun, apparently moving along the ecliptic, crosses the celestial equator from the south to the north. When this occurs, spring begins in the northern hemisphere. Our calendar is adjusted by using leap years so that March 21st continues to coincide with the sun's passage across the vernal equinox. Similarly, on or about September 21st, the sun crosses the celestial equator from the north to the south and spring begins in the southern hemisphere and autumn in the northern hemisphere.

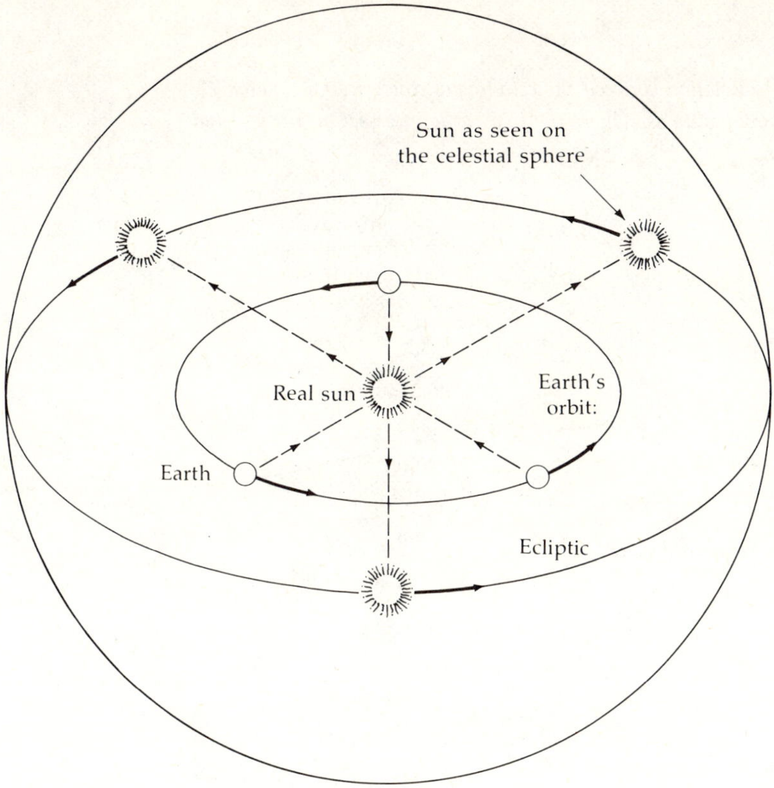

5.14 As the earth revolves around the sun, the sun appears by projection to move along a great circle on the celestial sphere. This circle is called the "ecliptic." It is both the apparent yearly path of the sun and the intersection of the plane of the earth's orbit with the celestial sphere.

The midpoints between these two intersections are called the *winter* and *summer solstices.* At the solstices, the sun is at the farthest points possible from the celestial equator. On or about June 21st, when summer begins in the northern hemisphere, it is as far north in declination as it can get—namely 23½° (Fig. 5.16). At that time we see it high in the sky at noon (about 71° above the horizon if our latitude is about 42°). On or about December 21st, the sun is at its lowest point in the sky at noon, at −23½° declination (or about 25° above the horizon). Is it any wonder that early civilizations, upon seeing the sun appear lower and lower in the sky each approaching winter, had many rituals and ceremonies at the time of the winter solstice to persuade the gods to "bring the sun back up again." Nearly all cultures in the northern latitudes had ceremonies in December that involved the lighting of fires in the symbolic attempt to entice back the sun, the ultimate source of light.

5.15 Relation between the ecliptic and the celestial equator. Equinox points are where the two paths intersect on the celestial sphere.

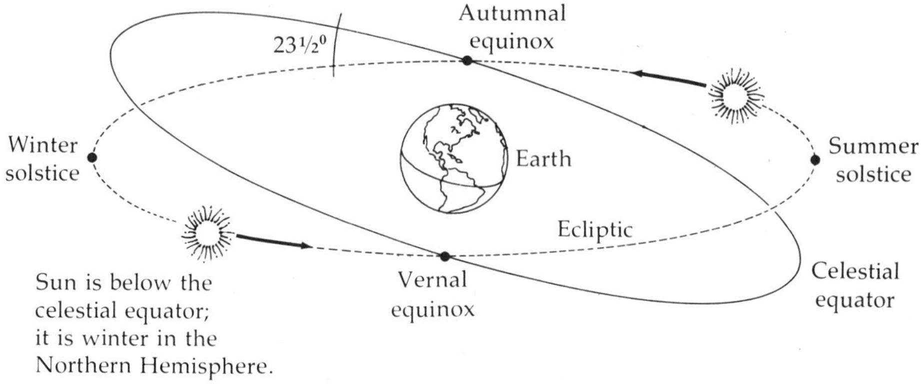

5.16 The sun's position in the sky at noon depends upon the season.

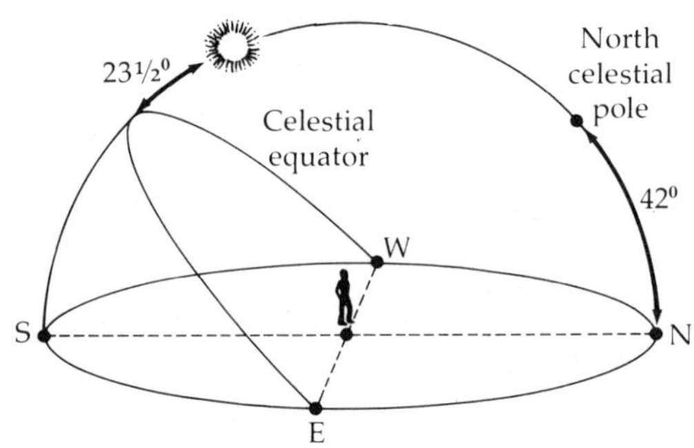

a. Sun's highest position at summer solstice.

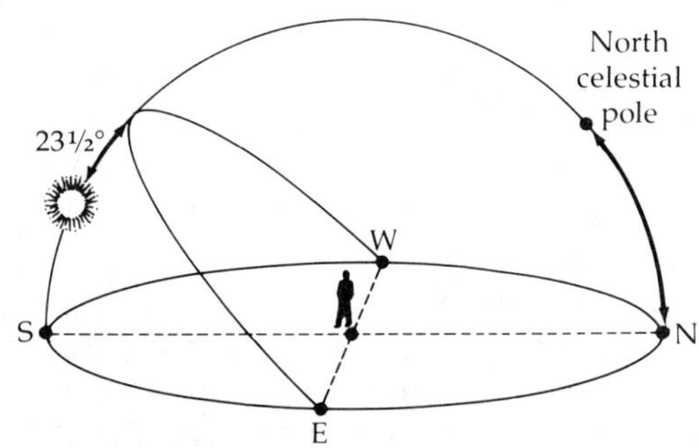

b. Sun's highest position at winter solstice.

The Precession of the Equinoxes

The positions of the equinoxes and solstices on the celestial sphere are not stationary. The sun and the moon exert gravitational forces on the earth, tugging on the earth's axis to bring the plane of the earth's equator in to the same plane as the ecliptic. The net effect is to cause the earth to act like a gyroscope (Fig. 5.17). Very slowly the earth's axis describes a large circle, causing the celestial poles to make similar circles on the celestial sphere. Each revolution takes 26,000 years. This motion, which is called *precession,* slowly changes the orientation of the earth's axis with respect to the celestial sphere. The circles that the north and south celestial poles make on the celestial sphere have radii of 23½°. Figure 5.18 shows this path on the celestial sphere for the northern sky and indicates the points in the sky that will be the north celestial pole in the future.

As the poles shift on the celestial sphere, the celestial equator, which is always 90° from the poles, also changes its position. Consequently, its intersections with the ecliptic, the equinoxes, change their positions as well. They move slowly westward on the ecliptic. This is called the *precession of the equinoxes*. Since right ascension and declination measurements depend upon the position of the celestial equator and the vernal equinox,

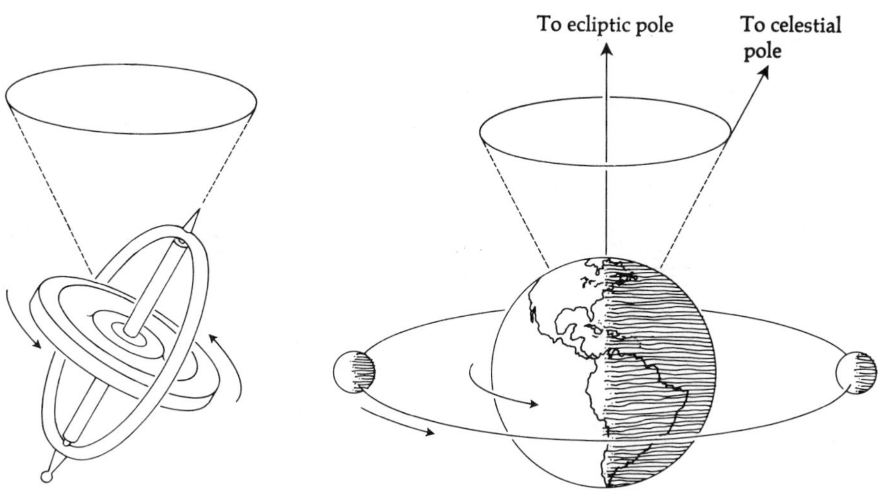

5.17 *One of the earth's motions is similar to that of a gyroscope. The pull of the moon and sun to bring the plane of the earth's equator into the same plane as the ecliptic combined with the earth's rotation produces the conical precession of the earth's axis.*

5.18 Path of the north celestial pole due to the precession of the equinoxes.

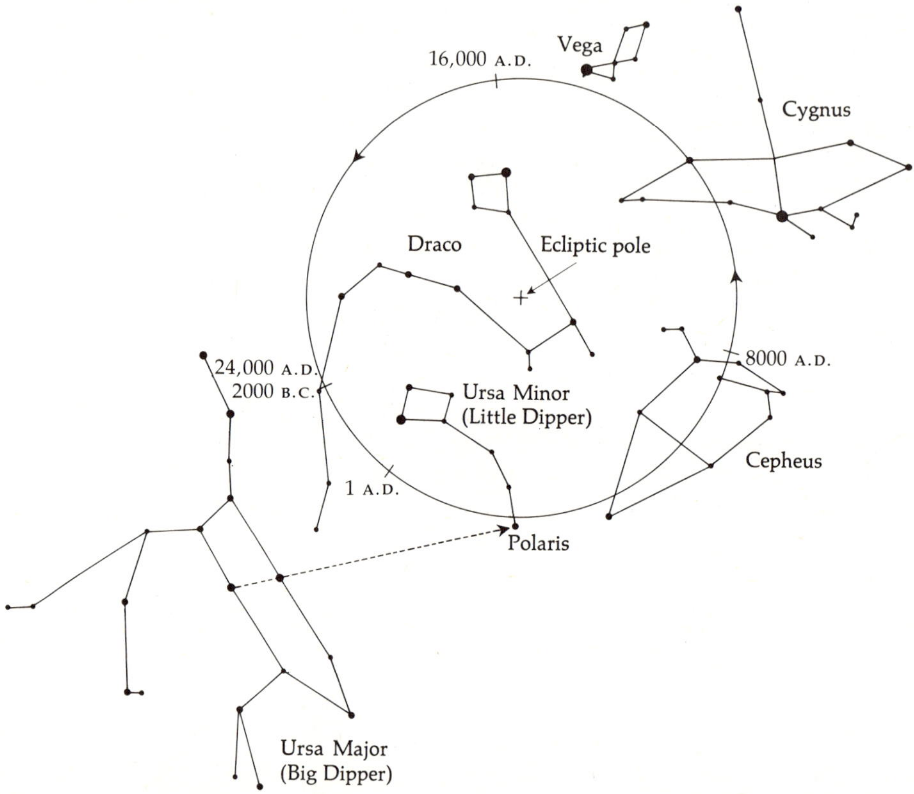

the measured positions of the stars slowly change over the years. Large catalogues are published each year giving the precise right ascension and declination of the stars for that year. However, although they are important in astronomical research, precession is not very noticeable on a day-to-day or even year-to-year basis. The North Star will continue to be near the north celestial pole for centuries to come.

However, because the vernal equinox also changes its position on the celestial sphere over time, it is not where it was 2000 years ago. When the astrology that is practiced today was formulated, the vernal equinox was in the constellation of Aries. This is still considered the first constellation by astrologists. But precession has slowly moved the vernal equinox through Pisces and it is now approaching the constellation Aquarius—hence, the term "the age of Aquarius." The vernal equinox will slowly move through Aquarius during the next 2000 years. If your birthday is between March

21st and April 21st, you were born when the sun was between the constellations of Pisces and Aquarius, but according to modern and ancient astrology, you are considered an Aries.

The Names of the Stars

Early civilizations, besides naming groups of stars or constellations, also identified individual bright stars with names that were meaningful to their cultures. Many of the names we use today are of Arabic origin (Aldebaran, Algol, Deneb) rather than Greek or Latin. The Arab astronomers bridged the gap between the ancient civilizations of Greece and Rome and the Renaissance. They were the last important star namers and therefore their identifications have lasted into modern times.

However, with the beginning of the Renaissance, more scientific methods of identifying the stars were developed. In 1603, the German astronomer, J. Bayer, besides depicting the constellations and their associated legendary figures in his handsomely illustrated *Uranometria*, also established a new system of stellar identification. He assigned Greek letters to the major or brighter stars. In general—but not without exception—he assigned *alpha* (the first letter in the Greek alphabet) to the brightest star in each constellation, *beta* to the next brightest, and so forth. Then he added the genitive or possessive form of the constellation's name to this designation. Thus, Sirius became Alpha Canis Majoris (the brightest star in the constellation Canis Major), and Algol was listed as Beta Persei (the second brightest star in the constellation Perseus).

Later, in the eighteenth century, John Flamsteed, the first Astronomer Royal of England, developed another method of identifying stars. He disregarded brightness entirely and instead assigned a number to each star in a constellation according to the order in which it crossed the meridian each night. Thus Arcturus, which was Alpha Boötes, also became "16 Boötes." Although neither Bayer's nor Flamsteed's methods of classification included all the stars that are visible to the naked eye, their designations are still much in use today.

All these bright stars as well as many others are now included in the *Yale Catalogue of Bright Stars*, which uses still another method to identify them. They are numbered without regard to their constellation in order of their passing the meridian, and are referred to by their *HR number*.[2] Thus, Arcturus is also HR 5340.

---

[2]HR refers to the *Harvard Revised Photometry*. The numbers are identical with those in the Yale catalogue.

With the help of the early telescopes, many more stars were ultimately catalogued. The two basic catalogues still very much in use are the *Bonner Durchmusterung* and the *Henry Draper Catalogue*. Although the method of listing each star differs, both catalogues ignore the constellation designations and rely solely on numerical classification. One speaks of a star's BD or HD number. The star Arcturus is therefore also referred to as BD +19 2777 or HD 124897.

Extensive as these catalogues may be, however, only about four-thousandths of one percent of the stars in our galaxy are catalogued. The very great majority of stars visible through modern telescopes remain un-

5.19 A small section of the Milky Way in the constellation Sagittarius. The dark sections are caused by interstellar dust obscuring our view of the stars beyond (Lick Observatory photograph).

named and unnumbered. Comprehensive atlases of the sky have been made, with detailed photographs of every section, but the stars they show are far too numerous and too densely packed to be individually listed (see Fig. 5.19). These stars are identified from the photographs, and their positions are determined in relation to other stars.

Today, not only stars are catalogued but also nebulae, galaxies, radio, x-ray sources, and so forth. Each group of objects, as soon as enough of them are discovered, gets catalogued and identified by its own system of numbers and letters. In each case, their positions are given in the right ascension-declination coordinate system which uses an imaginary hollow sphere as its point of reference. And on this celestial sphere, these are points of light or other electromagnetic radiation that are being studied.

Discussion Questions

1. What uses have constellations had in astronomy in the past? How are they used today?

2. Some students confuse constellations and galaxies because they are both described at times as "groups of stars." Define these two "groups of stars" in terms of size, number of stars, where they are seen, where they really are, etc.

3. The sun's position in the sky relative to the stars changes a little each day, so that it appears to move through all the zodiacal constellations each year. This is not the same apparent motion that we are aware of each day as the sun rises in the east, crosses the sky, and sets in the west. What causes each of these two apparent motions?

4. The star Alpha Orionis (Betelgeuse) and the star Beta Orionis (Rigel) are listed respectively in the Bright Star Catalogue as #2061 and #1713. Which star is the brighter one? Which star will cross the meridian first each night?

5. A star is overhead at midnight in the spring months. When will it be overhead in summer? In autumn? In winter?

6. Where would an observer on the earth have to be in order to be able to see all the stars on the celestial sphere sometime during the year? Where would he see only half the celestial sphere during a 24-hour period?

7. Figure 5.2 was made by taking a long exposure of the sky with the camera pointed at the north celestial pole. What would such a photograph look like if the camera were pointed at the celestial equator?

8. Why are the apparent nightly swing of the constellations around the north celestial pole or the rising and setting of the sun not proofs that the earth rotates on its axis?

9. The altitude of the North Star (Polaris) approximately corresponds to the latitude of the observer. If Polaris is at the zenith, what is the latitude of the observer? Where would he or she be on the earth? If Polaris is not visible in the night sky, where would the observer be located on the earth?

10. If you were on the equator, where would you see Polaris? On what date(s) would the sun be directly overhead? On June 21st, would the sun be in the northern sky, the southern sky, or directly overhead?

6 *Stellar Distances and Motions*

We know today that the stars are not really luminous markers on a crystalline sphere. Even the young child who may cling to this illusion quickly learns its falsity and comes to understand the reality of the infinite depths of space. We must therefore look at the stars not as points of light but as physical bodies.

The Brightness of the Stars

When you look at the night sky, it is obvious that some stars appear brighter than others. And as we have seen, the very brightest were named and watched closely by many ancient peoples. In the second century B.C., the Greek astronomer Hipparchus noted that the stars could be grouped into six convenient classes of brightness. He designated the very brightest ones as stars of the first magnitude. Those that were a detectable step dimmer were termed stars of the second magnitude, and so on. He classified the faintest stars visible to the naked eye as sixth magnitude.

In 140 A.D., the Greek astronomer Ptolemy incorporated this classification into his *Almagest,* a thirteen-book work on astronomy which became the authoritative source throughout the Middle Ages. Undoubtedly because of this, Hipparchus' system of stellar magnitudes is still used today, although it is now based on precise measurements rather than simple visual estimates. It was these brightness measurements that were used to measure the Cepheid variations which eventually led to the determination of the distances to the distant galaxies (see Chap. 2).

On a clear, moonless night, study the stars carefully. Can you divide them into classes of brightness? You probably won't be able to see stars of the 5th or 6th magnitude unless your location is far from the city lights and atmospheric pollution. Use the star charts to identify the stars and constellations. Then check your results with the list of bright stars in Appendix IV.

Modern astronomers using photometers (which are essentially very accurate exposure meters) discovered that the light ratio between these magnitude steps was approximately the same, and that each was equal to about 2.5 times more radiant energy than the next lower step.[1] That means that a first-magnitude star radiates 2½ times more energy in the visible region of the spectrum than a second-magnitude star and is 2½ × 2½ or $(2\frac{1}{2})^2$ times more luminous than a third-magnitude star. It is therefore $(2\frac{1}{2})^5$ or nearly 100 times more luminous than a sixth-magnitude star. Since this was the case, astronomers defined the magnitude system so that a difference of five magnitudes was exactly equal to a ratio of 100. In other words, although a first-magnitude star appears to the eye as merely five steps removed from a sixth-magnitude star, it actually sends us 100 times as much energy. A single magnitude step is then the 5th root of 100 or 2.512. This means that a fourth-magnitude star radiates 2.512 times more energy than a fifth-magnitude star.

When more precise measurements were made, it was found that some of the stars were actually brighter than those of the first magnitude. Rather than reclassify the entire system, these brighter stars were given stellar magnitude designations of 0.0 and −1.0, thus extending the scale in that direction. The planet Venus can reach a −4.4 magnitude at its brightest, which is more than 100 times brighter than a first-magnitude

6.1 *Stellar magnitude scale.*

Magnitude difference	*Brightness ratio*	*Magnitude difference*	*Brightness ratio*
0.00	1 : 1	7.50	1000 : 1
0.50	1.6 : 1	10.00	10^4 : 1
1.00	2.5 : 1	12.50	10^5 : 1
1.50	4.0 : 1	15.00	10^6 : 1
2.00	6.3 : 1	17.50	10^7 : 1
2.50	10.0 : 1	20.00	10^8 : 1
3.00	16.0 : 1	22.50	10^9 : 1
3.50	25.0 : 1	25.00	10^{10} : 1
4.00	40.0 : 1	27.50	10^{11} : 1
4.50	63.0 : 1	30.00	10^{12} : 1
5.00	100.0 : 1		

[1]This brightness ratio, which the eye detects as basic steps or divisions of brightness, is due to what is known in physiology as Fechner's law. If the stimulus we receive increases in geometric progression, (2), $(2)^2$, $(2)^3$, $(2)^4$, then the sensation we perceive increases in arithmetic progression, (1,2,3,4).

star. The full moon shines with a magnitude of −12.5, while the sun is −26.7.

Although this classification was originally applied only to stars visible to the naked eye, it has been extended to include the fainter stars visible only through a telescope. However, the same ratio of light intensity has been kept throughout the scale, i.e., just as a first-magnitude star is 100 times brighter than a sixth-magnitude star, so a ninth-magnitude star is 100 times brighter than a 14th-magnitude star.

Note that a difference of 10 magnitudes indicates 100 × 100 or 10,000 times difference in luminosity or energy output (see Fig. 6.1). The 200-inch telescope at Mount Palomar, the biggest optical telescope in the United States, can detect stars of the 23rd magnitude. These stars appear about two billion times fainter than the first-magnitude stars. Telescopes on earth can see nothing fainter than that because of our own atmosphere. The night sky is not black but has a residual glow to it which drowns out fainter stars. A telescope on the moon or in space could detect star considerably fainter than those detectable from earth.

The magnitudes referred to above are measures of how bright the stars *appear* to us and are therefore called *apparent magnitudes*. When Henrietta Leavitt at Harvard was first measuring the brightness of the Cepheid stars in the Magellanic clouds, she could only use apparent magnitudes since she did not know their distance from us. Since the stars in our own galaxy are not all at the same distances from us, their apparent magnitudes tell us nothing of their true luminosities. Only by knowing its distance can we determine if a bright star is really bright or just appears that way because it is very close.

For example, if our sun were about 32½ light years away from us (instead of 8⅓ light minutes), it would only appear as bright as about a fifth-magnitude star. The distance of 32½ light years (or more accurately 32.6 light years) was used in this example because astronomers use this distance to calibrate the actual light intensity of the stars (see p. 101). The intrinsic brightness of a star or other celestial object is called its *absolute magnitude*. This is defined as the apparent magnitude a star would have if it

---

Locate the stars Arcturus and Spica in the night sky. They are visible in the spring and summer. (See the suggestion for finding these two stars on page 00.) Arcturus is 36 light years away and has an apparent magnitude of −0.06. Spica is 261 light years away and has an apparent magnitude of +0.9. What would you conclude about how these two stars compare in real (absolute) brightness?

---

were 32.6 light years from us. In this way intrinsic stellar brightnesses can be compared. The same scale of 5 magnitudes equaling a luminosity 100 times as great is used. The sun's absolute magnitude is +4.8, a relatively faint star. Figure 6.2 shows a comparison of absolute and apparent magnitudes for several celestial objects.[2]

6.2 A comparison of the apparent and absolute magnitudes of several celestial objects.

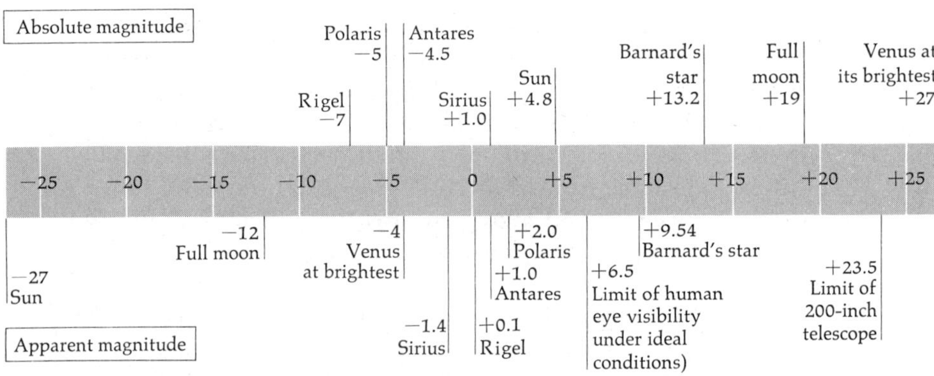

The Distances of the Stars

In order to determine the absolute magnitude of a star or other celestial object, its distance must be determined. We saw in Chapter 2 how the Cepheids were used as distance determiners to the galaxies. But before this could be done, the distances to the Cepheids in our own galaxy had to be

---

[2]The sun is about 1/60,000 of a light year away. The standard distance of 32.6 light years is almost 2 million (2×10^6) times more distant. According to the inverse square law, the sun, if put that far away, would appear $(2 \times 10^6)^2$ or (4×10^{12}) times dimmer. Using Figure 6.2, we find that a brightness ratio of 4×10^{12} is equal to a magnitude difference of 31.5. The sun's apparent magnitude is −26.7. The difference between the absolute and apparent magnitude equals 31.5. Of course, this is the difference in magnitude that the sun would have if it were 32.6 light years away rather than in its real position.

$$M - (-26.7) = 31.5$$
$$M = 31.5 - 26.7 = +4.8 \quad (M = \text{absolute magnitude})$$

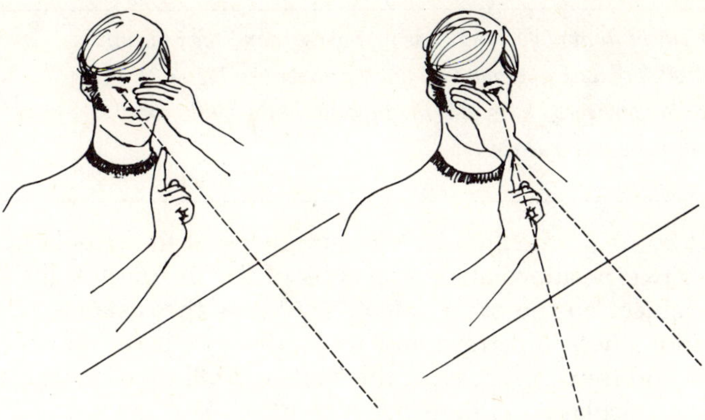

6.3 A simple example of the parallactic shift: Hold your finger in front of you and close your left eye. Note where the finger appears against the background wall. Now open the left eye and close the right one and again sight the finger against the background. The apparent angular shift is the "parallactic shift."

determined. Once this was done, their true or absolute magnitude could be obtained and that information could be used to establish the actual period-luminosity relationship.

For nearby stars, the method of distance determination is very similar to that used by surveyors for obtaining distances across rivers and other terrain that cannot be easily crossed. If you perform the simple experiment suggested on page 98, you will become aware of the apparent shift of the position of your finger against the background. This is called a *parallactic shift.* As shown in Figure 6.4, the distance from your face to your finger can

6.4 If angles 1 and 2 and the distance between Eye 1 and Eye 2 are known, the distance between A *and* B *can be computed using simple trigonometry.*

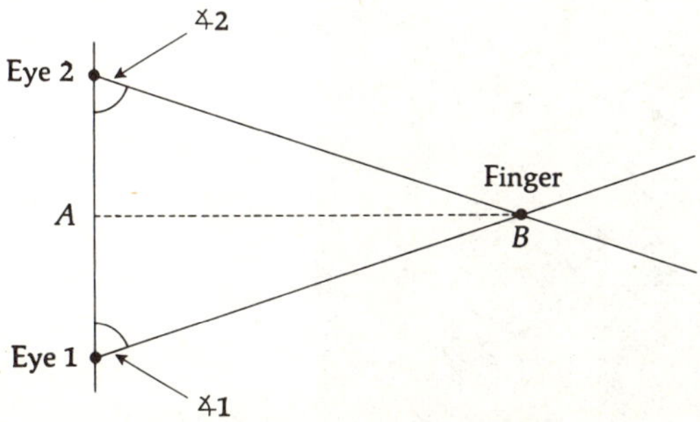

Hold your finger out in front of you and close first one eye and then the other. Note the apparent shift of the background behind the finger (see Fig. 6.3). Bring your finger closer to your eyes and repeat the process. Is the shift (parallax) greater or less when the finger is closer?

be easily calculated by using the angular measurements of the parallactic shift. Surveyors use two positions along one bank of the river and sight a tree or some other object on the other side. They know the distance between the two positions (which corresponds to the distance between your eyes in the exercise) and they can measure the angles used in the sightings across the river. The calculation is then not difficult.

In astronomy, a nearby star is sighted against the background of far more distant stars. The two sighting positions are on nearly opposite sides of the earth's orbit (see Fig. 6.5). Photographs of the star are taken against the background stars approximately every six months for several years. The star's position relative to the other stars in the field is measured with great precision. More than just two photographs must be taken and measured because (as we will discuss in the next section) the star has its own "proper" motion which must be separated from the tiny annual wobble from side to side that represents its parallactic shift (see Fig. 6.6). The distance from one side of the earth's orbit to the other, corresponding to the

6.5 *To obtain the parallax of a relatively nearby star, it is sighted against the background of much more distant stars from nearly opposite sides of the earth's orbit.*

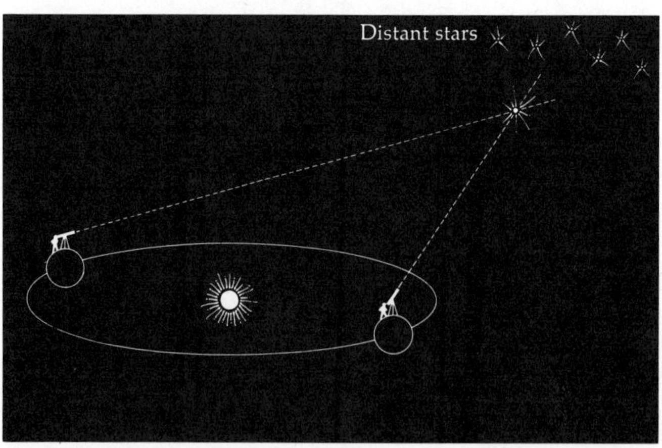

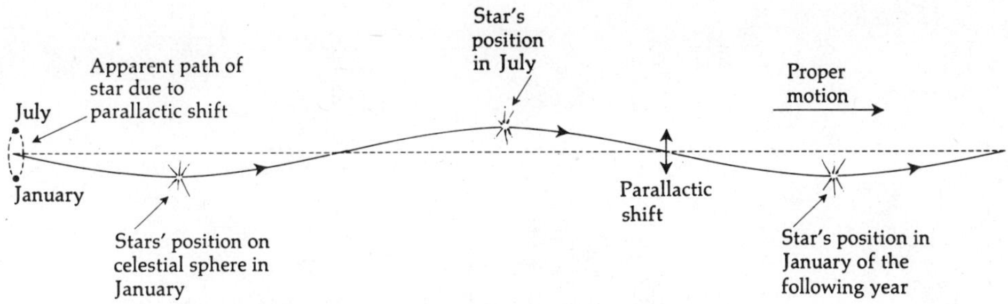

6.6 The combination of a star's proper motion through space and its parallactic shift. These two motions create the apparent wobbly path of the star on the celestial sphere (schematic drawing, greatly exaggerated).

distance between your eyes in the exercise, is known (2 AU) and therefore the distance to the star can be calculated.

However, the parallax of a star is defined as the angle subtended by the *radius* of the earth's orbit as viewed from the star (Fig. 6.7). Thus, if we draw a triangle between the earth, the star, and the sun, the angle at the star is the parallax of the star. We obtain this angle by using an equal angle which can be measured from the earth (Fig. 6.8). The results are actually one-half of the amount of total angular shift, since we are using only one-half of the base line (which is 2 AU) in our computation.

The parallactic shifts that are measured are very small angles because of the great distances to the stars. In fact, pre-telescope astronomers (Fig. 6.9) such as Tycho Brahe were well aware of the principle of this method of measurement, and used the fact that they could not detect any parallactic shift as proof that the earth was stationary and did not revolve

6.7 The parallax of a star is the angle subtended by the radius of the earth's orbit as viewed from the star. Drawing is greatly exaggerated.

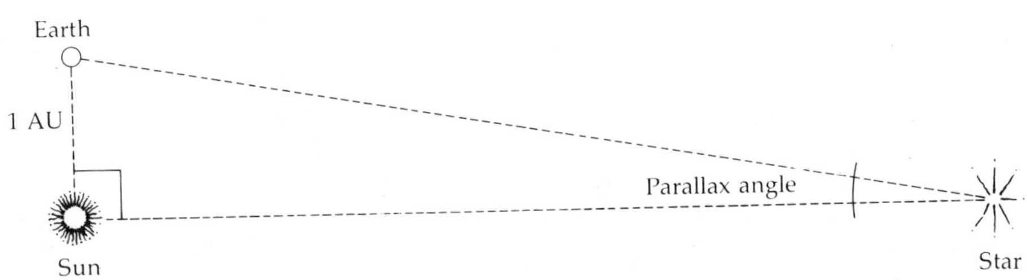

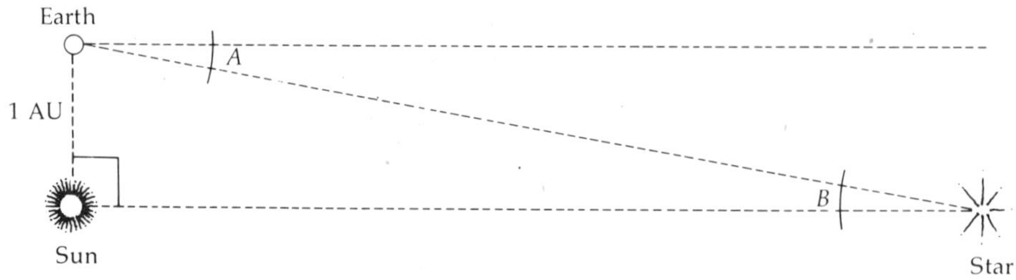

6.8 Angle A *is equal to the parallax angle* B *but can be measured from the earth and is therefore the one used to obtain the stellar parallax.*

6.9 Frontispiece of Peter Apianus' Instrument Buch *published in 1533. The woodcut illustrates the many early methods used to sight and measure the stars on the celestial sphere.*

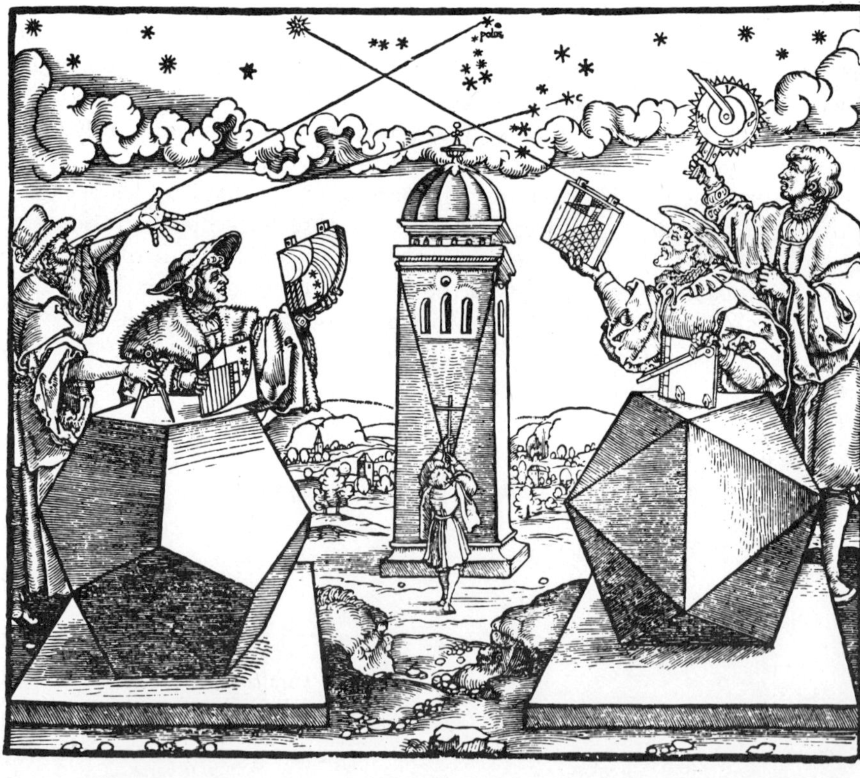

No star is close enough to have a parallax of even one second of arc. To get an idea of the size of such an angle, try drawing an angle of one degree (use a protractor). How far from the apex of the angle must you extend the sides before you can measure a separation of one millimeter? How far would it take to obtain the same separation with an angle of one second of arc (one second is 1/3600 of a degree)?

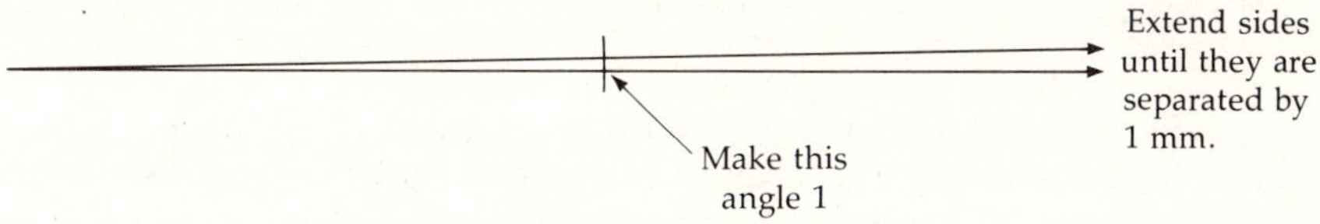

around the sun. Actually, these early astronomers' instruments were not precise enough to record such tiny angles which are no larger than the angle subtended by a penny about three miles away! (Fig. 6.10)

We now know that no star is close enough to have a parallax of even one second of arc (1″—1/3600 of a degree of a circle). If there were such a star, it would be at a distance of 3.26 light years or 206,265 Astronomical Units. This distance has been defined as one *parsec* (i.e., *par*allax of one *sec*ond—see Fig. 6.11) and is the distance at which an object has a parallax of one second of arc. It is a very widely used yardstick in astronomy, which is why the distance of 32.6 light years (10 parsecs) was selected as the reference point for absolute magnitude measurement.

6.10 *Even the largest stellar parallax is no larger than the angle subtended by a penny 3 miles away. At that distance the angle is 1/5000°.*

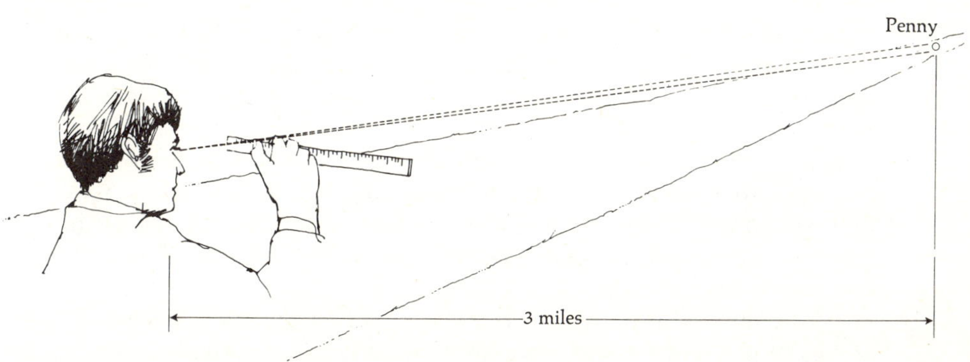

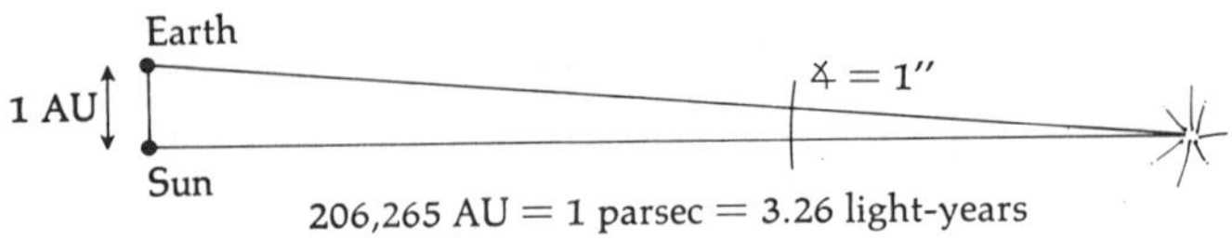

6.11 Hypothetical star with a parallax of one second of arc (1″) would be one parsec away from us.

The distance in parsecs equals the reciprocal of the parallax. For example, if a star has a parallax of ¼ second of arc (¼″), its distance from us is 4 parsecs. The largest parallax belongs to the star Proxima Centauri, which is a close companion to Alpha Centauri. Its parallax is 0.76 seconds of arc which is equivalent to 1.3 parsecs or 4.29 light years. It is the closest star to the earth (other than the sun, of course).

Because the angles involved are so very small, it was not until the mid-nineteenth century that the first parallax measurements were possible. Because trigonometry is used in the computation, this method became known as "trigonometric parallax." Obviously, as the distance increases, the parallactic angle becomes smaller, and there is a point at which even our most precise instruments can no longer measure accurately. The smallest reliable trigonometric parallaxes are of the order of one one-hundredth of a second of arc, which corresponds to 100 parsecs or 326 light years. The distances to stars farther away must be obtained by other means, but these other means are based on the accuracy of the trigonometric parallax measurements, which are thus basic to the determination of stellar distances.

The Motions of the Stars

The parallactic shift of a star is due solely to our own motion around the sun; it has nothing to do with the real motion of the star. However, this real motion can also be detected and measured. To do this, the two components of this motion are treated separately and then combined. The star is moving either toward or away from us (the radial direction) and it is also moving across our line of sight (the tangential direction) (see Fig. 6.12). The first component is the radial velocity, which is obtained by measuring the Doppler effect on the stellar spectral lines (see Chap. 3). This is directly measured in kilometers per second.

The other component of the star's motion through space is at right angles to the radial velocity motion and is called the *proper motion* of the

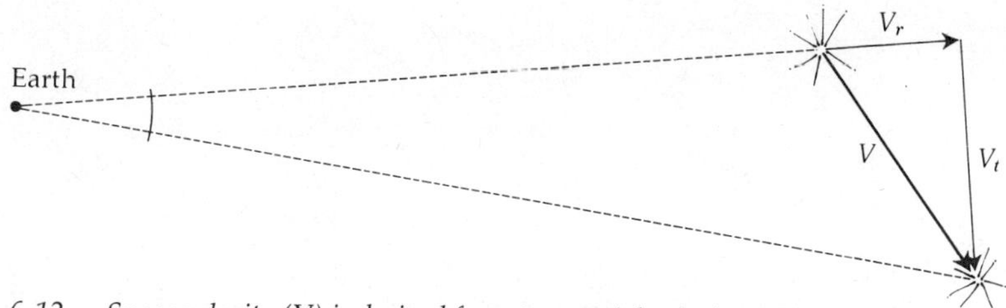

6.12 *Space velocity* (V) *is derived from tangential velocity* (Vt) *and radial velocity* (Vr).

star. This is the observed angular motion of the star as it crosses the celestial sphere, and is measured in seconds of arc per year. Because the stars are so very distant, even for the very fastest stars we cannot detect this motion for dozens of years, and then only with the most precise measuring instruments. No wonder the ancient civilizations referred to them as "fixed" stars. It was not until 1718 that the English astronomer Edmund Halley discovered that the stars do in fact change their places on the celestial sphere. He compared the position of the stars that he had measured at Greenwich, England, with those in the catalogues of the ancient astronomers Hipparchus and Ptolemy and found that some had changed by minute amounts over nearly 1500 years.

Today such work is done with the use of photographs taken many years apart. The motion of nearby stars is measured with reference to stars more distant whose proper motion is so small as to be undetectable.

This motion across our line of sight can only be expressed in terms of its annual angular change (i.e., seconds of arc on the celestial sphere) unless we know the distance to the object being studied. A nearby star that appears to move across the sky the same amount as a more distant one is actually moving through space much more slowly (Fig. 6.13*a*). We see this same effect when observing a single-engine airplane and a 747 jet airliner in the sky. Both may appear to move across the sky at the same rate, but we know that the single-engine airplane is much closer to us and is moving much slower.

If a star's distance is known, then we can translate its proper motion into a velocity. This is called its *tangential velocity.* Given a specific tangential velocity, the farther away a star, the smaller its proper motion (Fig. 6.13*b*). Thus, whereas radial velocity can be measured directly by the motion of the star itself, tangential velocity requires that both the star's proper motion and its distance be known.

The largest proper motion known is over 10 seconds of arc per year. This is the angular change of Barnard's Star, the second closest star to the sun (5.9 light years away). It was named after its discoverer, the American

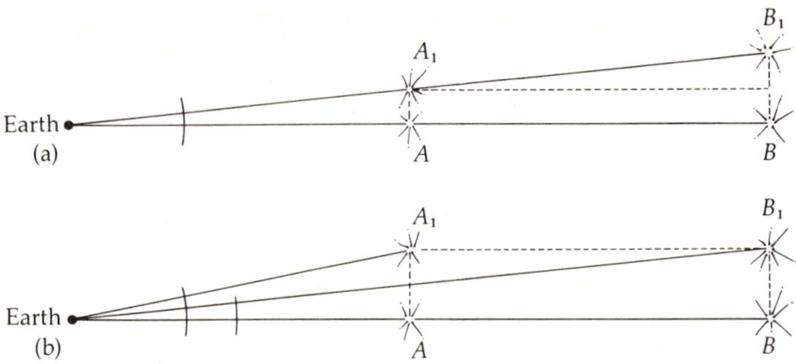

6.13 The relationship between proper motion, tangential velocity and distance. (a) Star A *and Star* B *both have the same proper motion, i.e., they have moved the same number of seconds of arc on the sky in a given year. (The same angular shift is seen.) However, Star* B *has had to move twice as far as* A *to effect this angular shift because it is twice as far away. (b) Star* A *and Star* B *both have the same tangential velocity, i.e., they have both moved the same real distance in space across our line of sight. However, since Star* A *is one-half the distance from the earth that Star* B *is, it will have twice the proper motion as Star* B.

astronomer E. E. Barnard, who located it in 1916. Although it is very close to us, it is so faint that it can only be seen with a telescope. Its relatively large proper motion causes it to move on the celestial sphere the diameter of the full moon every 180 years.

Most of the naked-eye stars have proper motions that are much less than one-tenth of a second of arc per year, yet this amount of motion is usually much greater than the star's parallactic shift each year. Since the stars all move in different directions, the forms of the constellations will eventually change. However, it will be centuries before we become aware of this. Figure 6.14 illustrates how the familiar shape of the Big Dipper will be affected by the motion of its seven stars.

If we can obtain both components of a star's motion through space, namely its radial velocity and its tangential velocity, we can recombine them to determine the star's *space velocity* or real motion relative to the sun (Fig. 6.12). However, because of the very precise measurements required, this can only be done accurately for those stars that are relatively nearby and in our immediate part of the galaxy, i.e., within several hundred light years' distance. We know, for example, that Barnard's Star is approaching

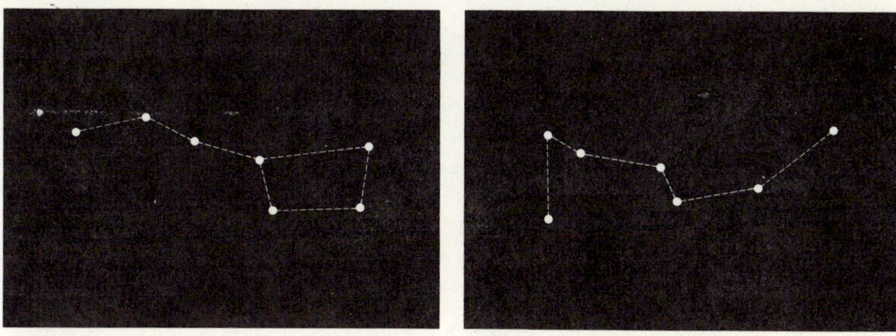

Today 100,000 years from now

6.14 As a result of the individual proper motion of each of the seven stars in the Big Dipper, its present shape (a) will change so that in 100,000 years the constellation will appear as shown in (b).

us at 108 kilometers per second and that in about 10,000 years it will be the closest star to our sun. It will still be invisible to the naked eye but will have a proper motion of over 25 seconds of arc per year.

The Rotation of the Galaxy

Although the space velocity of the more distant stars is difficult if not impossible to obtain, we can still measure their radial velocities. From an analysis of these, we can establish a fairly accurate picture of the rotation of the galaxy. It has been found that stars farther from the galactic center than the sun move around the center at velocities slower than the sun whereas those nearer to the center of the galaxy are moving at faster rates relative to the center. The sun and its neighboring stars revolve around the galactic center at about the same speed, namely 240 kilometers (150 miles) per second (see Fig. 6.15).

Therefore, the galaxy does not rotate like a solid wheel. If it did, the stars farther away from the center would revolve more rapidly than those closer in. The evidence shows that the opposite is true. The stars nearer the center revolve faster than those farther out. (This is also true of the planets in their motions around the sun and for the same reason: see page 261, Kepler's third law).

As each star revolves around the galactic center, it also describes its own individual orbit rather than following concentric circles. Therefore,

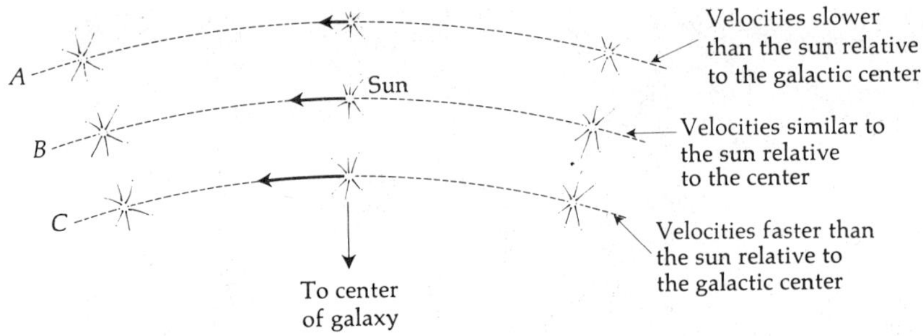

6.15 *Radial velocity studies have been used to calculate galactic rotation. Stars on track C (closer to the galactic center) travel faster than the sun (track B) which in turn overtakes stars on track A (farther out).*

The earth revolves around the sun at 30 kilometers per second. At the same time the sun revolves around the center of the galaxy at 240 kilometers per second. To orbit the sun takes one year but to orbit the galaxy takes 200 million years (a cosmic year). How much of the sun's orbit around the galaxy is covered in one earth year? In terms of degrees of a circle, what does this represent in curvature? Do you think that, over the course of the year, this deviation from a straight line is more or less than the deviation of the center line of a typical straight highway over the course of a kilometer (as compared with the curvature of the earth)? What would be the path of the earth as seen from outer space?

---

although all the stars near the sun are revolving at basically the same rate, there is still some relative motion between them as well. Relative to the nearby stars, the sun moves at a slow 19 kilometers (12 miles) per second. All of this is much like a car race in which dozens of cars are speeding rapidly around an oval track more or less together. As they race, they also change their positions relative to each other, but at a much slower rate. The driver of any one car will be aware of this slower motion as well as the much more rapid motion around the track. The former is relative to the cars near him; the latter is relative to the center of the racetrack.

Therefore, when we refer to how fast we are moving, we must specify our reference point. We move relative to some other object which is also moving. As the earth revolves around the sun, the sun in its turn revolves around the galactic center. But whereas the earth takes one year (earth time) for its revolution, the sun takes about 200 million years to complete one orbit.

Binary Stars

Besides being members of a galaxy and revolving about its center, almost all stars are also part of a double or a multistar system. Only about 15% are single stars like the sun, but because of the great distances to the stars, practically all of them appear to the naked eye as single stars. With the aid of telescopes, however, almost half of the stars are found to be in pairs or double star systems. These are called *binaries* or *binary stars*.

The two stars actually revolve around each other in space—or more accurately, each revolves about the common center of mass of the system, constantly balancing its distance from the other with its own mass (Fig. 6.16). Two children of different weights playing on a seesaw do basi-

---

Obtain the weight of two small objects such as a screw driver and a roll of mending tape. Attach each to either end of a yardstick and then find the point on the yardstick at which it balances evenly on your finger or some other narrow object. Unless the two objects are of exactly the same weight (which is a measure of mass), that point will not be at the center of the yardstick. What is the relationship between the weight of each object and its distance from the balance point (the fulcrum)?

6.16 *In a binary system, the two bodies have elliptical orbits around their common center of mass. In this example, the mass ratio of the two stars is 3:1, and their distances from the center of mass at any instant is in the ratio of 1:3.*

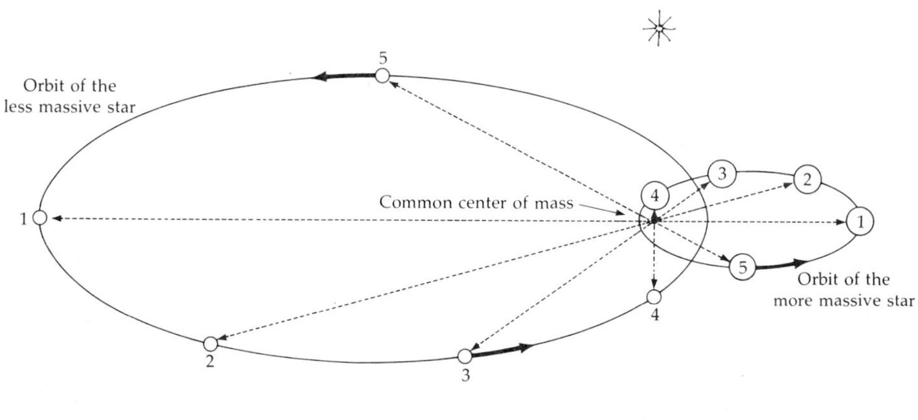

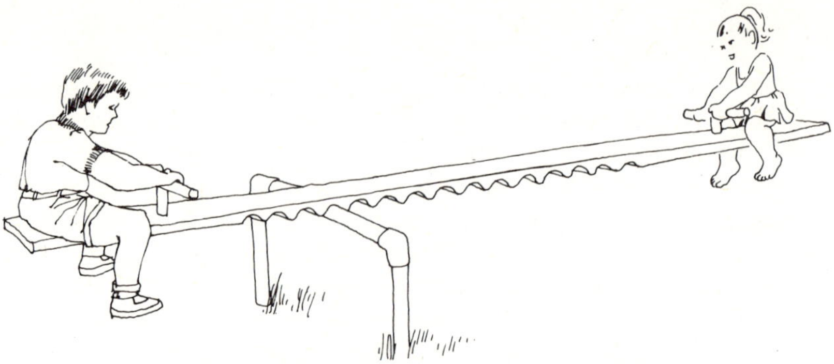

6.17 Two children of different weights can seesaw successfully if their distance from the center of mass (the fulcrum) compensates for their weight difference. Similarly, the sizes of the orbits of binary stars depend upon the relative mass of the two stars.

cally the same thing. The heavier (more massive) child must sit closer to the center post than his smaller companion if the balance is to work properly (Fig. 6.17).

The two orbits of the binary members can be plotted after extensive observations. Since the point where the common center of mass lies is not

6.18 Because the center of mass is not a luminous point, it is easier to consider the brighter star as stationary with the dimmer star revolving around it.

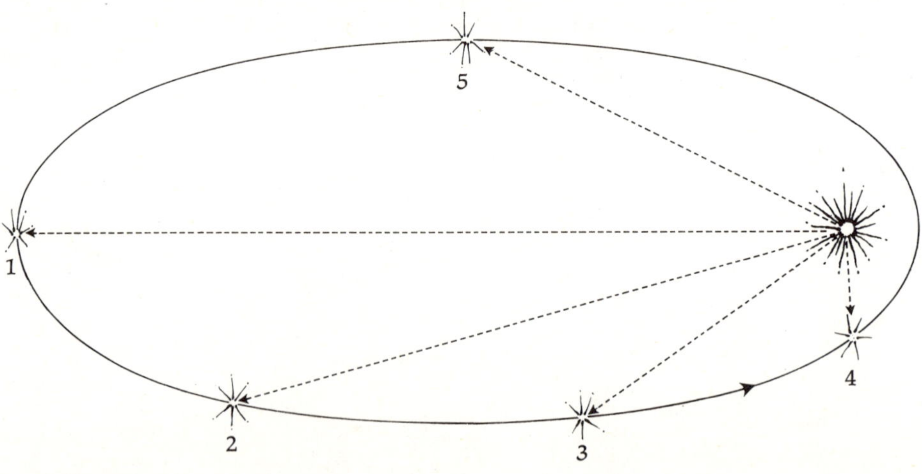

luminous, it is easier to consider the brighter star as stationary and the dimmer star as revolving around it (Fig. 6.18).

Binaries which can be observed through a telescope as two separate stars are called *visual binaries*. In order to be visible as separate stars, these components must often be hundreds of Astronomical Units apart. (Let us remember, however, that the sun's nearest neighboring star is about 270,000 Astronomical Units away.) The orbital speed of visual binaries is relatively slow and therefore their change in position relative to each other may take many years to detect. However, over 64,000 visual binaries are known and catalogued. One that is easy to see with a pair of binoculars is Eta Cassiopeiae (see Fig. 6.19). It takes the two stars in this system 480 years to orbit each other.

There are some binaries in which the two stars are so close together that they cannot be separated even by the most powerful telescopes. The only method of determining their orbits is by measuring their radial velocity changes as they approach and recede from us. Because this requires the measurement of spectral line shift, these systems are called *spectroscopic binaries*. Figure 6.20 shows us how the radial velocity will vary in such a

---

Locate Mizar and Alcor in the night sky. These are the middle stars in the handle of the Big Dipper. Can you separate them without a telescope? With a telescope or a good pair of binoculars separate the components of Mizar. Make a sketch of the stars you can see in the telescope's field of view.

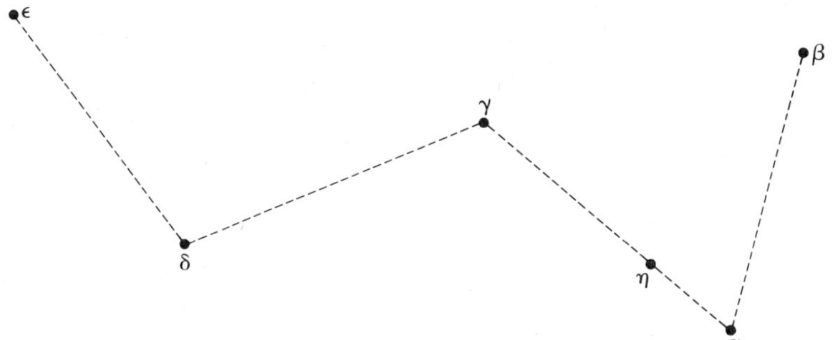

6.19 Eta Cassiopeiae (η) in the constellation of Cassiopeia is a visual binary. Both stars can be seen separately with a pair of binoculars.

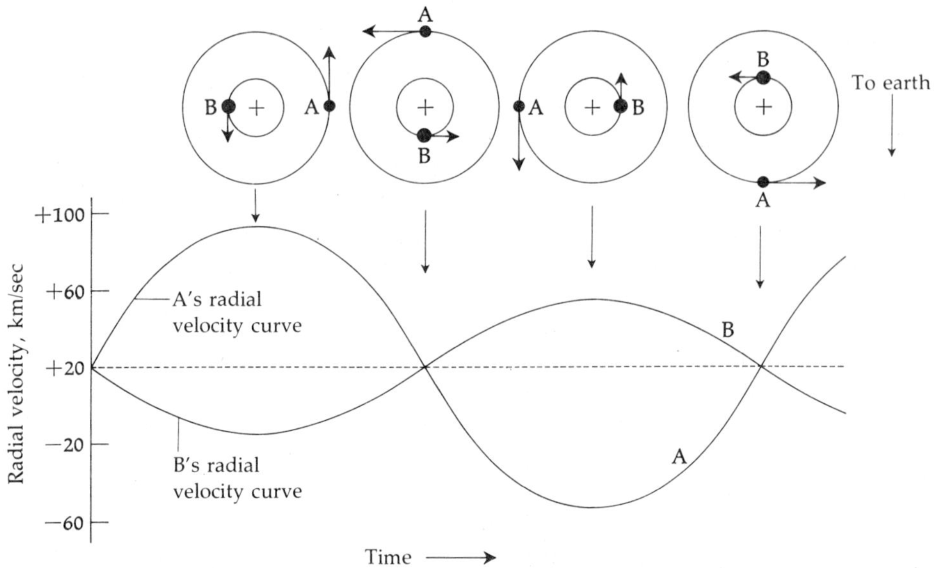

6.20 Hypothetical spectroscopic binary system. The radial velocity of each star in the system varies as both stars revolve about their common center of mass, which is indicated by a small +.

system. Perhaps the most famous spectroscopic binary is the middle star in the handle of the Big Dipper. On a clear night, sharp eyes can see two stars in this location, Mizar and Alcor. Legend has it that the ancient Arabs used this pair of stars as an eye test. But they did not know that Mizar itself is really four stars—two stars whose period is about 20,000 years but each of which is itself a spectroscopic binary. The spectroscopic binaries complete one orbit in a matter of days or months. These four stars of Mizar are in turn orbited by Alcor in a period of 10 million years (see Fig. 6.21).

With visual binaries, we can plot the orbits simply by observing the changes in the positions of the stars with time. But sometimes the orbit of a binary system is so oriented toward us that, as the stars revolve around each other, they periodically eclipse one another—each in turn passes in front of its partner in our line of sight. This causes the binary system's light to vary depending upon which star is in front at any particular time. Figure 6.22 shows how the various positions of the two stars will affect the amount of light that reaches us. This is called an *eclipsing binary.* Besides measuring the light variations, we can also determine the radial velocity of each star as it approaches us and then recedes. In fact, more can be learned from this type of binary than from any other. The combination of the *light*

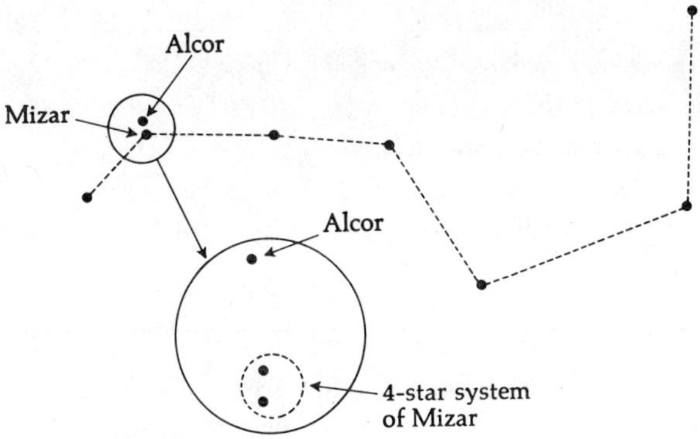

6.21 *Through a telescope, Mizar and Alcor appear as shown in the larger circle. Each of the two components of Mizar (dashed circle) is a spectroscopic binary. Thus Mizar is actually a 4-star system.*

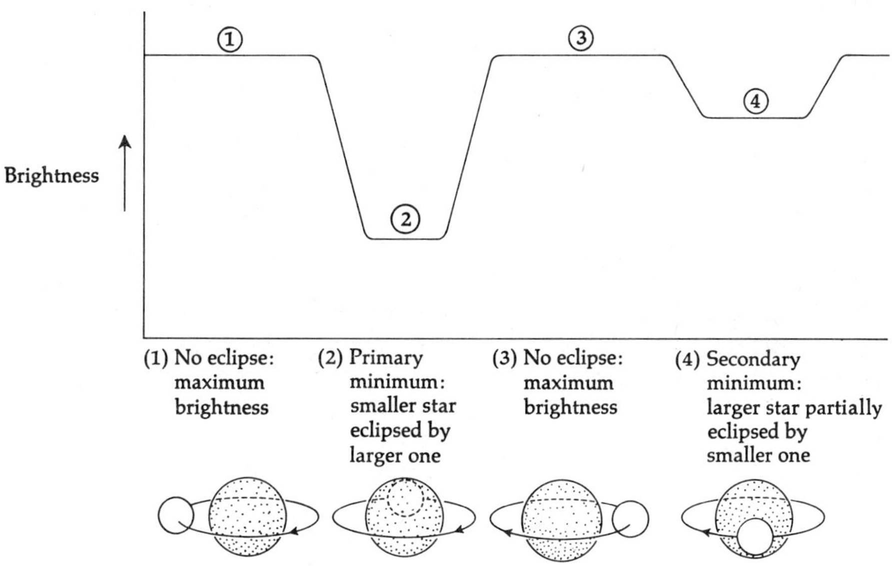

6.22 *Light curve of a hypothetical eclipsing binary star system. In this case, a common one, the smaller star is the brighter one.*

curves and the *radial velocity curves* enable us to determine not only the size of the stellar orbits but the linear dimensions of the stars themselves.

The closer the two stars are in a binary system, the shorter their *orbital periods*, much like the stars close to the galactic center which revolve faster

---

Obtain two flashlights, preferably of different diameter and brightness. Using a standard light meter, measure the amount of light received at a set distance when the two lights are held side by side. Now measure the light received at the same distance when first one light and then the other is placed in front of its companion, eclipsing it. Chart the light differences (light curve) similar to Fig. 6.22.

---

than those farther out. There are some binaries whose periods are measured in minutes. At the other extreme, some have periods so long that they have not completed even one since observations of binaries began. Proxima Centauri, which revolves around Alpha Centauri and is at present the closest star to our sun, takes about a million years to complete one orbit.

Determination of Stellar Mass

The relationship between the period of a binary star and the closeness of its two members leads us to valuable information which cannot be obtained from single stars, namely *stellar mass*. In 1615, Johannes Kepler first discovered this relationship in his study of planetary orbits (see page 260). But he was concerned with the periods of planets orbiting the sun and therefore did not have to consider the relative mass of the two objects involved, namely the sun and one of the planets. The mass of the sun is so much greater than any of the planets that its mass is in effect the only important one. Kepler found that for the planets, the squares of their periods[3] *(p)* was proportional to the cubes of their distances from the sun *(d)*. If we put these quantities in earth years and AUs, the relationship can be written as: $p^2 = d^3$.

With binary stars in which each member has a large percentage of the total mass of the system, their respective masses must be considered in order to find the relationship between the period and distance. The formula for binary stars must read: $p^2(m_1 + m_2) = d^3$, where m_1 and m_2 are the respective masses of the two stars. (Note that m_1 and m_2 were also in the sun-planet formula, but if we put them into solar masses, they would have

---

[3]The period of a planet is the time it takes to orbit the sun.

added up to only a minute fraction more than 1.0 and therefore would not have effected the rest of the calculations.)

In a binary system, the period can be observed. If the distance to the system is known, the distance between the two member stars (*d*) can be expressed in actual linear terms (AUs, kilometers, or miles) rather than in angular degrees of separation as measured on the celestial sphere. Thus, the total mass of the system can be calculated.

The ratio of the masses of the two stars, like the children on the seesaw, is inversely proportional to the distance of each star from the center of mass. In other words, the less massive the star, the farther away it will be from the center of mass of the system. In Figure 6.16, the distances of the two stars from their common center of mass is three to one. The ratio of their masses therefore is one to three. The more massive star has three times more mass than its companion and at any point in its orbit, it is ⅓rd as far away from the system's center of mass as the less massive one.

Actually, relatively few stars have had their masses determined in this direct manner. Although there are many binary systems, only a small percentage have had their distances and periods measured accurately. From these, however, another very important relationship has been discovered, namely the mass-luminosity relationship. For most stars, the more massive it is, the greater its intrinsic brightness (see Fig. 6.23).

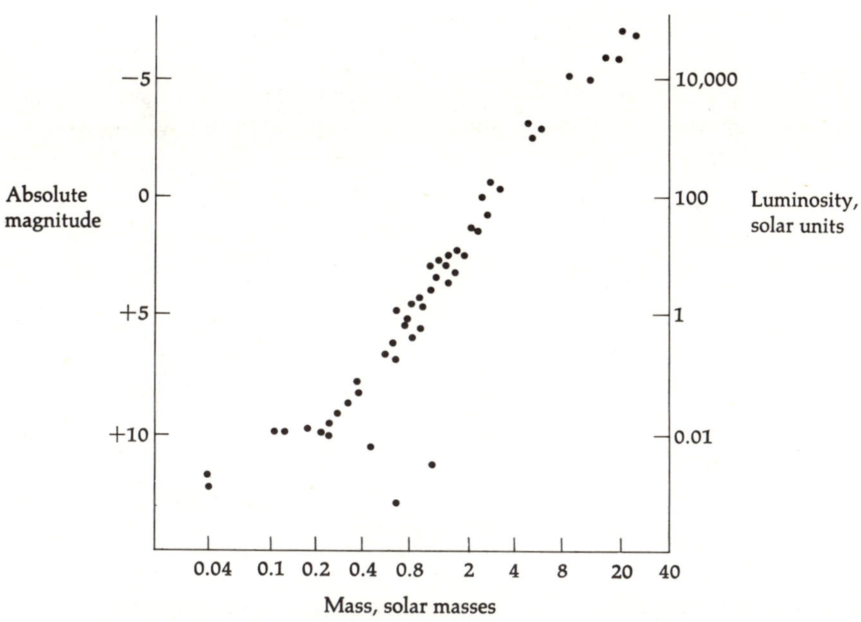

6.23 *The mass-luminosity relationship.*

Roughly, luminosity increases somewhat more than the cube of the mass. This means that a star with two solar masses will have a luminosity about eight times as great as the sun.

Once this relationship was established, the masses of stars of known luminosity could be closely estimated. Note from Figure 6.22 that most stars range in mass from about one-tenth to 50 times that of the sun whereas their luminosities range from less than 1/100th to over one million times that of the sun.

Thus, from studies of binary stars astronomers were able to obtain the masses of the component stars, and subsequently the mass-luminosity law. From the latter, the masses of nonbinary stars could be estimated. In the next chapter, we will discover how very important the knowledge of stellar mass is in the study of stars as physical bodies and in the study of stellar evolution.

Discussion Questions

1. How was the inability to detect parallax of stars used by early astronomers as an argument that the earth was stationary?

2. If astronomers had made their observations from Mars rather than from the earth, which of the following units of distance would still have the same value? Why?

    a. Parsec
    b. Light year
    c. Astronomical Unit

3. Two stars have the same proper motion but different parallaxes. Which star has the greater tangential velocity? Draw a diagram showing this.

4. Radial velocity studies have been used to determine galactic rotation. In Figure 6.15, what would have been the velocity of each star as seen from the earth (approaching, receding, or neither)?

5. What is the difference between apparent and absolute magnitude?

6. Define:

    a. Radial velocity
    b. Proper motion
    c. Tangential velocity
    d. Space velocity

7. Stars *A* and *B* have the same apparent magnitude. Star *A*'s absolute magnitude is +5, whereas star *B*'s is +10. Which star is intrinsically brighter? How many times brighter is it? (Answer in terms of brightness, not magnitude difference.) Which star is farthest from the earth? How much farther is it from the earth than the other?

8. If one star appears 40 times brighter than another, what is the difference in their apparent magnitudes?

9. The full moon has an apparent magnitude of about −12. How much brighter does it appear than Sirius, whose apparent magnitude is −1.58?

10. What is the distance in parsecs of a star that has a parallax of 0″.1? of 0″.5?

11. Star *A* has one-half the parallax of star *B*. Which star is closer to us? If they both have the same apparent magnitude, what can be said about their absolute magnitude?

12. Describe the various types of binary stars.

13. What is a "light curve"?

14. If the two stars in Figure 6.16 have masses of 12 and 4 solar masses, and their average distance from each other is 4 AU, what is their period of revolution?

7 Stars as Physical Bodies

In the mid-nineteenth century, August Comte stated, "There are some things of which the human race must forever remain in ignorance, for example the chemical constitution of the heavenly bodies."[1] It is understandable that Comte and others held such despairing views, since the complex nature of electromagnetic radiation had not yet been discovered. This serves to remind us that knowledge of the true physical nature of the stars has been only a relatively recent discovery.

Atomic Energy Levels

Electromagnetic radiation from the stars has provided us with nearly all of our information about them. Therefore, in order to study the stars, we must learn more about this radiation. Since light is its most familiar form, we will concentrate our attention on it, although we must remember that what we find to be true of light is also true for the other regions of the spectrum.

We speak of light as a wave of specific length when we measure the radial velocity of receding galaxies. We also speak of light as particles or *photons* when we are measuring the amount of light received on a photograph. Red light, as we have seen, has a longer wavelength and lower frequency than blue light. It also is composed of photons having less energy than those of blue light. A given hue of light (or light of a specific wavelength) is composed of identical photons, all having the same energy or *quanta*. The intensity of the light depends upon the number of photons reaching your eye or the recording instrument per second. In the mathematical discipline called *quantum mechanics*, these two concepts are combined.

[1]Comte, August, *Cours de Philosophie Positive,* Paris: J. B. Bailliere et Fils, 1864.

In order to understand where these photons originate and why they may have different amounts of energy, we must look at the atom itself. Although it has many components it is still considered the main building block of matter, since when an atom of any element is divided, it loses its chemical identity. It becomes subatomic particles but can no longer be called hydrogen, neon, iron, or whatever it was before being split.

A simple way of visualizing the atom is to think of it as having relatively tiny electrons orbiting a much larger nucleus made up of protons and neutrons. The electron is a negatively charged particle while the proton is positively charged. The neutron, as its name implies, has no charge. Of course, the atom is far more complicated than this model, but for our discussion this model will be sufficient.

The simplest atom is that of the element hydrogen (H). It can be thought of as one electron orbiting a nucleus containing just one proton. Helium (He), the next simplest element, has two electrons orbiting a nucleus of two protons and two neutrons. Because helium has four major particles in its nucleus, it is about four times heavier than the hydrogen atom. (The mass of the electron is so negligible compared to even the single proton that it can often be ignored.) The heavier and more complicated elements have larger nuclei (i.e., more protons and neutrons) and a greater number of orbiting electrons (Fig. 7.1).

In this simple model of the atom, there is an enormous amount of empty space, similar to what we find in interstellar and intergalactic space. The distance from the typical atomic nucleus to its electrons is more than 10,000 times the size of the nucleus itself. Our densest solids here on earth

7.1 *Schematic illustration of the atoms of three elements. Note that neither the size of the particles (protons, neutrons, or electrons) nor the size of the orbits is drawn to scale.*

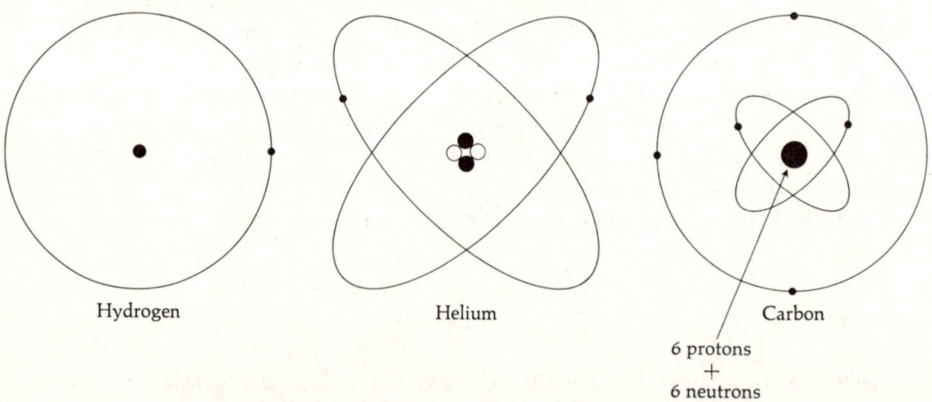

are therefore almost completely "empty." An iron bar is composed of iron atoms, each occupying its own volume of space at any given moment. The empty space within each atom is considered an integral part of the atom itself. Normally no other particle can enter or stay in this space without changing the make-up and identity of the atom itself.

Make a diagram comparing the size of a typical atom and its nucleus with the size of the solar system (out to Pluto) and the sun.

In the empty space of each atom the electrons revolve around the nucleus in specific orbits. There are only certain physically permitted sizes of these orbits that each electron can have at any given time. The larger the orbit, the more energy the atom as a whole possesses. We can speak of the electron orbits as representing various energy states or "levels" of the atom. Therefore, the atoms of each chemical element can have only specific energy states, which correspond to the sizes of the electron orbits permitted.

In cool hydrogen, for example, when the atom is in its normal state, the single electron is at the lowest possible energy level or orbit, which is called the *ground state*. If a photon of light comes along which has the exact amount of energy necessary to raise the electron to the next highest "allowed" energy level, the photon will be absorbed by the atom. This added energy causes the electron to jump to this correspondingly higher orbit. We say that the atom is then in an *excited state*. If the photon's energy is not the exact amount required, it will pass right through the atom and not affect the electron.

The excited electron stays in this new orbit (or at its higher energy level) only a very short time (about one-hundred-millionth of a second) and then jumps back down to the lower level, thereby emitting the same amount of energy that it just absorbed. A photon identical to the one absorbed is released. Similarly, if an incoming photon has the precise energy needed to raise the electron to still higher orbits, this photon may also be absorbed (Fig. 7.2). Then the electron may either jump down directly to the lowest level or it may cascade down, stopping momentarily at various levels (Fig. 7.3). Thus, each electron has several specifically allowed orbits or energy levels which are attained when the atom gains or loses energy. Thus there are a specific number of discrete jumps that can be made up or down.

Photons with more energy than is necessary to raise the electron to the highest allowed levels will tear the electron free from its bond to the nucleus. The atom is then *ionized*. In elements other than hydrogen, which

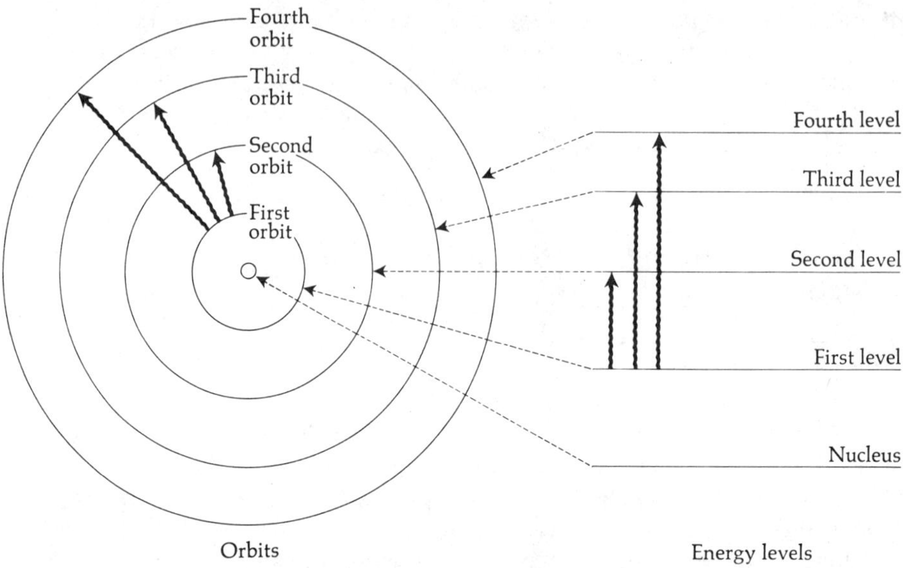

7.2 The "permitted" higher orbits to which an electron may be raised can also be thought of as energy levels. For the sake of simplicity, the diagram shows only three of the possible upward jumps an electron can make from the first orbit or energy level. (The drawing does not represent any specific atom, it is a schematic diagram of a hypothetical atom.)

has only one electron to begin with, still more energy is required to remove the second electron, and of course, still more yet to remove the third, fourth, fifth, and so on. A hydrogen atom can be ionized only once; an oxygen atom, with eight electrons, can be ionized eight times. The free electrons move about the gas as free particles until they are again captured by an ionized atom, usually by colliding with it. This capture causes the emission of one or more photons, depending upon the number of jumps the electron takes as it goes down to a lower level or orbit. Obviously, the more diffuse the gas involved (i.e., the fewer the actual particles of matter per cubic centimeter) the fewer the chances of a collision and therefore the longer the atom will stay in its ionized state.

This jumping down from different higher levels to a specific lower level produces a series of photons ranging from large-jump, high-energy photons to small-jump, low-energy photons. Figure 7.4 shows the several series for the hydrogen atom (named after the various men who first studied them). The large high-energy jumps emit photons with shorter

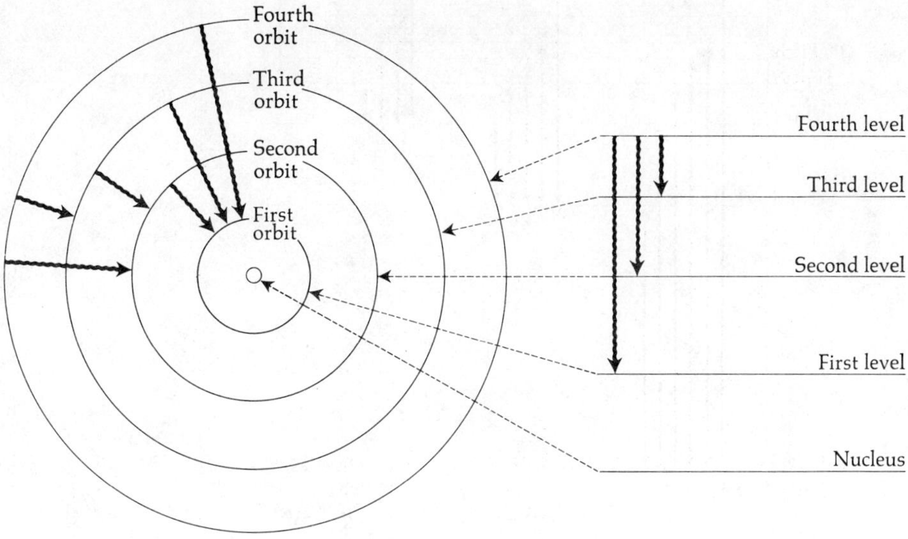

7.3 Once raised from the first level to any higher "permitted" level, the electron jumps back down to the lower levels. It sometimes makes a series of jumps rather than one large one. With each jump, a photon of energy is emitted. (The drawing does not represent any specific atom, it is a schematic diagram of a hypothetical atom.)

wavelengths, whereas the small low-energy jumps produce photons with longer wavelengths. Thus the energy carried by an emitted photon is clearly the *difference* in energy between the starting and ending levels of the electron jumps. The kind of electromagnetic radiation that we receive is determined by the size of the jump and the level to which the photon jumps. Thus, if an electron jumps from the fifth to the second energy level of hydrogen, a photon in visible blue light will be emitted. If the jump is only from the third to the second level, red light results.

In a heated hydrogen cloud that contains billions of hydrogen atoms, there will be a certain percentage of atoms whose electron is found at each level. The hotter the gas, the greater the percentage of atoms that will have their electrons located at the higher levels. The hydrogen gas will be visible only if it is hot enough to have a large percentage of its electrons at the second level. Then, as the atoms absorb and emit energy, the electrons will jump up from and return back to this second level, producing the *Balmer series* of hydrogen (Fig. 7.5).

Cooler hydrogen, which has most of its electrons at the lowest level, absorbs incoming photons with energies that correspond to jumps from

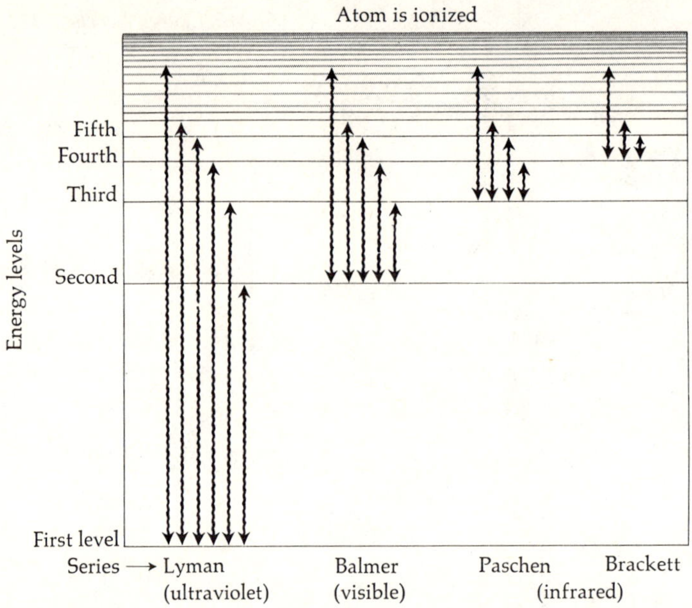

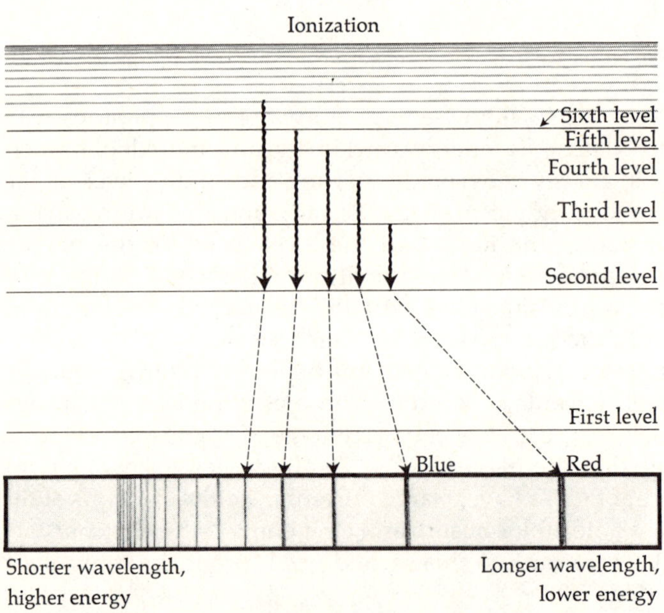

7.5 *Illustration of how the Balmer series of electron jumps creates the hydrogen emission spectrum in the visible wavelength region.*

◄7.4 *Energy level diagram for hydrogen. Arrows pointing upward represent electrons jumping from their base positions to a higher level as their atoms absorb specific amounts of energy. Downward arrows represent the emission of these energies as the electrons return to their original levels. The type of electromagnetic energy absorbed and re-emitted is indicated below the diagram.*

this lowest level to higher levels. When the electrons jump back to this lowest level, they emit the ultraviolet *Lyman series.* Hot hydrogen, with its electrons excited to various higher levels, absorbs and emits photons that correspond to the other hydrogen series. As Figure 7.4 illustrates, the jumps are shorter, corresponding to infrared photons. In any given hydrogen cloud, all these series can be emitted. The strength of each series depends upon the temperature of the cloud.

Spectral Lines

Like hydrogen, every chemical element has its own unique pattern of permitted transitions between energy levels. The series of lines thus produced are characteristic of that element. To distinguish these specific colors (or wavelengths) a *spectroscope*[2] is necessary, because our eyes cannot separate these distinct stimuli the way our ears can recognize distinct tones in music.

If we simply had pure hydrogen emitting energy in the visible region of the electromagnetic spectrum, our spectrogram would have a series of bright, distinct lines on it, each representing specific electron jumps in the Balmer series (Fig. 7.5). If our picture were in color, the lines would range from blue-violet at the shorter wavelength end of the spectrum to green, yellow, orange, and finally to red at the longer wavelength end. These are called *emission lines*. Since every element has its own specific set of emission lines in the various regions of the spectrum, it can be identified by these unique "fingerprints."

However, emission lines are produced only when the atoms are quite free to move about, as in a gas cloud. When they are "locked in place" as in an iron bar or in the tungsten filament of a lightbulb, the energy levels

---

[2]If we attach a camera to the spectroscope rather than just observe visually, the instrument is called a *spectrograph*. The photograph produced is a *spectrogram*.

become grossly disturbed and jumbled together. The heating of such matter, as when an electric current is passed through the filament, produces a mixture of all colors instead of just one and is perceived as "integrated white light." When such light is viewed through a prism, the white light is spread out into its many wavelengths and the familiar rainbow effect is seen—a continuous gradation of color from deep red to violet. Light of all wavelengths is represented. This is called a *continuous spectrum* (see Fig. 7.6).

Although a star is entirely gaseous, its extremely high temperatures cause its atoms to emit light much like a solid. Energy levels overlap each other and spectral lines blend, causing a continuous spectrum to be produced. We normally see this radiation as integrated white light, but through a spectroscope it can be spread out into its colorful array.

On practically all stellar spectra, superimposed on the continuous spectrum, are dark lines (Fig. 7.6). These are caused by the stellar atmosphere, which is composed of gases not unlike the main body of the star but much cooler and under much less pressure. The atoms in this gas move

7.6 *A light bulb will produce a continuous spectrum. If the light is passed through a cloud of gas first, the atoms in the gas absorb some of the light rays and re-emit them at the same wavelengths in all directions. As viewed from direction A, the spectrum has been depleted in these specific wavelengths. However, as viewed from* B, *only the light re-emitted by the gas cloud is seen, and this produces an emission line spectrum. (Adapted from G. Abell,* Exploration of the Universe, *2nd ed. New York: Holt, Rinehart & Winston, 1969.)*

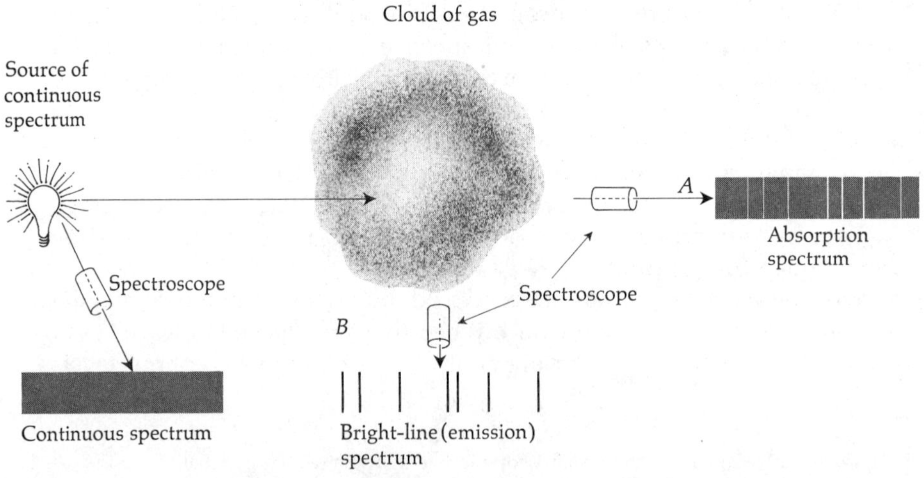

freely and are able to absorb and emit those energies that correspond to their characteristic wavelengths. They absorb radiation from the main body of the star but only those same specific wavelengths that they can emit. Almost immediately they re-emit this energy, at one or more wavelengths. However, because the emission is scattered in all directions, the number of photons of these specific energies or wavelengths originally traveling in our line of sight is much less. The dark lines that result on the spectrum are called *absorption lines* (see Fig. 7.7). Their patterns are identical with those that the atoms would exhibit if they were emitting energy rather than absorbing it. Only those wavelengths or energies that can be absorbed by the atoms present in the stellar atmosphere are effected. The rest of the stellar radiation passes right through the stellar atmosphere and is recorded as part of the continuous spectrum.

The same process of absorption and emission of energy takes place constantly throughout the body of a star as the photons slowly work their way from the central core where they are generated outward to the surface. It is only near the surface that we can record and analyze such activity, since the stellar interior is hidden from our view. The absorption lines in particular have given us a tremendous amount of information about the stars. The shift of these lines in stellar spectra is used to measure the Doppler effect and hence the radial velocities of the stars.

But the information that can be obtained from stellar spectra contains far more than just radial velocities. Once the absorption lines were identified in laboratory studies with the levels of excitation of specific elements,

7.7 Schematic drawing of continuous, bright-line (emission), and absorption spectra.

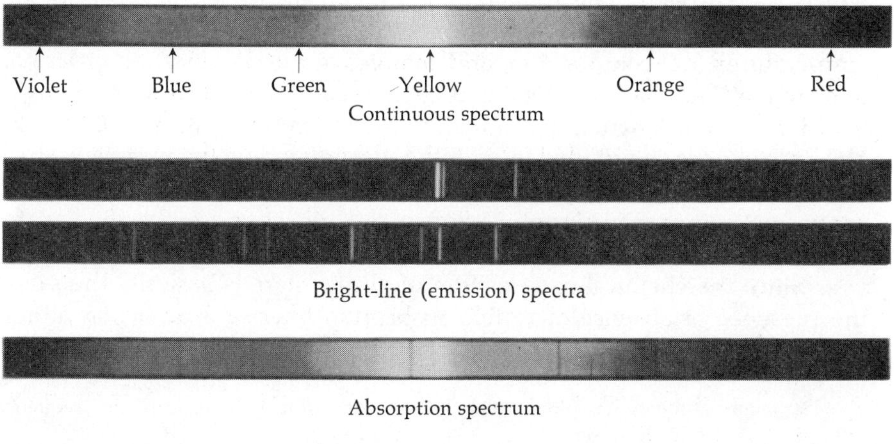

the presence of these elements could be assumed and the temperatures of the stellar surfaces could be calculated. Note that the presence of an absorption line of a certain element is absolute proof that that element is present in the stellar atmosphere, but the line's absence does not necessarily indicate the absence of the element. The electrons may be occupying energy levels that give lines in the nonvisible parts of the electromagnetic spectrum.

Spectral Classification of the Stars

By the latter half of the nineteenth century many stellar spectrograms had been taken. Virtually all of the spectra were found to have absorption lines, although there were wide differences in the number of lines and their appearances (see Fig. 7.8). In the 1860's, an Italian astronomer, Father A. Secchi, made the first attempt at classifying these spectra, but the work that really established the classification system used today was begun in 1886 at Harvard University. The great majority of the stellar spectral classifications were made by Annie J. Cannon. She classified 400,000 stars of the northern and southern hemisphere according to their spectral characteristics. The resulting *Henry Draper Catalogue* (named as a memorial to this famous astronomer) has since been greatly expanded, but the basic classification system is still used.

Originally the spectral differences were thought to be due to the chemical composition of the stars, and spectral classes simply followed the letters of the alphabet. Later it was found that most stars are basically similar in their composition, having at least 96% hydrogen and helium, most of which is hydrogen. The remaining four percent or less consists of the heavier elements. Spectral differences were found to depend upon surface temperatures, not composition, and changes in the classification order had to be made. The classes, which had been already named, were rearranged in order of decreasing temperature, and thus became: O, B, A, F, G, K, and M.[3] Each class is subdivided into tenths, designated by the numbers 0 to 9, because the differences between the classes were also found to be on a continuum rather than sharply delineated. Thus, stars are classified as, for example, K2, A5, M3, etc. The sun is classified as a G2 star (Fig. 7.9).

Since the chemical composition of all the stars is basically the same, the presence or absence of visible absorption lines of a particular atomic

---

[3]Astronomy students have long used the following mnemonic to learn this sequence: "Oh, Be A Fine Girl, Kiss Me."

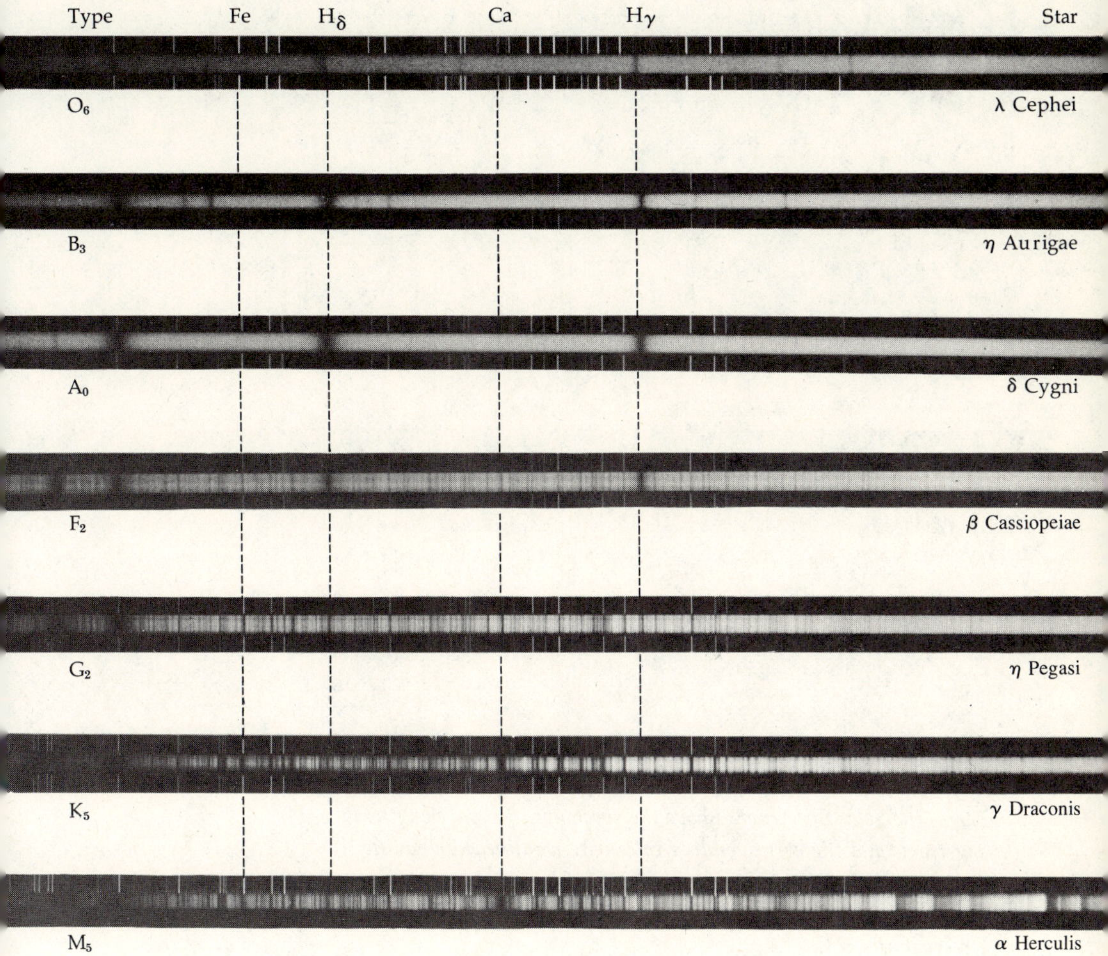

7.8 Examples of the principal types of stellar spectra (photograph from the Hale Observatories). Note the progressive changes in the absorption lines from class to class. Compare the absorption lines of hydrogen (H_Δ and H_δ) calcium (Ca), and iron (Fe) which have been marked.

series reflects the physical state of the atmosphere rather than its composition. The temperature of the atmosphere must be such that the atoms of a given element are able to absorb the characteristic wavelengths which lie in the observable region of the spectrum. As discussed on page 121, in order to produce absorption lines of hydrogen in the Balmer series (which is the only one in the visible region of the spectrum), the electrons must first be in

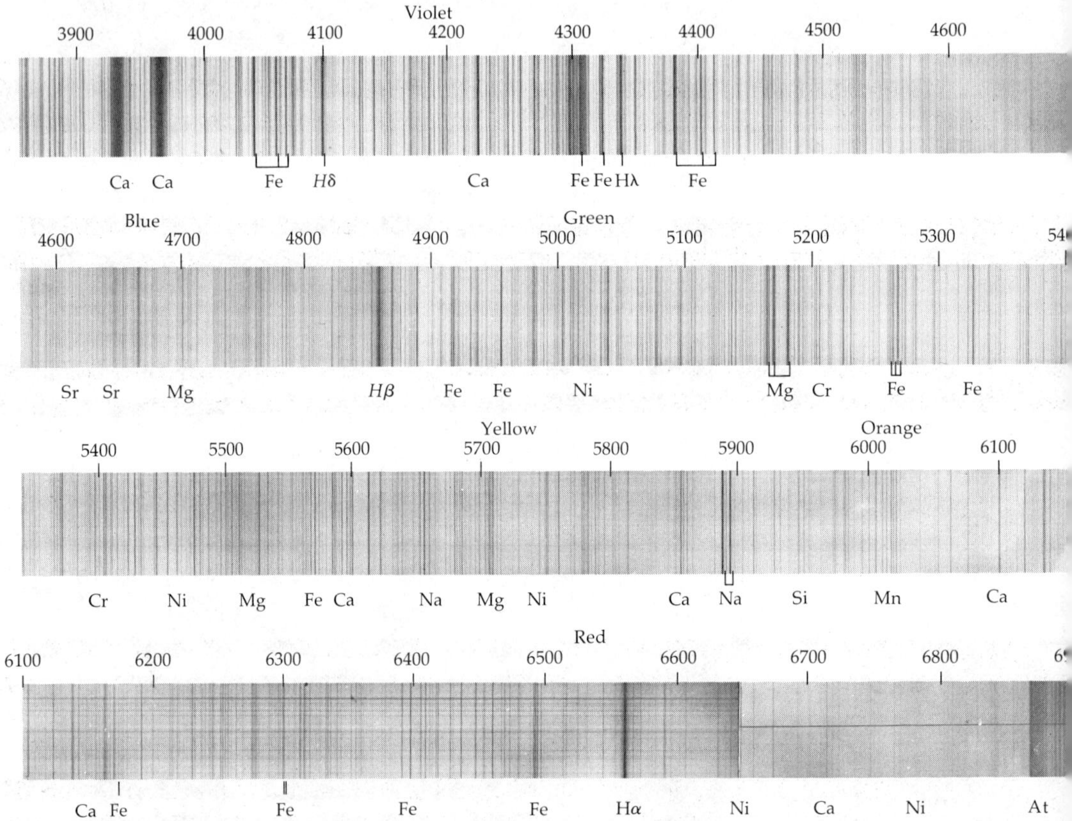

7.9 The Solar Spectrum. Above the spectrum, the wavelengths in angstroms and the corresponding colors are identified. Below the spectrum, several spectral lines are identified by the element present in the photosphere that produced them (courtesy Hale Observatories).

the second orbit or energy level. If the star is too cool or too hot, most electrons will be in the first, third, fourth, or fifth orbit and will therefore produce spectral lines at shorter or longer wavelengths than those which produce visible radiation. At about 10,000 K,[4] a majority of the hydrogen

---

[4]The Kelvin (K) or absolute temperature scale is the one most often used in astronomy. Its "zero point" is the lowest possible temperature, *absolute zero,* when all molecular motion ceases. Its scale is the same as the Centrigrade (C) scale, with 100° between the freezing and boiling points of water.

$$\begin{aligned} 0\ \text{K} &= -273^\circ\ \text{C} = -459^\circ\ \text{F absolute zero} \\ +273\ \text{K} &= 0^\circ\ \text{C} = +\ 32^\circ\ \text{F water freezes} \\ +373\ \text{K} &= +100^\circ\ \text{C} = +212^\circ\ \text{F water boils} \end{aligned}$$

Note that the degree sign (°) is not used when using the Kelvin scale.

atoms will have their single electron in the second orbit, and therefore a star with that surface temperature will show the strongest hydrogen absorption lines. These are the A-type stars (see Fig. 7.8). In hotter stars these lines weaken because most of the electrons of the hydrogen atoms are at higher orbits and radiate in a different region of the electromagnetic spectrum. This does not mean there is any less hydrogen in the star, but rather that the atoms are incapable of absorbing the series that results from the second level.

In cooler stars, the hydrogen lines also get very weak or virtually disappear. In the sun, with a surface temperature of about 6000 K, only one out of 100 million hydrogen atoms has its electron in the second orbit. That is enough, however, to produce very weak hydrogen lines in the solar spectrum.

As is shown in Figure 7.8, the hotter stars (O and B) have very few absorption lines of any element in their spectra. At these high temperatures, most of the atoms are emitting energy in the far ultraviolet regions of the spectrum. The cooler stars (especially M-type) show many lines, particularly those of the metals, because at that temperature (about 3000 K) the absorption lines of metallic gases happen to fall in the visual region. Figure 7.8 shows the spectral type and typical spectrum for each of the classes.

The Hertzsprung-Russell Diagram

The spectrum of a star and its luminosity (or absolute magnitude) are two of the most obvious ways in which stars differ from each other. In 1913, two astronomers, each working independently in his own country, found that there was a definite relationship between these two variables. Ejnar Hertzsprung of Denmark and Henry Norris Russell of the United States expressed this relationship graphically, and it is called, in their honor, the Hertzsprung-Russell diagram (or the H-R diagram). The study of this diagram, which is probably one of the most important in astronomy, has stimulated much of the research in astrophysics (Fig. 7.10).

Note first in Figure 7.10 that practically all of the stars fall into one of a very few specific areas. Over 90% of all stars are on the narrow band that slants downward across the diagram. This is known as the *main sequence*. It curves diagonally from the upper left to the lower right, covering the entire range of spectral classes. Its existence demonstrates the close linkage between luminosity and temperature for stars on the main sequence. The most luminous stars on the main sequence band are hotter and bluish-white in color. The dimmest are the coolest and are reddish. Our sun, with its predominantly yellow color, falls about midway.

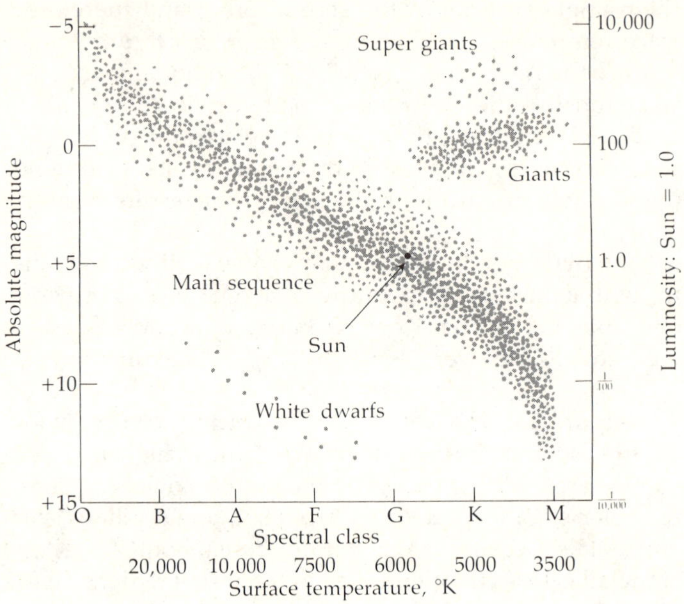

7.10 *The Hertzsprung-Russell diagram (schematic).*

A star's place on the main sequence is determined by its mass, which is the measure of the actual amount of matter that is in it. The blue-white O and B-type stars, which are the most luminous, are also the most massive, while the red M-type stars, the faintest on the main sequence, are the least massive. This is the *mass-luminosity relationship*, already diagramed in Figure 6.23 and discussed on page 113. This relationship applies only to those stars on the main sequence (see Fig. 7.11).

Obtain the height and weight of each person in the class. Make a graph of height versus weight. The data from each person will be represented by a dot on the graph. Draw a line down the middle of the dots, indicating the typical student for any given height or weight. This could be called a "Main Sequence" since it represents a relationship for the people of your age in general. However, remember that the sequence holds true for the average person but not necessarily for any specific person.

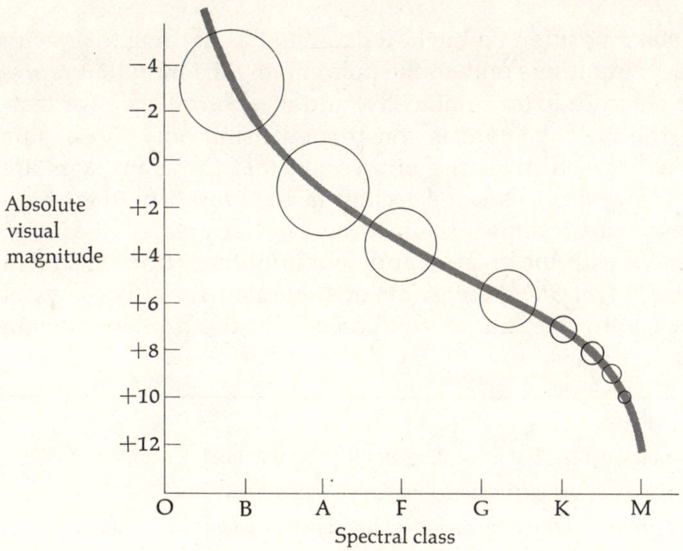

7.11 The main sequence of the H-R diagram showing the relationship between position on the main sequence and mass of star. Note that these are all main sequence stars compared with the giants and supergiants.

The second most prominent region of the H-R diagram lies above and to the right of the main sequence. This is called the *giant branch*. Although it contains many fewer stars than the main sequence, this branch is the only other well-defined sequence. Above it is found a scattering of "bright" giant and supergiant stars.

A very few stars fall into the lower left-hand corner of the diagram. When the first one was plotted there, it was thought to be an error. Actually, these are very important parts of the graph. The stars plotted here are *white dwarfs*, which are relatively hot, small, white stars with very low luminosities. A good example is Sirius B, the small companion to the bright star Sirius.[5]

---

[5]Sirius exhibited an unusual wobble in its path through space which could not be attributed to its parallectic shift (see Chap. 6). An invisible companion was predicted in 1844, but it was not seen until 1862. Alvan Clark, a famous American telescope maker, was testing the quality of a new 18-inch telescope which was eventually to be at Dearborn Observatory, Northwestern University. He noticed a tiny speck of light close to Sirius, almost lost in the brilliance of the larger star. Careful computations and measurements determined that this speck was indeed the long-searched-for companion of the star that appears brightest in the sky.

Note that a star's position on the H-R diagram has nothing to do with its position in space, but refers only to the point on the diagram that represents its absolute magnitude (or luminosity) and its spectral class (or temperature.) When the H-R diagram is constructed using only those stars visible to the naked eye, it gives the impression that the giant stars are quite numerous. However, this is a selection factor caused by their high intrinsic brightness, which allows them to be seen at greater distances. Over 50% of the stars with the greatest apparent brightness (those listed in the *Yale Catalogue of Bright Stars*) are giants or supergiants. Yet very few of these are found close to the sun, as compared with the numerous main sequence stars.

Appendices IV and V list the brightest stars and the nearest stars. Make an H-R diagram using these two groups of stars (use two different colors to differentiate between the two groups). Why is an H-R diagram constructed by using only the nearby stars or only the brightest stars misleading?

Actually, the vast majority of stars are less luminous than the sun. The relatively few stars which are more luminous contribute the major part of the starlight that we see with our naked eye, despite the fact that more than 90% of the stars in our own part of space are below the sun on the main sequence and are therefore less luminous, whereas less than 1% of these nearby stars are on the giant or supergiant branches. The rest of the neighborhood stars fall either on the upper main sequence or on other regions of the H-R diagram. Figure 7.12 illustrates this distribution of stars on the "ladder of luminosity," which graphically shows the hierarchy of luminosity and stellar mass.

On the H-R diagram you will note that there are several places where there are stars of the same spectral class but with widely different luminosities or those with the same luminosities but different spectral

---

Once again locate the stars Arcturus and Spica in the sky (see Chap.6). Can you detect any difference in their colors? Arcturus should appear somewhat orange; it is a K-type star. Spica will appear bluish; it is a B-type star. Using their absolute magnitudes as found in the listing in Appendix IV, locate them on the H-R diagram. What kind of stars are they? How big are they compared to the sun? Locate some other bright stars and then check their spectral types and locations on the H-R diagram.

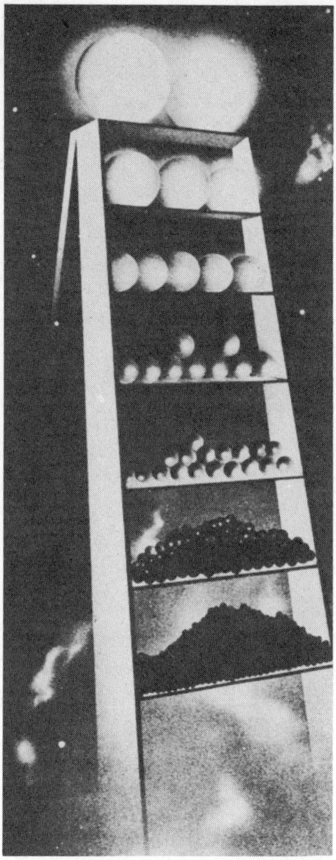

7.12 The Ladder of Luminosity. Most stars are on the lower part of the main sequence. A very small percentage of all stars are giants or supergiants. (From J. A. Hynek, Challenge of the Universe. New York: Scholastic Press, 1965.)

classes. For example, the star Capella and the sun are both G-type stars, yet Capella is some 100 times more luminous (about five magnitudes brighter). Since they both have the same surface temperature, this must mean that Capella has a radiating surface 100 times larger than the sun. Since surface area is proportional to the square of the radius of a spherical object, Capella must therefore have a radius ten times the radius of the sun (see Fig. 7.13).

The relationship between surface temperature, luminosity, and size may be easier to understand with the following example. Imagine that $\frac{1}{8}$th of a mile away a solitary candle is burning. Nearby but at the same distance a cluster of 1000 identical candles is also burning. Even though each candle of the 1000 has the same temperature as the single candle standing alone, the cluster will obviously be far more luminous. In the

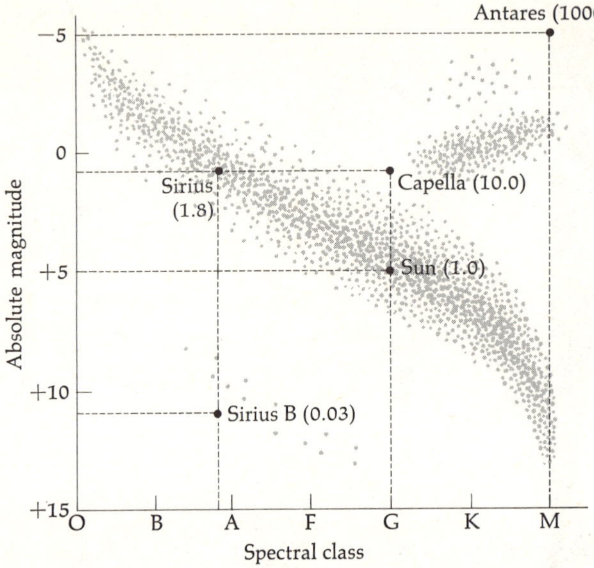

7.13 The comparison of various stars on the H-R diagram. The number in parentheses indicates the approximate diameter of the star. (Sun = 1.0)

same way, a star can have the same surface temperature per square centimeter as another star, but if it has a bigger radius, it will have a greater luminosity.

In Figure 7.13 you will also note that Capella has the same absolute magnitude as Sirius, but Capella is relatively much cooler. Since a cooler object emits less energy per unit area than a hotter one, Capella must also be much larger than Sirius. Comparing the size of Capella to either Sirius or the sun gives the reason for the name "giant." The term "supergiant" is reserved for stars whose radii may be several hundred times that of the sun. One such star is Antares. Even with a surface temperature only about one-half that of the sun, it is about 10,000 times more luminous. If Antares were substituted for the sun in our solar system, its atmosphere would extend out to beyond the orbit of Mars!

At the other extreme are the white dwarfs such as Sirius B, which is about the size of the earth. However, its mass is nearer to that of the sun. That means that its density (actual amount of matter per volume of space) is about 50,000 times that of water. Some white dwarfs are as small as the planet Mercury (about ⅓rd the size of the earth) and have densities up to one million times that of water. A golf ball of such matter on the earth (if it were possible to get it here) would weigh several tons!

Thus, from a star's luminosity and surface temperature, its size can be determined. Whether it is a giant or a main sequence star can also be deduced from its spectrum. For example, the spectral lines due to ionized atoms are stronger in the spectra of giant and supergiant stars than they are in the spectra of main sequence stars. This indicates that the atmospheric pressure of larger stars is lower than that of smaller stars. Low pressure means that the gas is less dense—there are fewer particles per volume of space. The particles of matter that are present have less chance of contact with each other and therefore an atom can exist for a longer period of time in an ionized state before it is neutralized by capturing an electron. The longer the atom stays ionized, the more chances it will have to absorb and emit energy at the ionized level, thus strengthening those spectral lines.

Thus, atmospheric pressure is very closely related to the size and therefore to the luminosity of stars of a given spectral class. Once the absolute magnitude (luminosity) is determined, the star's distance can be obtained (the apparent magnitude is, of course, obtained by observation). This method of distance determination is called *spectroscopic parallax*. It can be used for stars far more distant than those measured by trigonometric parallax.

Total Energy Output

Although we use the magnitude scale for much of the research in astronomy, it is arbitrary and relative. To have any real meaning, it must be put in terms of the actual quantity of energy output. What does +4.8 magnitude mean in terms of the sun's actual candlepower or energy? What is the wattage it is actually producing?

By definition, a watt equals 10^7 ergs/second, an *erg* being an extremely minute amount of energy—approximately the amount expended by a mosquito flapping its wings just once! A 60-watt bulb gives off 60×10^7 ergs/second. In other words, a watt is a rate of expending energy.

Imagine a huge sphere around the sun, with a radius the size of the earth's orbit. Any photon of energy from the sun would have to pass through this sphere (although a minute amount of energy would be absorbed by Mercury and Venus). If the amount of energy that reaches one square centimeter on the earth could be accurately measured (since the earth would be on the imaginary sphere at all times), then by elementary geometry the amount reaching the entire sphere could be calculated.

The Smithsonian Astrophysical Observatory has made just such measurements over a period of years. It has been found that 1.36×10^6 ergs/second are received from the sun by one square centimeter on the earth (as measured just outside the earth's atmosphere). This is called the *Solar Constant*. If we compute the surface area of our imaginary sphere and multiply this figure by 1.36×10^6 ergs/second, the total output of the sun's energy comes to 4×10^{33} ergs/second. This is the energy expended by 400 billion billion million-watt bulbs in one second. The earth actually intercepts only about one two-billionth of this energy. The great majority of it keeps going out into space.

Since a star's real luminosity is directly related to its energy output, every star of absolute magnitude +4.8 has an energy output of 4×10^{33} ergs/second. A star of absolute magnitude 0.0, which is approximately five magnitudes brighter and therefore about 100 times more luminous, will have an energy output of $(4 \times 10^{33}) \times (100)$ or (4×10^{35}) ergs/second. Thus, the energy output of all stars can be calibrated if we know their absolute magnitude.

Sizes of the Stars

Although the radii of some stars in a binary system have been calculated from observations of their light curves and radial velocity curves, for most stars another method must be employed to determine their sizes. We start with several laws of radiation, namely Planck's, Wien's, and Stefan-Boltzmann's, all named after the physicists who first discovered them. They all incorporate the concept of a "perfect radiator" or *blackbody*. Such an idealized object will absorb all of the radiation it receives, reflect none, and its temperature will depend entirely upon this radiation. The amount and distribution of the energy it re-emits can then be precisely calculated from the three radiation laws.

The hot gases found in stars closely approximate blackbodies in that they are also very opaque, that is, they absorb practically all the radiation they receive. This is why it takes a million years or more for energy created in a star's core to make its way to the surface. As we will see in Chapter 14, it is only in the outer 200 kilometers of the sun in a layer called the photosphere that the photons of energy are finally free to escape into space.

According to Planck's law, a blackbody of any temperature will emit radiation at all wavelengths, but not in equal amounts. The hotter the blackbody, the more it emits at all wavelengths, but it emits the greatest proportion of its energy at shorter wavelengths than would a cooler blackbody. This is why hot stars appear blue (shorter wavelength) and cool stars appear red (longer wavelength). Figure 7.14 shows three different "Planck curves" for blackbodies of different temperatures. The energy distribution is schematically plotted against the wavelength. Our sun has a surface

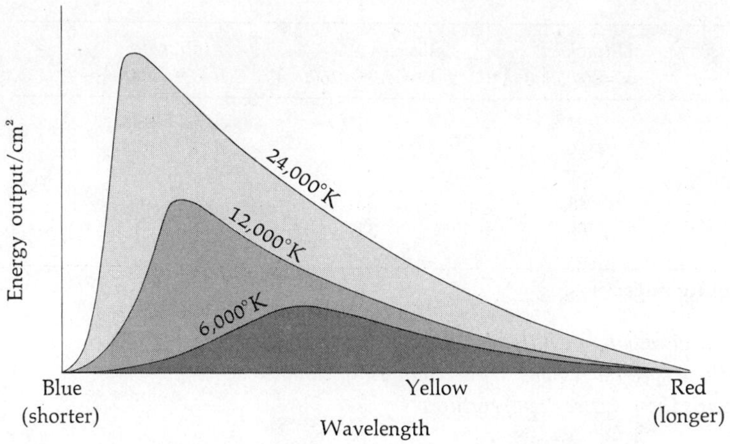

7.14 The energy distribution of three blackbodies of different temperatures according to Planck's law. This diagram is schematic. If drawn to scale, the 12,000° curve should have 16 times as much total area under its curve as the 6000° curve and similarly, the 24,000° curve should have 16 times as much total area as the 12,000° curve, or 256 times as much as the 6000° curve.

temperature of about 6000 K, whereas a hot, blue star would be about 20,000 K.

Wien's law states that the wavelength at which a blackbody emits the most energy is inversely proportional to its temperature. The higher the temperature, the shorter the wavelength at which the maximum amount of energy is emitted. In Figure 7.14, this peak is quite obvious for each of the Planck curves represented. At low temperatures, this peak of emitted energy is not in the visible region but rather in the infrared.[6] Conversely, a substance at a temperature of many thousands of degrees will radiate most of its energy in the ultraviolet region (see Fig. 7.15).

The total amount of energy per square centimeter from the entire electromagnetic spectrum is represented by the total area under the curve. According to the Stefan-Boltzmann law, this energy is proportional to the temperature of the blackbody raised to its fourth power. Thus, a star with

---

[6]It is interesting to note that a healthy person with a temperature of 98.6° F will radiate in the far infrared region. This is why infrared devices like "snooperscopes" can "see" people and objects in complete darkness.

Temperature, °K	*Ultraviolet* λ = 0–4000 Å*	*Visible* λ = 4000–7600 Å	*Infrared* λ = 7600 Å—∞
1,800	0%	1%	99%
2,300	0	3	97
5,000	7	38	55
10,000	48	36	16
20,000	86	11	3

*Å is symbol for angstroms.

7.15 The percent of radiation emitted by "blackbodies" of various temperatures in three spectral regions. This relationship can also be applied to stars with these surface temperatures.

twice the temperature of our sun radiates 16 (2^4) times as much energy per square centimeter.

The total energy that a star radiates is equal to the amount radiated per square centimeter multiplied by the surface area. Earlier in this chapter we found that we could determine the total energy or luminosity of a star by comparing its absolute magnitude with that of the sun's, using the solar constant as a basis for our calculations. From the Stefan-Boltzmann law, given the temperature of the surface of the star, we can calculate the amount of energy emitted per square centimeter. The total energy divided by the energy per square centimeter will give us the number of centimeters of surface area and from that we can obtain the stellar radius.

To estimate the surface temperature of the star most simply, we note its predominant color. Arcturus, for example, is yellowish-orange; Vega and Rigel are bluish-white. Of course, these stars emit energy in all wavelengths, but as is stated in Wien's law, their predominant color (or wavelength) is determined by their temperature.

For a more precise temperature determination, the stellar spectrum is used. Stellar classifications are based on surface temperature and it is from these that we get our best data. Once we know the stellar temperature, we can determine the amount of energy it emits per square centimeter, using the Stefan-Boltzmann law. From that we can then calculate the size of the star.

We find that stars vary greatly in size, from approximately the size of the earth to that of the entire orbit of Mars or more (Fig. 7.16). The sun, with a diameter of 1,391,700 kilometers (864,400 miles) is about midway between these two extremes. However, this is not the average value for all stars, since there is a great preponderance of small, faint stars which are visible only through a telescope.

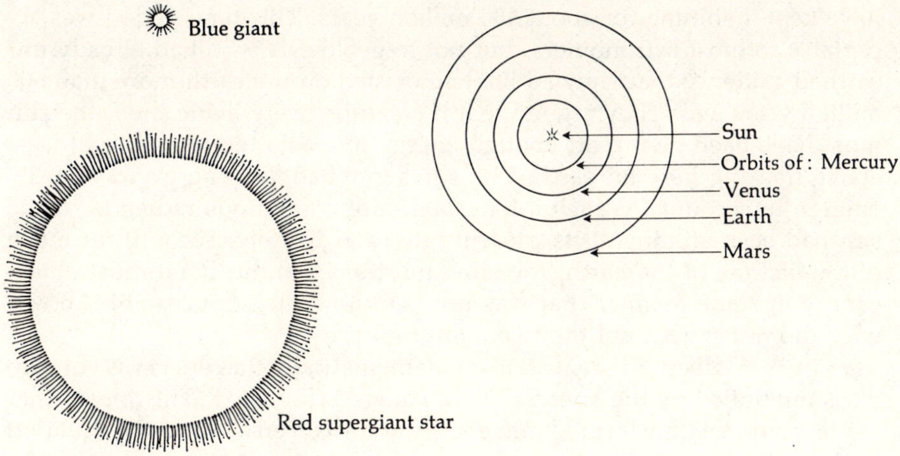

7.16 A comparison of some stellar sizes with the orbits of the four innermost planets.

Because the total energy output (which we measure by absolute magnitude) depends upon both the size and the energy output per square centimeter, a larger star is not always the brighter one. As we will see in the next chapter, there are tremendously large cool, red stars whose luminosity is no greater than some much smaller, hot blue stars.

Creation of Stellar Energy

All the stars radiate enormous amounts of energy. We see some of this energy each clear night. We see and feel some of the energy of our own star—the sun—every day. Where does this energy come from? How is it created? How long has the sun been shining?

In the seventeenth century, Archbishop James Ussher of Ireland calculated that the world was created at 9:00 A.M., Sunday, October 23, 4004 B.C. Although the theologians did not dispute this date on scientific grounds, the scientific world rejected it because the time scale was absurdly short and unrealistic.

Later, in the nineteenth centry, the physicists Helmholtz and Kelvin theorized that if the sun got its energy by slowly contracting, thus transforming its gravitational energy into radiant energy, this process would have kept it shining for about 100 million years. This time period was acceptable to most astronomers, but not to geologists who had already unearthed evidence that showed life had existed on the earth more than 500 million years ago. They argued that if creatures were living then, the sun must have been giving off enough energy to create temperatures at least above freezing here on earth. The sun's gravitational energy was hardly enough to account for such a long period of continuous radiation. If the sun had been shining at its present rate (4×10^{33} ergs/second) for much of the lifetime of the earth, the sun (and therefore the stars) must obtain energy in some manner that was not yet known to science. This knowledge did not come until the twentieth century.

In 1905 Albert Einstein showed mathematically that energy is equal to mass multiplied by the speed of light squared ($E = mc^2$). This meant that matter could be transformed into energy and vice versa. Since the equation contains the term "c^2", even a very minute quantity of matter would be transformed into a large amount of energy. One gram of matter (about $\frac{1}{28}$th of an ounce) is a concentrated form of 9×10^{20} ergs or about 90 billion kilowatts. In other words, one ounce of matter completely transformed into energy could operate 2800 modern cars for an entire hour!

If we apply the formula $E = mc^2$ to the sun we get:

$$4 \times 10^{33} \text{ ergs/sec} = m \text{ (grams)} \times (3 \times 10^{10} \text{ cm/sec})^2$$

Solving this for the mass m, we find that it equals 4.5×10^{12} grams, or 4.5 million tons. That means that every second the sun uses about 4½ million tons of its mass to produce 4×10^{33} ergs of energy. By terrestrial standards, this loss seems incredibly great, but the sun's mass (2×10^{27} tons) is not appreciably decreased by this rate of attrition, even in the course of several billions of years.

It was not until the late 1930's that the mechanism that causes this transformation of stellar matter into energy became known. Although there are several peripheral mechanisms that can operate, especially in the early and late stages of a star's life, the two nuclear mechanisms that are believed to furnish the nuclear energy by which stars shine are the *proton-proton reaction* (p-p) and the *carbon cycle.* Both take place deep in the stellar interior. The carbon cycle is particularly effective in the more massive stars, whereas the p-p reaction is primarily responsible for the energy production in stars the mass of our sun or smaller.

In the p-p reaction, four hydrogen nuclei are converted into one helium nucleus, as shown in Figure 7.17. In the carbon cycle, carbon acts as a catalyst for the conversion of hydrogen into helium. Note that only the nucleus of the atom is involved in these nuclear conversions. At the ex-

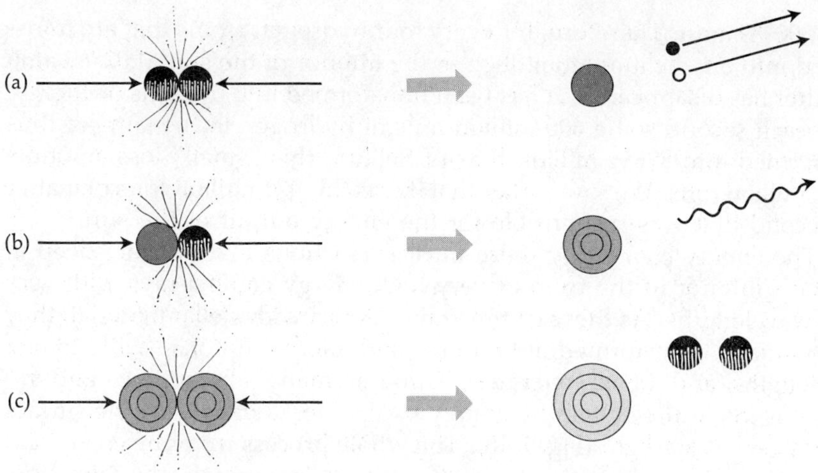

7.17 The proton-proton reaction. (a) Two protons (hydrogen nuclei) collide, producing a deuterium (heavy hydrogen) nucleus plus a positron and neutrino (which are two tiny particles). (b) The deuterium nucleus collides with another proton, producing a light helium nucleus (2 protons and 1 neutron) plus a photon (of gamma ray energy). (c) Two light helium nuclei (both formed by the above processes) collide, producing a regular helium nucleus (2 protons and 2 neutrons) plus two protons which are now free to start the process all over again.

tremely high temperatures found in stellar interiors, the forces holding the electrons to the nucleus have long since been overcome and the atoms have been "stripped" of all their electrons.

We sometimes say that the sun and stars are "burning." This is really a misnomer. The kind of burning that takes place in our fireplaces is simply the result of orbital changes in the outer electrons of the atoms involved. These ordinary chemical reactions do not create changes in the atomic nucleus. When changes in the nucleus occur, they are nuclear reactions, involving far greater forces. Nuclear reactions are sometimes referred to as "nuclear burning" but the two types of "burning" must not be confused. One is the type of burning that occurs in a bonfire, whereas the other is the type of "burning" that occurs in an H-bomb.

Careful measurements in atomic mass units[7] have shown that four hydrogen nuclei weigh 4.03252 amu, whereas one helium nucleus weighs

---

[7]The *atomic mass unit* (amu) is a unit of weight used in atomic physics and is based on the weight of one oxygen atom.

only 4.00389 amu. Therefore, for every four hydrogen atoms that are transformed into one helium atom deep in the interior of the sun, 0.02863 amu of matter has disappeared. It has been transformed into photons of energy. Since each second some 600 million tons of hydrogen in the sun are thus transformed into 595½ million tons of helium, this "small" loss amounts to 4½ million tons. We saw earlier that it was this 4½ million tons of matter per second that was responsible for the energy output of the sun.

The energy created by these nuclear reactions first appears deep in the star's interior in the form of very high energy gamma rays with very short wavelengths. As these gamma rays interact with stellar material, they are gradually transformed into more and more photons, with longer wavelengths and lower energies. Through many absorptions and re-emissions in all directions, the energy finally works its way to the surface and appears as starlight (Fig. 7.18). The whole process from the creation of the energy to the time it finally reaches the stellar surface may take more than a million years. It is this impeding of energy or radiation within the stellar interior that allows us to use the term *opaque* to describe the stellar gases.

Theoretical calculations show that the p-p reaction and the carbon cycle mechanisms operate only at very high temperatures. The p-p reaction requires five million degrees, whereas the carbon cycle requires 15 million degrees or more. These are the temperatures indicated for the deep inte-

7.18 Energy is created as gamma rays in stellar interiors finally work themselves to the surface in the form of many light rays with much less energy.

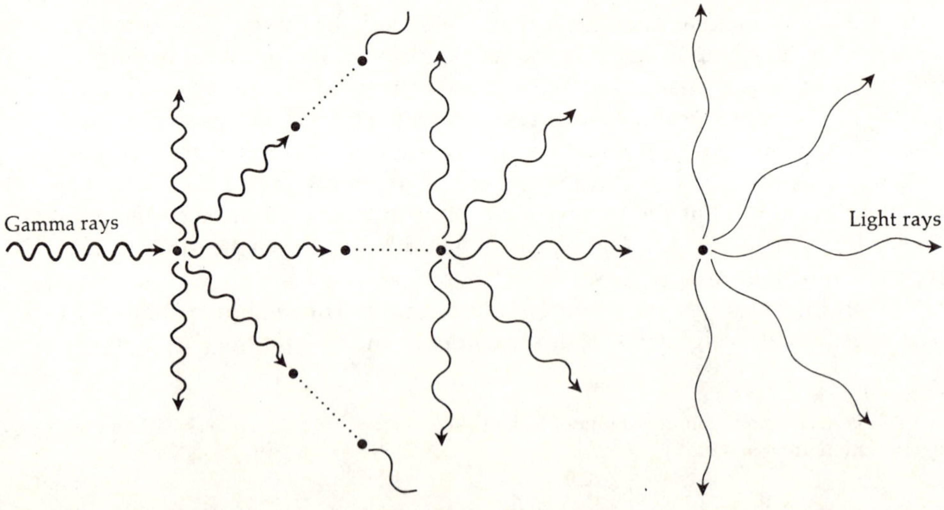

riors of stars. Such calculations also show that these mechanisms will produce enough energy at the proper rate to account for the outflow of energy from the star's surface. Although these mechanisms describe only the conversion of hydrogen to helium, heavier elements also are created in a similar manner, but at much higher temperatures. However, such temperatures are found in stellar interiors at much later stages of a star's evolution. The conversion of hydrogen to helium is the most important process that occurs during most of the star's life.

Discussion Questions

1. If the stars exhibited only continuous spectra, how would this limit our knowledge of their physical nature?

2. As late as the 1940's, some schoolchildren were still being taught that "matter can be neither created nor destroyed." Discuss how this concept has changed in the last 35 years.

3. Since the atmosphere of a star absorbs and almost immediately re-emits light, why are there absorption lines at all?

4. What is the difference between emission and absorption lines?

5. Why does a star shine? How is the energy created? Where is it created? How does it get to the surface of the star?

6. In "$E = mc^2$, what does each letter stand for? What is the meaning of this formula?

7. Why can we describe the interior stellar gases as "opaque"?

8. What is the difference between the Balmer and Lyman series of hydrogen lines?

9. The greater the distance the electron jumps when it is excited, the ______ (longer, shorter) the wavelength and the ______ (higher, lower) the energy.

10. What type of spectrum would you get from the light emitted by a neon bulb?

11. List the steps in determining a star's size by the use of the laws of radiation.

12. Star *A* has a surface temperature of 5000 K, whereas star *B*'s temperature is 20,000 K. Which star emits more radiation per square centimeter of surface area? Which star emits a greater percentage of its radiation in the visible region of the spectrum? In which part of the spectrum would the other star emit the most energy?

13. An average home uses about 10,000 kilowatt hours per year. What equivalent amount of matter, if completely transformed into energy, would supply this home with the year's energy? How many such homes could be run for a year on one ounce of matter if it were completely transformed? (1 kilowatt hour = 3.6×10^{13} ergs; 1 ounce = 28 grams.)

14. Why does the surface temperature of a star play the major role in determining the appearance of its spectrum?

15. Given a star's apparent magnitude, its trigonometric parallax, and its spectral type, how would you go about placing it on the H-R diagram?

16. Why does the presence of a specific absorption line of a chemical element in a star prove that that element is present, whereas its absence does not prove the absence of the element?

17. Consider two stars: one is an A-type main sequence star and the other is a K-type supergiant.

    a. Which star has the hotter surface temperature?
    b. Which one is brighter?
    c. Which one is larger?
    d. What color is each one?
    e. Which one will show the stronger hydrogen lines in its spectrum?
    f. Locate each on the H-R diagram.

18. What are the differences in the spectrum of an A-type star and an M-type star if both are main sequence stars?

8 *The Life Cycle of a Star*

Deep in the stellar interiors, hydrogen is transformed into helium by the process of nuclear "burning." The energy that is produced by this process we see ultimately as starlight. A star's light is measured in terms of its luminosity or brightness. We know that a star that is brighter than the sun has a greater total energy output. This means it is transforming more hydrogen into helium per second than the sun is. The mass of most stars ranges from one-tenth to fifty times that of the sun. Luminosities, however, range from as little as 1/100th to over one million times that of the sun.

A star that is, say, 1000 times more luminous than the sun will use up its hydrogen "fuel" 1000 times more rapidly. If it were the same mass as the sun, it would use it up in 1/1000th of the time taken by the sun. It is, of course, a more massive star, but even if it is ten times more massive, it will still use up its fuel in 10/1000th or 1/100th of the time taken by the sun. On the other hand, a star that is only one-tenth the solar mass and 1/100th as bright will taken ten times longer to transform its hydrogen into helium. As an everyday example of this, a small campfire with a modest fuel supply can last much longer than a blazing bonfire even though the bonfire has many times as much fuel.

Star Clusters

The relationship between a star's mass and its rate of energy expenditure is clearly seen when we study star clusters. These are groups of stars moving together through the galaxy in relative proximity (Fig. 8.1). Binary and multi-star systems are also groups of stars, but the clusters we are referring to consist of a few hundred or more stars held together by their mutual gravitational bond. They move together through space like a flock of geese.

The most common type of cluster is the *open cluster* such as the Hyades or Pleiades (Fig. 8.2). Hundreds of open clusters have been found near the plane of the galaxy, usually in or near the spiral arms. Open clusters are loosely formed with no special distinguishing shape.

8.1 *Open star cluster in the constellation Cancer. The size of the star images as well as the "points" on many of them are caused by the long exposure and photographic process (photograph from the Hale Observatories).*

8.2 *The Pleiades cluster, an open star cluster. Note the nebulosity still surrounding each star (photograph from the Lick Observatory).*

Locate the Pleiades in the fall or winter night sky. If possible, view this group of stars with both the naked eye and with a small telescope or good pair of binoculars. In each case, how many stars can you distinguish?

We may assume that for any given group or cluster, the stars in it were formed from the same mass of interstellar gas and dust and are therefore all approximately the same age. If this is so, then their differences cannot be a factor of age. If we plot the stars in each cluster on an H-R diagram, we notice very definite differences from cluster to cluster (Fig. 8.3). Comparing them (Fig. 8.4), we note that the length of the main sequence varies markedly. In some clusters, there are obviously no longer any O and B-type main sequence stars (H & χ Persei), whereas in others, even A-type stars are missing (Hyades).

We also note in the diagrams that the clusters without upper main-sequence stars have well-defined giant sequences. It would seem likely that these giant stars were once on the main sequence as upper main-sequence stars. The theory of stellar evolution states that this is indeed true. These upper main-sequence stars were more massive and therefore used up their hydrogen in the deep interior core more quickly than the less massive, lower main-sequence stars. When this happened the star

8.3 *H-R diagrams of several open clusters.*

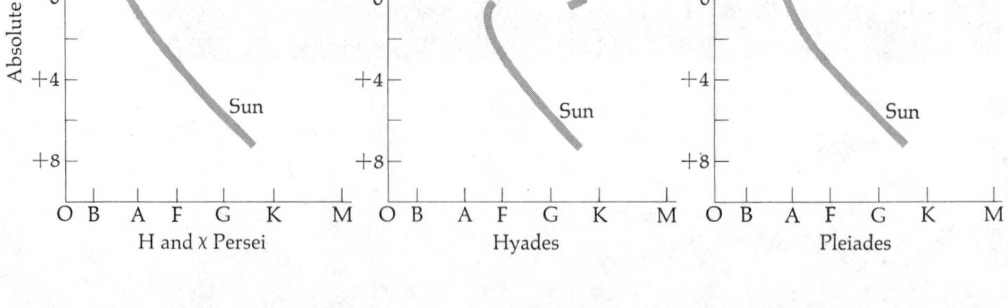

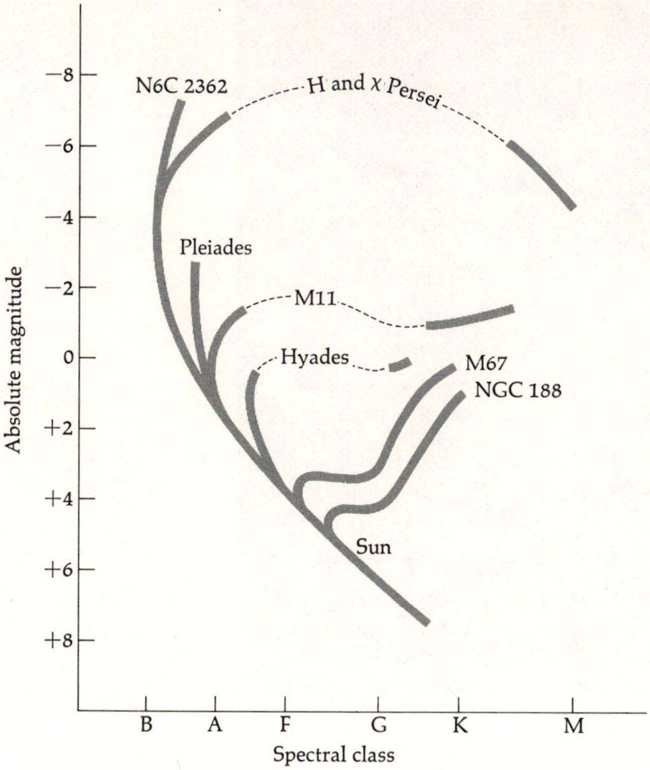

8.4 *The H-R diagram for several open clusters superimposed upon each other. (After Sandage,* Astrophys. J. *CXXV (1957) 436.)*

changed, both internally and externally. In doing so, it became a red giant, and accordingly is plotted on the giant branch of the H-R diagram.

The main sequence of these clusters can be thought of as "peeling off" to eventually form the giant sequence. The older the cluster, the more the "peeling off" has progressed. Since all the stars in a cluster are the same age, the amount of "peeling" gives us a clue as to the age of the cluster. The Hyades cluster must be older than H &χ Persei, since more of its stars have left the main sequence.

There is another type of star cluster, which is very different from the open clusters. These are the *globular clusters* (Fig. 8.5). Only about 125 are known in our galaxy and they tend to be distributed around the nucleus or center of the galaxy, forming a sort of spherical halo. None are near our sun and most are very far from the galactic plane. Figure 8.6 shows their dis-

8.5 A globular cluster.

8.6 Schematic distribution of globular clusters around the nucleus of the galaxy.

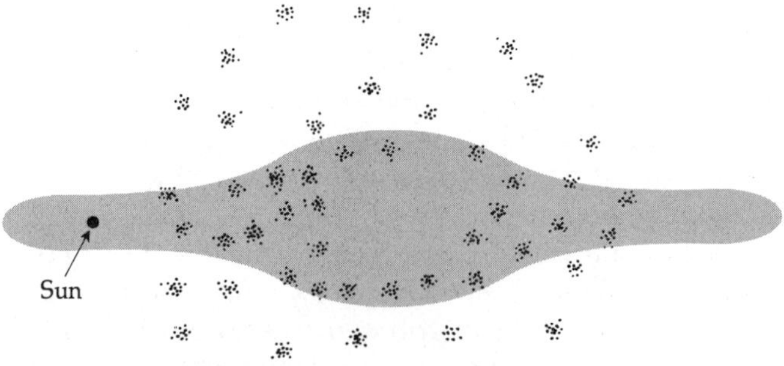

tribution, which was determined after the discovery that half of the known globular clusters in the constellations Scorpius and Sagittarius, which are in the direction of the galactic center. Similar clusters are found distributed spherically around the centers of other galaxies.

A globular cluster contains many thousands of stars relatively closely packed into a spherical shape of some 100 light years across. If we plot an H-R diagram of these stars, we see that the "peeling off" is much farther advanced than with the open clusters (Fig. 8.7), thus indicating their greater age. The upper main sequence is missing down to where the giant branch appears to join the very short lower main sequence. From what has been learned from the open clusters, these globular clusters are judged to be extremely old. Estimates of their age on the order of many billions of years have been made, as compared with the open clusters, whose ages range from perhaps one million to one billion years. In the globular cluster, there has been enough time for stars to evolve even beyond the normal giant stage; these are found in the "horizontal branch" which stretches across the diagram.

Thus, we see how very important the H-R diagram has been in determining the relative ages of star clusters and in giving the first clues of how stars evolve. When we compare the ages of stars we must take into consideration this rate of evolution. The more massive the star, the faster it

8.7 The H-R diagram of a globular cluster (schematic).

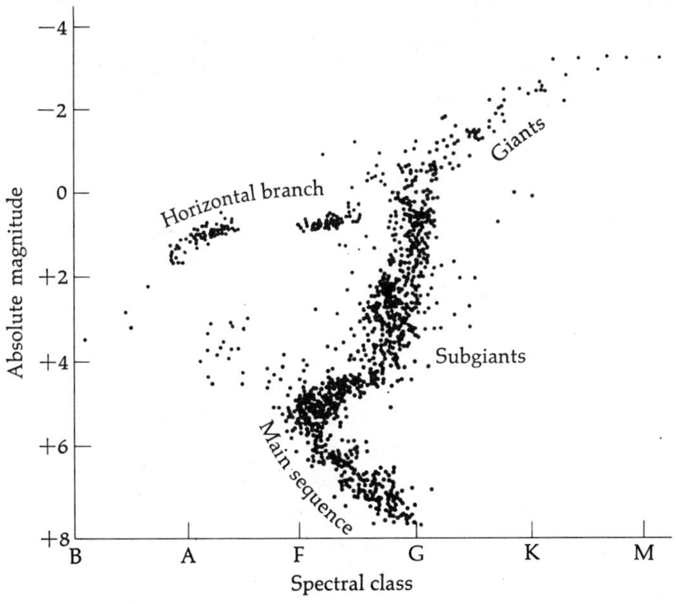

will evolve. A star can be chronologically young yet evolutionarily old, much in the same sense that a one-month-old mouse is "older" as far as its life cycle is concerned than a one-year-old elephant.

A very busy photographer takes pictures of every section of a crowded ball park on the last day of the World Series. From such "crowd scenes" how can he then go about determining the general characteristics of baseball fans? He can judge such things as size, sex, race, weight. From these, could he get some inkling as to how the human race ages? What possible factors would stand in the way of a true evaluation of such a study? (For example, the unlikelihood of any newborn baby being in the audience.)

Stellar Evolution

Stellar evolution is the term applied to the slow structural changes a star undergoes during its lifetime. It is the aging process of a star. Do not confuse the term "evolution" as it is used here with its use in the biological realm, where it refers to the changes in a species over an interval of time comprising the lifetime of countless sequential individuals.

Throughout our galaxy (and others as well) there is a large amount of gas (mainly hydrogen) and dust that lies between the stars. This matter is very unevenly distributed, sometimes forming into relatively dense clouds. However, even the densest clouds contain less matter per cubic centimeter than any vacuum attainable on earth. The amount of dust and gas is considerable only because the size of these clouds is so enormous.

Interstellar dust becomes visible to us when it reflects or obstructs light from the stars in its neighborhood. A good example of this is the Horsehead nebula in the constellation of Orion (Fig. 8.8). The center of our galaxy would shine as brightly as the full moon were it not for the dust in the plane of the galaxy obstructing our view. The "dark rift" that runs lengthwise down the Milky Way galaxy and in many others as well (Fig. 8.9) is caused by this dust. Even the naked eye can readily detect breaks along the relatively smooth whitish band in the Milky Way.

Interstellar gas can be detected optically when it is luminous or by radio observations when it is dark. It becomes luminous either when it reflects the light of nearby stars or when the ultraviolet light of much hotter

8.8 The Horsehead nebula in the constellation Orion. This is an example of a "dark nebula," a cloud of interstellar dust that obscures starlight (photograph from the Hale Observatories).

nearby stars causes it to *fluoresce* (Fig. 8.10). Fluorescence occurs when the ultraviolet light from a hot star excites the otherwise cool atoms in the cloud, raising their electrons to higher orbits, from which they then cascade downward, emitting energy in the visible region of the spectrum.

Most of the gas in space is cool hydrogen in its lowest energy state, which cannot be detected optically from the earth. However, such hydrogen does radiate a single *radio line* of a wavelength of 21 centimeters. This radio wavelength is about 400,000 times longer than the light waves of the Balmer lines. This low-energy radiation does not occur as a result of orbital jumps of electrons. The proton and electron of a hydrogen atom rotate or spin on their respective axes. If both are spinning in opposite directions (one clockwise and the other counterclockwise), the atom as a whole has a slightly lower energy than if the two motions are aligned. This alignment can occur when the electron in the hydrogen atom is raised to a very slightly excited state—not one of its regular orbits. This can happen when

8.9 A spiral galaxy in the constellation Andromeda photographed with a 60-inch telescope. The edge-on view of this galaxy shows the dust lanes down its center plane. Do not confuse this galaxy with the great Andromeda galaxy (Fig. 2.6) (photograph from the Hale Observatories).

the atom collides with another atom or with a free electron. Its electron "flips" over so that it is spinning in the same direction as the nucleus (the proton). When the atom loses this extra energy, it returns to its ground state, emitting a photon of 21-cm wavelength. This process is extremely rare and may take millions of years before the minute amount of energy is reemitted, but statistically there are enough atoms in the hydrogen cloud so that the 21-cm radiation is detectable.

As the interstellar clouds of dust and gas move through space, local concentrations of matter form. These may be caused by a shock wave or some other force that pushes the atoms and molecules closer together. Each of these concentrations of matter or *globules* is gravitationally bound,

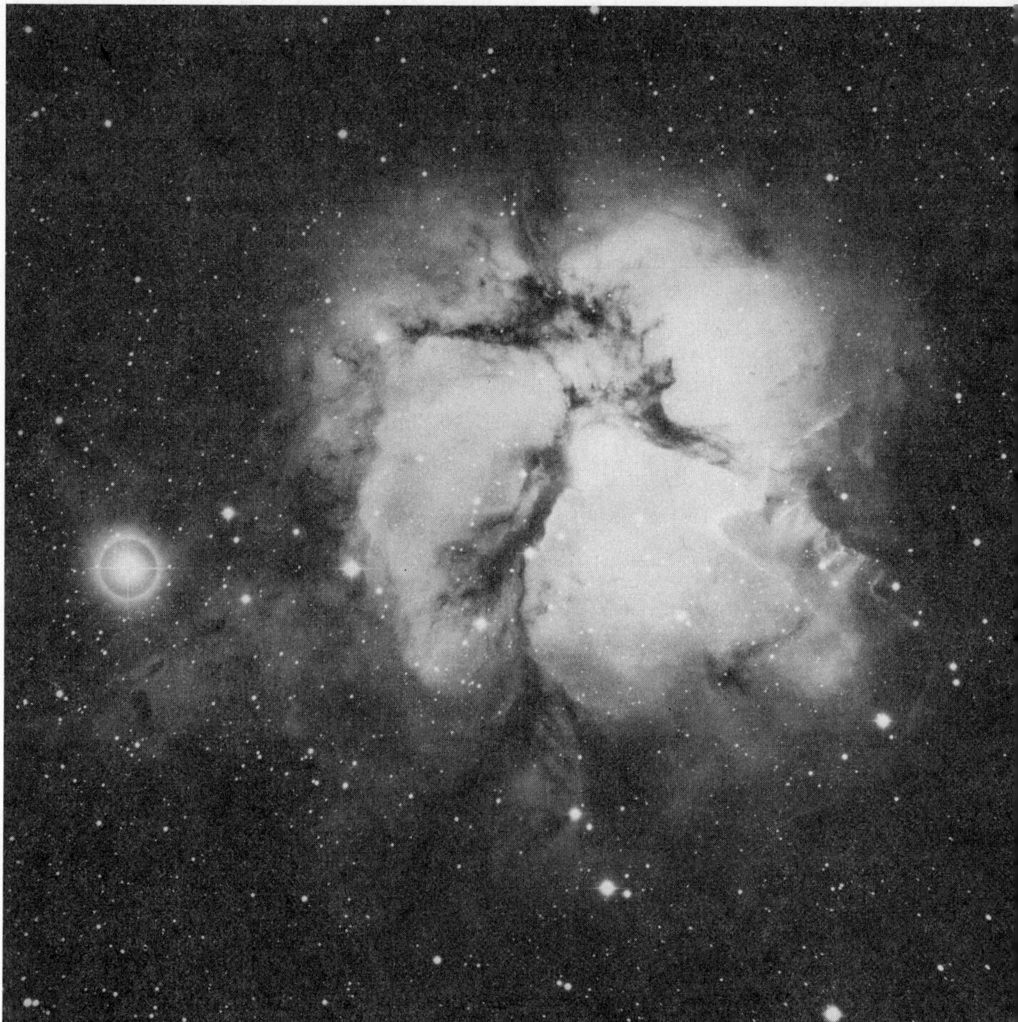

8.10 The "Trifid" Nebula (M22) in the constellation Sagittarius as photographed with the 200-inch telescope. This is an example of an "emission nebula," one that fluoresces (photograph from the Hale Observatories).

and in time is able to attract more material to itself, increasing its size and density. The more massive it becomes, the greater its gravitational attraction, causing still more matter from greater and greater distances to fall towards it, increasing its mass still further. Eventually it begins to contract under its own gravitational attraction, converting gravitational energy to heat energy. When the matter becomes dense enough to be opaque to the radiation of energy, the globule starts to heat up. A *protostar* is born.

Newborn stars do not shine at first in visible light. They radiate brightly in other regions of the spectrum where most regular stars appear very dimly or not at all. Clouds such as the Orion nebula (see Fig. 2.9) have been called "stellar nurseries" because newly-forming stars have been detected among their gases and dust particles, radiating in invisible wavelengths. In dense dark clouds such as the Coal Sack nebula, which lies near the Southern Cross constellation, thousands of globules of matter (or protostars) have been detected by their radio and infrared radiation.

As the protostar continues to contract and become more dense, especially in its central core, its internal temperature and pressure increases. There comes a point at which the gravitational contraction or "infall" of material is stopped by the heat pressures pushing outward. An equilibrium is achieved and a stable "star" is formed. The young star still continues, however, to derive its energy from gravitational attraction. This energy is carried to the surface by means of convection currents. The physicists Helmholtz and Kelvin believed a star derived its energy from gravitation throughout its entire life (see p. 139). We know now that such a method of producing energy could not account for the long period of time the stars continue to shine.

During this period of convective heat transfer, the star is actually brighter than a star of similar mass on the main sequence. It then gradually moves downward from a very high point on the H-R diagram as it contracts further and becomes less luminous. It moves along its *Hayashi line*, named after the Japanese astrophysicist C. Hayashi who first computed and plotted this part of the H-R diagram (Fig. 8.11). A star of less than about one-half the solar mass reaches the main sequence after this stage. A more massive star, after several thousand or million years of moving downward along its Hayashi line, reaches the point where its central core is so hot and dense that convection currents give way to radiation as the main method by which energy is transported. The star now starts to move horizontally from right to left across the H-R diagram as its internal temperature and pressure rise even further. It is getting hotter but not brighter.

Just before it reaches the main sequence, the temperature and pressure in the central core become great enough for nuclear processes to begin. This "ignition" may create enough of a shock wave within the star's cloud to start the formation of yet other stars. The protostar becomes a full-fledged star on the main sequence at a point determined by its mass, and it stays at that point for most of the remainder of its life. Once on the main sequence, the star's hydrogen starts to be converted into helium by the processes described on page 141. Millions of years may have been re-

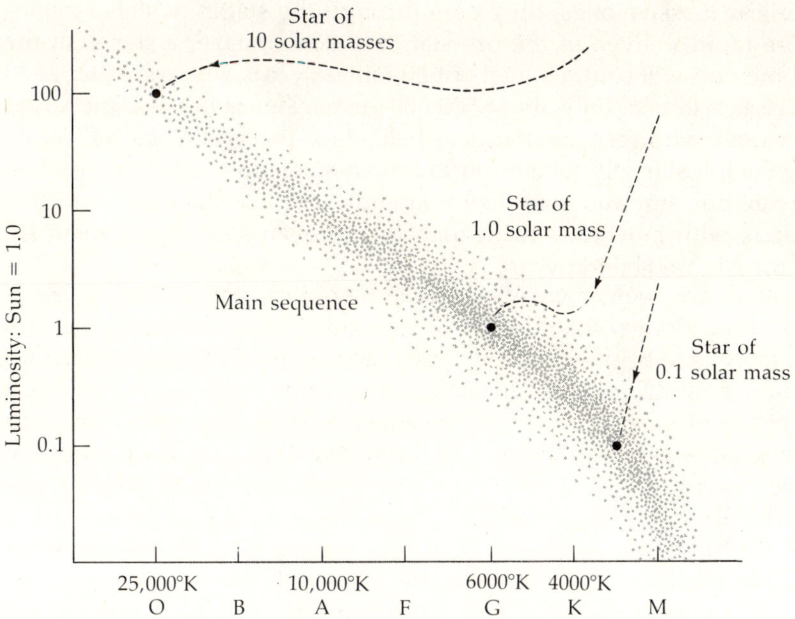

8.11 "Hayashi lines": Theoretical evolutionary tracks of pre–main-sequence stars on the H-R diagram.

quired for the star to pass through the pre–main-sequence stage but by comparison, a star with the mass of our sun will stay on the main sequence for many billions of years. The main sequence is thus the locus of stars in that stage of stellar evolution in which hydrogen is converted into helium in their deep interiors. Ninety-nine percent of the star's existence will be spent on the main sequence.

The more massive a star is, the greater its gravitational attraction. To counterbalance this inward pressure, the central core must produce a greater amount of energy than would be necessary in a less massive star. It has been calculated that the energy production of a star is roughly proportional to the cube of its mass. Thus, a star on the main sequence with twice the mass of the sun will produce about eight (2^3) times as much energy. Its surface area, however, will not be eight times greater at this stage of its evolution, and therefore more energy will have to be emitted per square centimeter than is necessary on the solar surface. This will give it a higher luminosity and put it higher on the main sequence than the sun. Similarly, because it uses less energy per second, a less massive star than the sun will be lower on the main sequence.

Since the more massive stars use up their fuel at a much faster rate than the less massive ones, they pass through the stages of stellar evolution more rapidly. Even in the pre-star contraction stage, a star with the mass of our sun will contract in about 50 million years whereas a star of 20 solar masses will take only about 500,000 years. This is because the larger mass creates a stronger gravitational field, i.e., the gases "fall in" more rapidly. Such a star will remain on the main sequence only a few million years while our sun and stars like it spend some 10 billion years in this stage. Stars with much less mass than the sun can stay on the main sequence for 30 or 40 billion years.

Because the cooler (less massive) fainter stars stay on the main sequence so much longer than the massive bright ones, and because they are more numerous to begin with, there are many more of them on the ladder of luminosity (see Fig. 7.12). In almost any volume of space, there is a much greater number of lower main-sequence stars than upper ones. In fact, if we could include all the stars in the galaxy, the upper main sequence would be a mere sprinkling of stars whereas the lower part would be solidly filled.[1]

As we discussed on page 140, in the nuclear process of converting hydrogen into helium, some of the matter is changed into energy. Actually this amounts only to 7/10ths of one percent of the original hydrogen and therefore the stellar mass remains virtually constant. Not until much later in the star's evolution will the mass change significantly.

Red Giant Stage

Eventually, after millions or billions of years (depending upon the mass of the star) the hydrogen in the star's core is depleted. It has been practically all converted into helium. Since helium requires hotter temperatures to "burn," the nuclear pressures pushing outward stop. The core begins to collapse from gravitational attraction. At the same time, the hydrogen surrounding the core continues to "burn," much like a prairie fire which burns around the edges of the already burned-out area. The helium it produces falls into the core, pulled there by the core's gravitational attraction. The core therefore becomes denser and hotter as it contracts and gains more

[1] Remember that when we refer to a star's position on the H-R diagram, this has nothing to do with its position in space. The H-R diagram is merely a way of plotting stellar variables in order to learn more about their physical nature. Moving on the H-R diagram (or staying in one position) has nothing to do with stellar motions through space.

matter. The star's energy now comes from this contraction of the core as well as from the hydrogen burning on the core's edges.

As the core contracts, the gravitational energy released causes the outer parts of the star to expand outward. Only the central core remains dense; the surrounding gases become enormously distended. With a much greater surface area, the energy distribution per square centimeter is far lower and the surface temperature therefore drops. This happens even though the star's total luminosity has become much greater than when the star was on the main sequence. The star has become a *red giant*.

The fact that such a short period of time is required for stars to pass from the main sequence to the red giant and supergiant branches is probably the reason why few, if any, stars are found in the transitional region of the H-R diagram (see Fig. 8.12). This empty space is called the *Hertzsprung gap*. Stars with greater mass move through this gap much faster than stars with less mass.

As a red giant or supergiant, the star's surface temperature may be only between 3000 K and 7000 K, but its size gives it a luminosity 100 to 10,000 times that of the sun. It has grown anywhere from 10 to 500 times the solar diameter. If it becomes a red supergiant, such as Antares, which has 50 times the sun's mass, its volume will be 64 million times greater than

8.12 The H-R diagram showing the "Hertzsprung gap." Very few stars are found in this region, and are believed to pass through it very quickly.

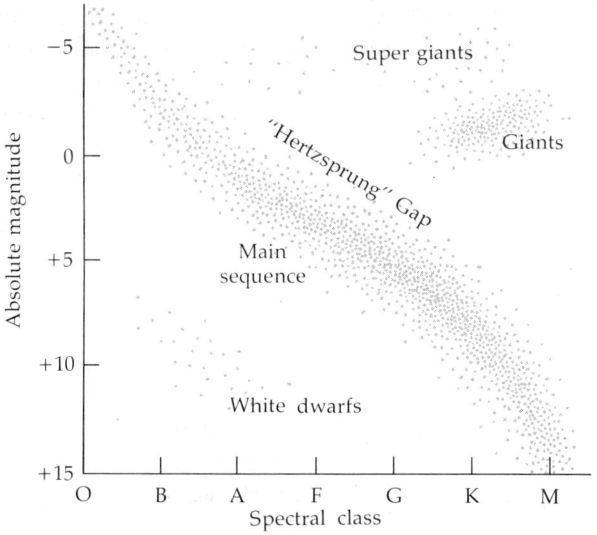

the sun's. But because it is so distended, its mean density will only be one ten-millionth that of the sun. Its atmosphere will actually be a better vacuum (i.e., with fewer particles of matter per unit volume) than can be produced here on earth in a laboratory!

The helium core of the red giant continues to contract under its own weight and its temperature continues to rise as the internal pressures become greater. At about 100 million degrees, nuclear reactions involving helium begin to be possible. The most important of these is the formation of a carbon nucleus by the combination of three helium nuclei. This is called the *triple-alpha process* because a helium nucleus is also known as an *alpha particle.* The start of the triple-alpha process is believed to be extremely rapid—so much so that its onset is called the *helium flash.*

With stars the mass of the sun or less, once the helium has been converted into carbon, the mass of the star is not great enough to create the internal pressures and temperatures necessary to continue nuclear "burning." The core collapses and the star enters its final evolutionary stage. Very gradually it contracts to the size of the earth or smaller, becoming a white dwarf. Because no new energy is created, the star slowly cools off, eventually to become a burned-out cold chunk of extremely dense matter.

With the more massive stars, however, still heavier elements are produced when the carbon formed by the triple-alpha process is bombarded by other helium nuclei. In fact, many of the chemical elements we know to exist in the universe are believed to have been created in the central cores of these red supergiant stars.

As each of these nuclear processes takes place, it produces more energy. This may expand the core and cause the outer regions to contract. The star will then move on the H-R diagram to the right as its surface temperature increases (with less surface area through which the energy can flow). Once the helium in the core is all transformed into carbon and other heavier elements, the star may again return to the red giant branch. A crossing of the H-R diagram may actually happen several times. Each time still heavier elements are created in the central core. However, elements heavier than iron, such as gold, uranium, or lead must wait until the later stages of the evolution of very massive stars for their creation.

Variable Stars

During most of a star's life it shines with a constant luminosity. In its old age, beyond the red giant stage, the star goes through an unstable stage when the intensity of its light can become noticeably variable in periods short enough for us to measure. It is thought that the particular type of stellar instability which is produced depends upon the mass and composition of the original star. Each star may take a different evolutionary path and which path a given star may take is not well understood.

There are many classes of unstable or variable stars, all with their own characteristics. Some expand and contract periodically, causing their luminosity to vary. Others throw off large quantities of their mass, either periodically or in one tremendous explosion. Because so many stars are in binary systems, exchange of matter from one star to the other is not an uncommon way of producing instability.

A star which periodically expands and contracts, causing its luminosity to vary, is called a *pulsating variable.* The characteristic differences of these types of stars are best illustrated by their *light curves.* These curves indicate how a star's brightness varies with time (Fig. 8.13). In Chapter 2 we saw how the variable light of the Cepheids enabled astronomers to find the distances to nearby galaxies. As the Cepheids expand and contract in size, their luminosity varies. A similar but much more common type of pulsating variable is the RR Lyrae type of star. These are actually a kind of Cepheid, but are smaller, denser, and pulsate more rapidly than the classical Cepheid. However, their light curves are similar. The RR Lyrae stars are found mainly in globular clusters and surrounding the nucleus of the galaxy, far from the galactic plane. Because of their similarity to the classical Cepheids, they were originally thought to be classical Cepheids in the globular clusters and were therefore originally used incorrectly for distance determination to these clusters. Now that their true luminosities have been determined and correlated with their periods of variability, their distances and the distances of the globular clusters in which they are found have been more accurately determined. RR Lyrae stars are considered to be among the oldest of stars.

A particularly fascinating group of pulsating variables are the *long-period variables,* so called because their periods can be even longer than a year. These are the most common type of variable. A notable example is Mira in the constellation Cetus. Mira (Omicron Ceti), has an average period of 330 days, and can vary in visible light from far below naked eye visibility to as bright as the 2nd magnitude. Like all long-period variables, Mira is a cool red giant. Its variability arises from relatively minor changes in its surface temperature as it expands and contracts, since the total energy that is emitted changes very little. However, because most of that extra energy is in visible light, we see a variation in luminosity of about 100 times or more.

Pulsating variables go through these fluctuations because the dense inner core and the thin outer layers of the star cannot reach the state of equilibrium that the star had in previous stages. The weight of the outer

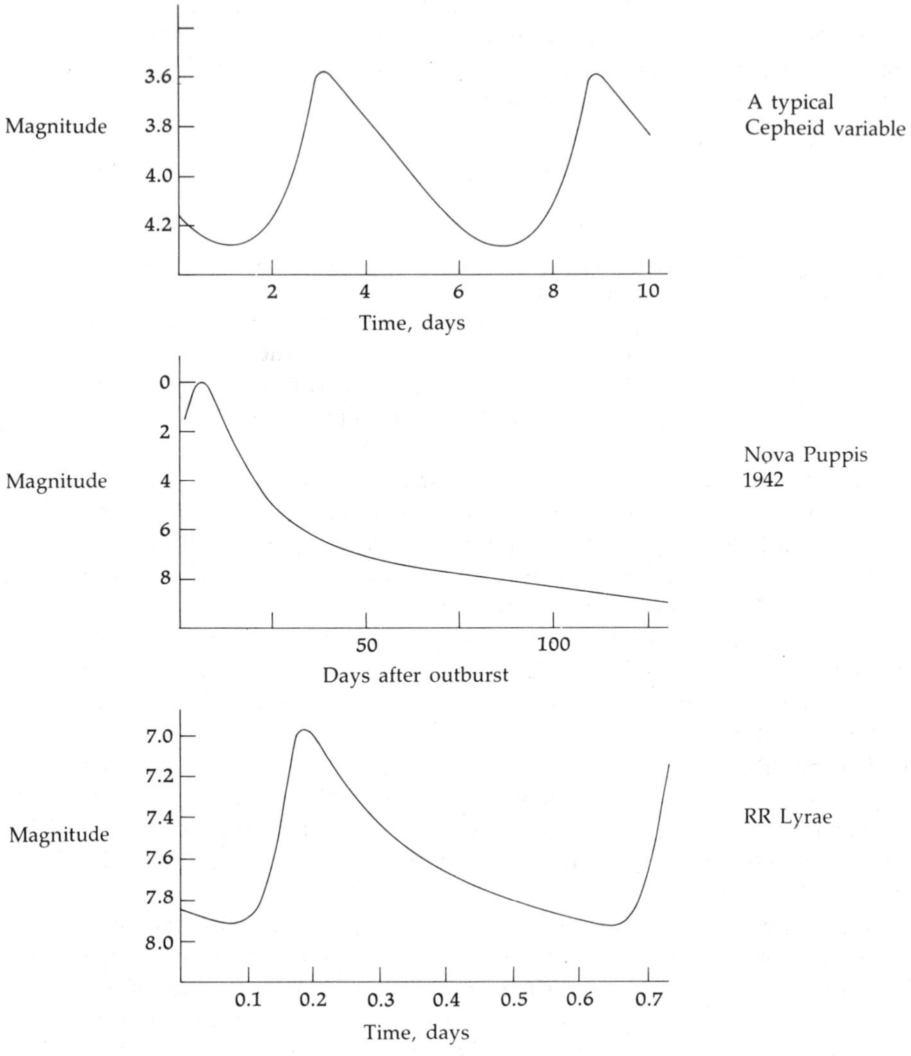

8.13 *The light curves of some variable stars.*

layers pushing inward must balance the pressures of the very hot gases pushing outward. If they do not balance, the star will be unstable and keep contracting and expanding in an attempt to equalize the pressures.

There are several groups of unstable stars that, instead of contracting and expanding, actually have extended atmospheres which are expanding gas shells. We learn of these conditions from studying the stellar spectra,

Here are the hypothetical light curves of two stars. How would you describe each? Which one brightens most rapidly? Which one dims most rapidly? What other differences do you notice?

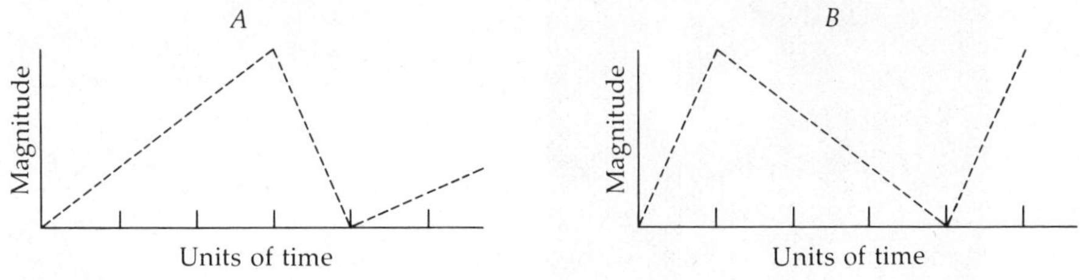

which show pecularities not found in normal stellar spectra. *Planetary nebulae* are examples of these stars. They have enormous expanding gas shells surrounding them which superficially resemble a planet when viewed through a telescope, hence the name. However, they have nothing at all to do with the planets. The gaseous shells are probably thousands of times larger than the entire solar system.

It is thought that the central stars in these planetary nebulae eject successive shells of gas into space, each shell containing perhaps 10% to 20% of the original mass of the star. The shells attain diameters many times larger than the original star, expanding outward at speeds of from 20 to 30 kilometers per second. There are only about 1000 planetary nebulae known but this may be due to the fact that the rapidly expanding shell soon becomes too thin to be seen. Since the shell is probably only visible for about 100,000 years, the thousand planetary nebulae that we do see allow us to surmise that a large fraction of all stars pass through this stage. One of the most famous, the Ring nebula in the constellation Lyra, is shown in Figure 8.14.

Draw and label an H-R diagram for stars in general. Indicate by a dotted line the probable future evolution of the sun.

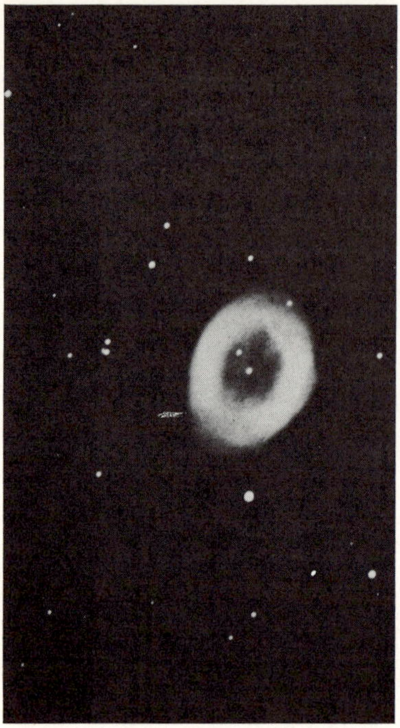

8.14 The Ring nebula, a planetary nebula in the constellation Lyra. Photographed with the 200-inch telescope (photograph from the Hale Observatories).

8.15 Possible evolutionary track of a star with one solar mass.

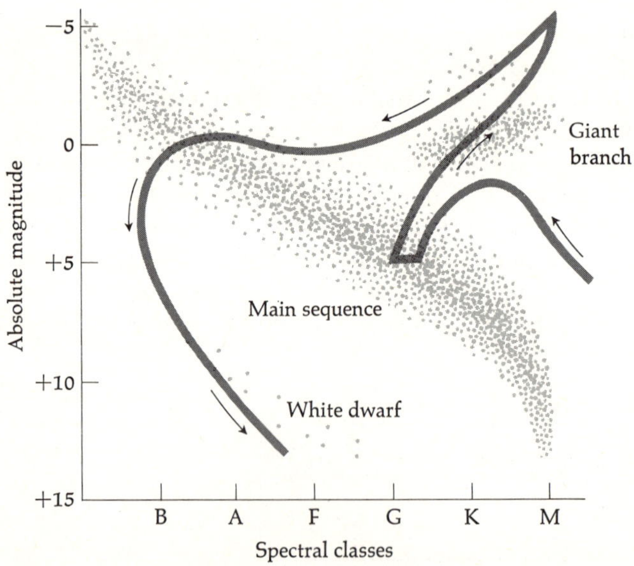

The central star of the planetary nebula is very hot, but with a very low luminosity, indicating its small size. The shedding of its matter by the ejection of gas shells may represent the star's final death throes before it completely depletes its remaining nuclear fuel and collapses into a white dwarf. This is probably the evolutionary path taken by stars with one solar mass or less (see Fig. 8.15).

Discussion Questions

1. If our sun was a star near the center of a globular cluster, what might be our view of the galaxy? What might it be if we were near the edge of the globular cluster?

2. How do astronomers determine the age of star clusters from study of their H-R diagrams?

3. What percentage of mass does a star lose while it is on the main sequence? What happens to this matter?

4. Star *A* is classed as a B8; Star *B* is classed as a F5. Both are main sequence stars. Which star will live longer? Why? How do we know this?

5. In an open star cluster, which of the following are about the same for all of the stars:

   a. parallax
   b. luminosity
   c. age
   d. spectral class
   e. size

6. What is stellar evolution? How does it differ from biological evolution?

7. How are stars created? Is it possible that some are being formed right now? What areas in the sky would astronomers study in order to find a "spawning ground"?

8. According to the theory of stellar evolution, stars move off the main sequence and become red giants at a certain stage in their development. Why then is there a gap between the main sequence and the giant branch? What is this gap called?

9. What are variable stars? Why are they considered unstable? Why are they unusual or peculiar stars?

10. What kind of star does the mass-luminosity relationship apply to? What is this relationship?

11. Star *A* has a mass 10 times that of the sun. Star *B* has a mass of only one solar mass. Which star will have the longer lifetime? Why?

12. Star *A* has a mass equal to 3 solar masses; how does its rate of energy output compare with that of the sun?

13. What is fluorescence?

14. What is the 21-cm line? What causes it?

15. How can the motion of nonluminous hydrogen clouds in space be detected?

9 *Final Stages of Stellar Evolution*

Until the late 1960's, it was thought that all stars in their late evolutionary stages eventually became white dwarfs. Calculations had shown that such a star could not have much more mass than the sun. Therefore, it was thought that somehow the more massive stars lost or ejected much of their matter before this final stage. It is now known that the more massive stars follow different evolutionary paths which do not require the loss of this mass.

White Dwarfs

The majority of stars end their existence as white dwarfs. When practically all the nuclear "burning" is over and no new energy can be created, the star slowly contracts, drawing the atomic particles closer and closer together. The density of the matter becomes about one million times that of water, far denser than any solid substance we know of. Lead, for example, is only 11 times denser than water. The nuclei and free electrons occupy much less space than is required by a normal atom with all its electrons revolving in their orbits. The free electrons are no longer able to move at random because they are so much closer to each other, but they are still moving at high speeds and exerting pressure in all directions. This pressure outward eventually stops the collapse of the star. But by this time, the star is about the size of the earth (but still with a mass on the order of one solar mass). We say that the stellar gases at this point have become *degenerate,* because they do not behave like normal gas, i.e., they do not obey the classical gas laws of physics. A golf ball of this degenerate matter would weigh several tons here on earth!

Because it is a very hot star (A-type) but also very small (low luminosity), the white dwarf is plotted near the lower left hand corner of the H-R diagram. It is thought that since the central stars of the planetary nebulae are also small and very hot, they may be associated with the white dwarf stage of evolution.

The ultimate size of the white dwarf depends upon its mass. The larger the mass, the greater the gravitational attraction inward and therefore the *smaller* the radius. This theoretical relationship was derived by the India-born physicist, S. Chandrasekhar, who also found that a white dwarf cannot have a mass appreciably greater than that of the sun. This mass-radius relationship is shown in Figure 9.1. For a white dwarf to have the radius of the earth, it can have only about one-half the mass of the sun.

At this stage the star's only energy comes from the heat of its core. Since no new energy is being created, it slowly cools off until, theoretically, it becomes completely dark. This cooling off period has been calculated to be many times longer than the accepted age of our galaxy, and thus we would not expect to find any "black dwarfs" in the Milky Way.

Many white dwarfs have been identified and studied, the first and most famous being Sirius B (see p. 131). There are almost certainly far more white dwarfs than are obvious in any star count. However, their extreme faintness makes them difficult to observe.

Novae

When a star is a member of a close binary system, its late evolutionary stages are influenced by its companion. If the two stars have different masses, the more massive one will evolve faster and reach its terminal stage first. If this happens to be a white dwarf, as it most often is, then

9.1 Theoretical mass-radius relationship of white dwarf stars. Note that the larger the mass, the smaller the radius.

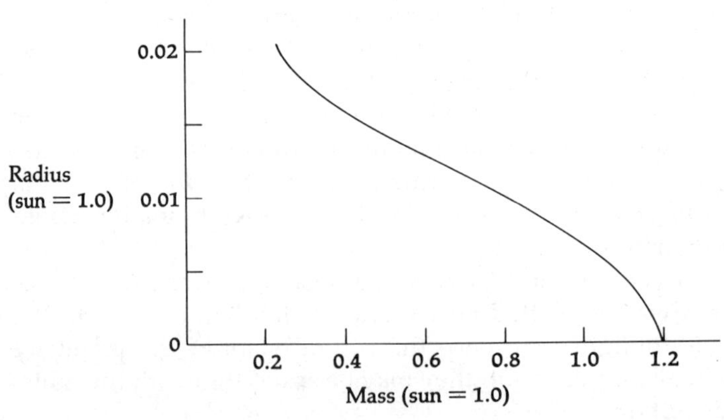

when the second (less massive) member later expands to become a red giant, matter can flow from its extended atmosphere to its white dwarf companion. This causes surface instability in the white dwarf, resulting in a sudden outburst of energy and an ejection of matter into space. This is probably one of the most famous types of *eruptive variables,* and is called a *nova.* The name comes from the original belief that the star was indeed new, that it had just been created. Such a concept is understandable, since previous to the eruption, no star had been seen in that location of the sky. And because the star, after erupting, only remains bright for a short period of time, the Chinese referred to novae as "guest stars." Figure 8.13 shows a typical light curve for a novae.

The white dwarf's luminosity explosively increases tens of thousands of times. But even at their brightest (absolute magnitudes of −6 to −9) most novae are still not visible to the naked eye because of their great distances from us. Therefore, only a few such phenomena have been recorded in past history. It is estimated, however, that each year several dozen such explosions occur in our galaxy alone. Figure 9.2, a photograph of Nova Herculis, illustrates a nova that was visible to the naked eye. Even seen through a telescope, however, a nova appears as all stars do, only as a bright point of light. In the photograph of Nova Herculis, the nova looks larger, but this is only a photographic effect.

As in the case of stars in general, spectroscopic analysis of a nova yields most of the information we have about their physical nature. Mea-

9.2 *Nova Hercules (photograph from the Lick Observatory).*

sures of radial velocity of the nova show that the material is being ejected at often many thousands of kilometers per second. This is roughly equivalent to moving from New York to San Francisco in one second! Some novae have been known to explode several times. This presumably occurs as additional material is transferred from the giant to the white dwarf.

Stars tend to evolve rapidly through these unstable, variable stages. On the H-R diagram these types of stars are generally found in areas not occupied by more common stars. Some of these locations are indicated on Figure 9.3. As can be seen, many of these stars are found in the gap between the red giants and later evolutionary stages. It is thought therefore that they represent various ways a star evolves toward its final stage.

Supernovae

The term"super" does not begin to convey the difference between a supernova and the much less spectacular nova. A comparable difference would

9.3 *The H-R diagram showing locations of some unusual stars.*

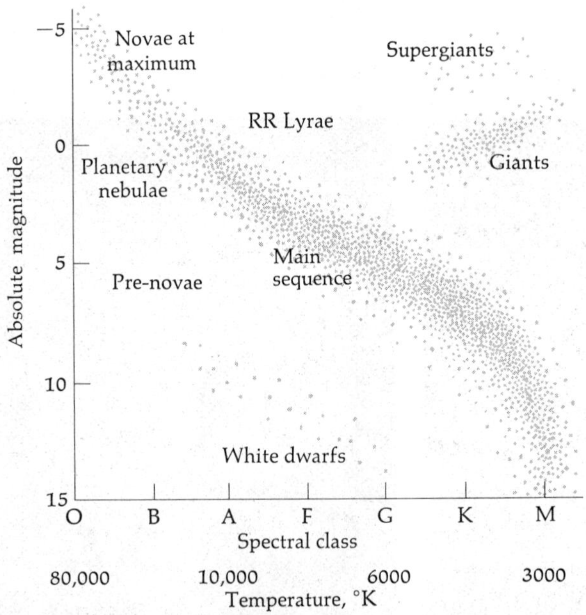

be between a stick of dynamite and a small firecracker. A nova flares to perhaps 10,000 times its normal luminosity, whereas a supernova flares up to hundreds of millions of times brighter than its original state, reaching absolute magnitudes of −14 to −18 or even brighter. For a few weeks, it may even rival in brilliance the entire galaxy in which it is located (Fig. 9.4).

Supernovae and novae are related only in that they are both explosive in nature. They are both eruptive variables, but the dynamics of the explosions and the masses of the stars involved are quire different. As we have seen, the explosive nature of a nova results from the inflow of matter from a companion star onto a white dwarf, thereby causing surface instability. The mass of the white dwarf can be no greater than about 1.2 solar masses.

A supernova involves a star of several solar masses or more. There is a rapid gravitational collapse that combines a tremendous implosion of the core with a cataclysmic ejection of the outer shell matter. The collapse is initiated when the material in the central core, because of complex nuclear reactions, can no longer withstand the tremendous gravitational pressures of the overlying layers. The star becomes more and more compact and the pressures increase. This collapse, rather than a gradual one like that of a white dwarf, is cataclysmic—like the crumbling of a tall building when the foundation gives way. Just as the building no longer has rooms, hallways, and other such spaces, so the stellar mass no longer preserves the spaces between the nuclei and electrons.

The gas in this core is degenerate, but the pressures are so much greater than in the white dwarf that they overcome the electron pressure and force the negatively-charged electrons and positively-charged protons even closer together until they finally combine and become neutrons. This is not the same process as the normal recombination of electrons and atomic nuclei to produce un-ionized atoms. Not until about 90% of the core is composed of neutrons does the strong nuclear force between these particles bring a halt to the contraction, and a *neutron star* is formed.

The energy released by this rapid gravitational collapse of the core causes a tremendous explosion of the overlying layers, which represent a major portion of the star's material. It is these gases hurtling outward at about 5000 kilometers per second that we see and define as a supernova. Although the supernova itself is only visible for a short period of time, its remnants may persist for many centuries. The Veil nebula in the constellation Cygnus (Fig. 9.5) is believed to be such a visible remnant. There have also been some radio sources that have been identified with filaments of gas that are believed to be remnants of prehistoric supernovae.

On the average, supernovae are observed to occur about once or twice a century in any given galaxy. The last two supernovae seen in our galaxy were the famous star seen by Tycho Brahe in 1572, and shortly thereafter, Kepler's star in 1604. When one occurs in our region of the galaxy it is a most spectacular object, shining brilliantly even in the daytime

(a)

(b)

(c)

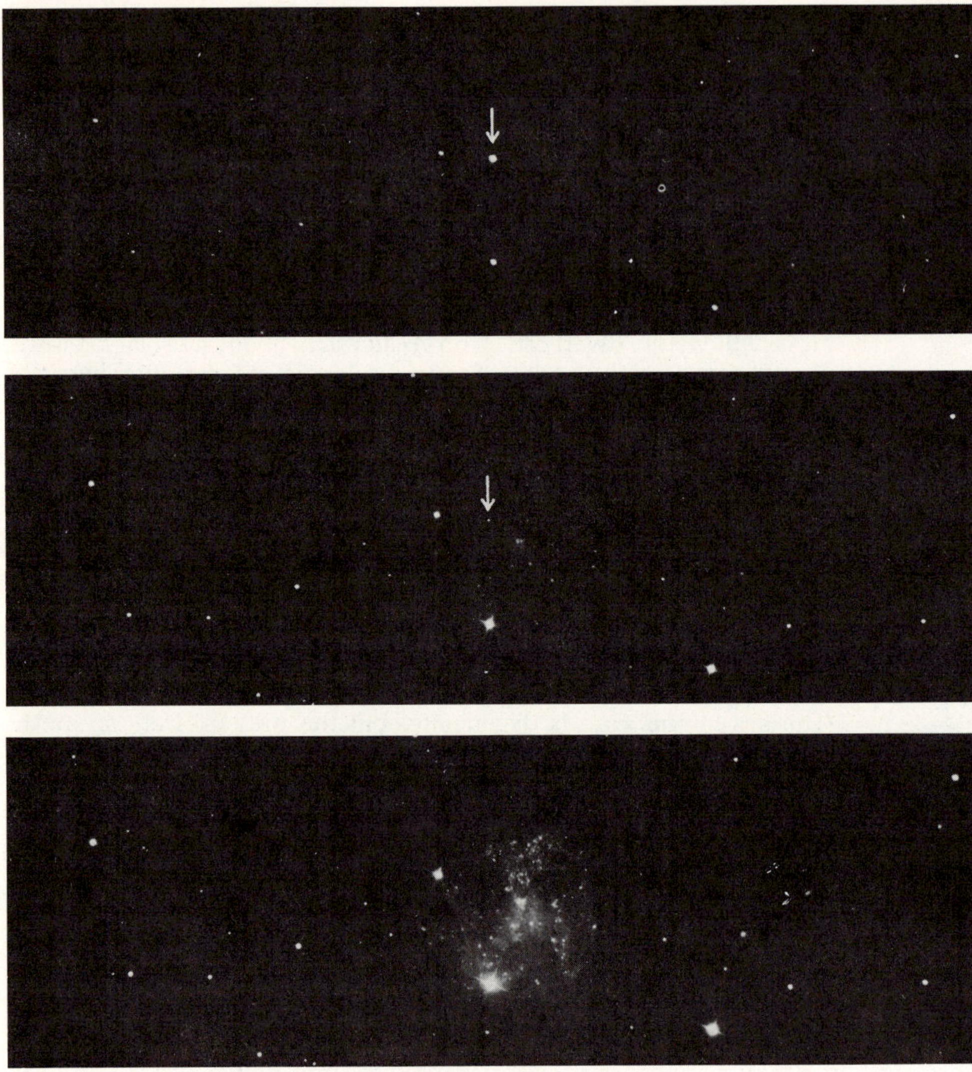

a) 1937 Aug. 23, exposure 20^m, maximum brightness
b) 1938 Nov. 24, exposure 45^m, faint
c) 1942 Jan. 19, exposure 85^m, too faint to observe

9.4 Supernova in a distant galaxy (IC 4182). Three views photographed with the 100-inch telescope (photograph from the Hale Observatories).

a) 1937 Aug. 23, exposure 20^m, maximum brightness
b) 1938 Nov. 24, exposure 45^m, faint
c) 1942 Jan. 19, exposure 85^m, too faint to observe

9.5 The Veil nebula in the constellation Cygnus, photographed with the 100-inch telescope. The circular object is a bright star which appears large on the photographic plate because of the long exposure time (photograph from the Hale Observatories).

sky. In addition to its visible brilliance, if it is a nearby star the earth receives a barrage of x-ray and gamma ray radiation which conceivably can have devastating effects. It has even been conjectured that supernovae in the past may have been responsible for mutation in life forms. The two supernovae mentioned above were sufficiently distant so as not to cause any radiation damage here on earth.

One cannot predict when the next supernova will occur. Statistically we are about due for another one somewhere in our galaxy. However, supernovae are important events in astronomy, and rather than wait for one to be detected in our own galaxy, regular searches for them in other galaxies are currently being conducted. For example, at the Corralitos Observatory operated in New Mexico by Northwestern University, a computerized program has been responsible for the detection and careful study of many extra-galactic supernovae. Although a rare event in any one galaxy, many are discovered each year because so many galaxies are under surveillance.

When an eruption of a star takes place, whether it be a supernova, a nova, or even a peaceful planetary nebula, matter is spewed forth into space. A small percentage of this material is no longer simply hydrogen or helium but has already been transformed into heavier elements. All of this matter—the hydrogen, helium, and heavier elements—may eventually be included in the dust and gas clouds that in time become new stars and even new solar systems. These stars will therefore have greater amounts of heavier elements than the stars formed earlier, since their formative material has already undergone some nuclear transformations.

This difference in chemical abundances can be detected in the stellar spectra and is especially obvious when we compare the spectra and H-R diagrams of the young open cluster stars versus the old globular cluster stars (see p. 151). Note in Figure 8.7 that on the H-R diagram of the globular cluster the giant branch slopes much more steeply upward to giants of much greater luminosity than those on the traditional H-R diagram or the H-R diagram of open clusters (Fig. 8.4). The globular cluster stars, which are much older than the open cluster stars, have a lower percentage of heavier elements. This makes them less opaque to the radiation that is working its way out toward the stellar surface. This results in a more luminous star than normal and therefore a distortion of the the usual H-R diagram. Actually, the percentage of heavier elements (everything except hydrogen and helium) is very small in all stars, but the difference between globular cluster stars and stars in general is enough to noticeably effect the slope of the giant branch.

Our sun is one of the newer stars in the galaxy and it is therefore altogether possible that its iron and phosphorus—as well as that in our own bodies—came from matter spewed forth from an ancient erruptive variable such as a supernova. So next time you eat a delicious steak, you might offer your compliments to a burned-out star!

The most famous supernova was the one observed by the Chinese in 1054 A.D. in the constellation Taurus. Today, we see the remnants of this explosion as the Crab nebula (see Fig. 9.6). From the rate of its current expansion and its diameter, it has been calculated that the explosion and subsequent expansion of the matter into space started some 900 years ago. (Note that it was 900 years ago that the first evidence of this explosion reached the earth. Actually, since the Crab nebula is about 5000 light years away, the explosion itself happened about 4000 years before Christ.) Long known as a powerful radio source and more recently as an x-ray source, the Crab nebula was also found to contain a very unusual object—a new type of star. This discovery confirmed the connection between supernovae and neutron stars.

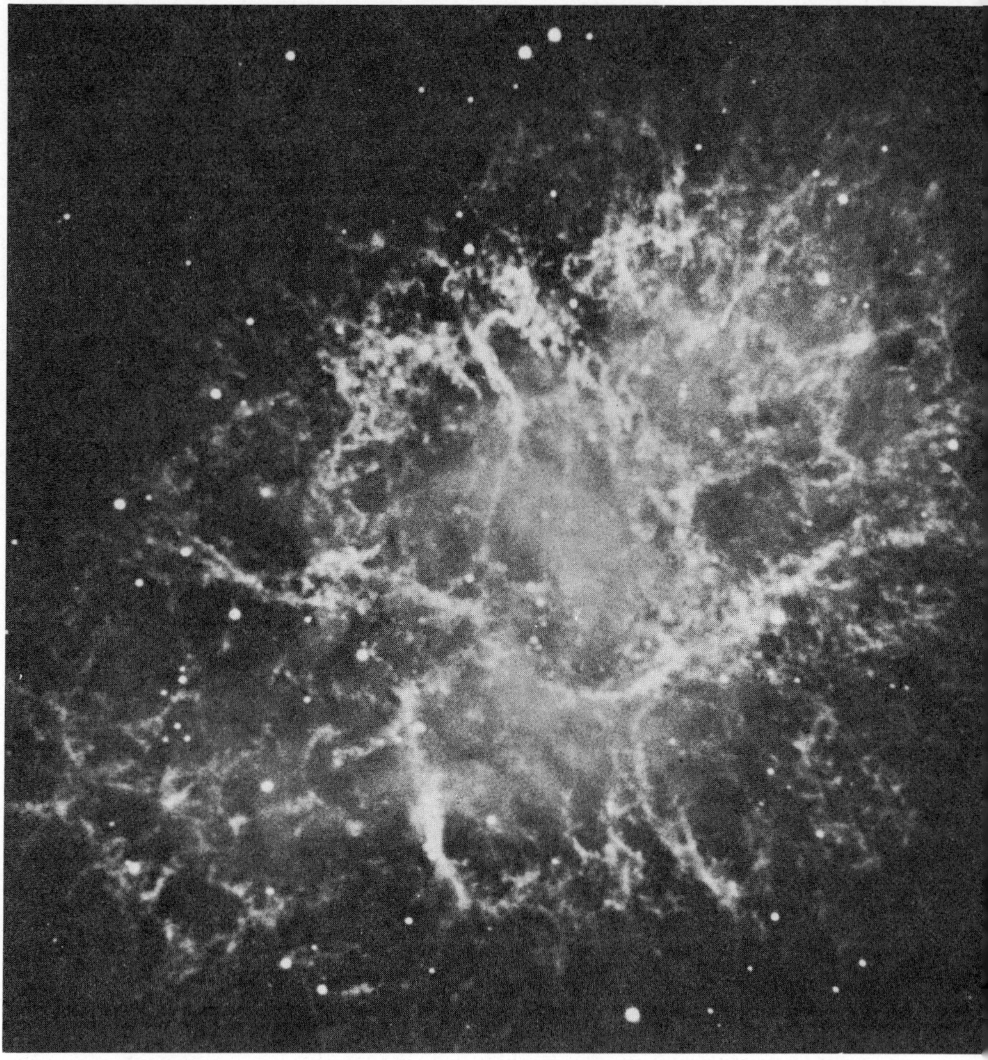

9.6 *The Crab Nebula (photograph from the Lick Observatory).*

The Neutron Star

Physicists and astronomers had hypothesized from purely physical principles the existence of a neutron star more than 30 years ago. It would be a star much denser than a white dwarf, a star in which one cubic inch of material would weigh a billion tons on earth! Whereas the white dwarf has a mass less than 1¼ solar masses, the neutron star would have a mass greater than this amount. A white dwarf collapses to about the size of the earth before the pressure of the electrons halts its collapse. A neutron star would reach diameters of about 15 to 20 kilometers. If all the material of the earth were packed into such density, the entire earth would be about the size of a marble!

Take a full sheet of newspaper (two pages unfolded) and cut out the largest possible circle. Let this represent the sun. On this circle draw an inner circle to represent the size of the sun when it becomes a white dwarf. Within this circle draw another circle representing the size of the sun if it could become a neutron star.

In August 1967, radio astronomers in Cambridge, England, working with a new specialized type of radio telescope, noticed something quite out of the ordinary. A large burst of radio energy came from a certain spot in the sky every 1⅓rd seconds. Actually, the precision of these bursts was even greater—they came every 1.3373013 seconds, which was far more precise than our best radio time signals. Special atomic clocks were necessary to make such time measurements, and only this new telescope was capable of detecting these short, quick changes in radio intensity.

The new radio source was called a *pulsar* because of the characteristic pulsating signal radiating from it. It was designated CP 1919, meaning "Cambridge pulsar at right ascension 19 hours 19 minutes." Rather quickly, more pulsars were discovered, each with a different period but all on the order of one second. Each burst of radio energy lasted only a fraction of a second. This *length* of each pulse must be distinguished from the *period* or interval between each pulse. It was the length of each pulse that gave astronomers information about the size of the pulsars.

Since no object as a whole can produce a pulse of energy shorter in length of time than it would take for that pulse to travel across the object, the shortness of the pulses from the pulsars indicated their relatively small size. If they were, say, a million miles across, the pulse would be observed to be over 5 seconds long (Fig. 9.8). The extremely short bursts of pulsars set an upper limit on their size of a few thousand miles. In actuality, they are much smaller, yet they continuously produce extremely strong radio pulses.

Thus far, only two of the more than one hundred known pulsars have

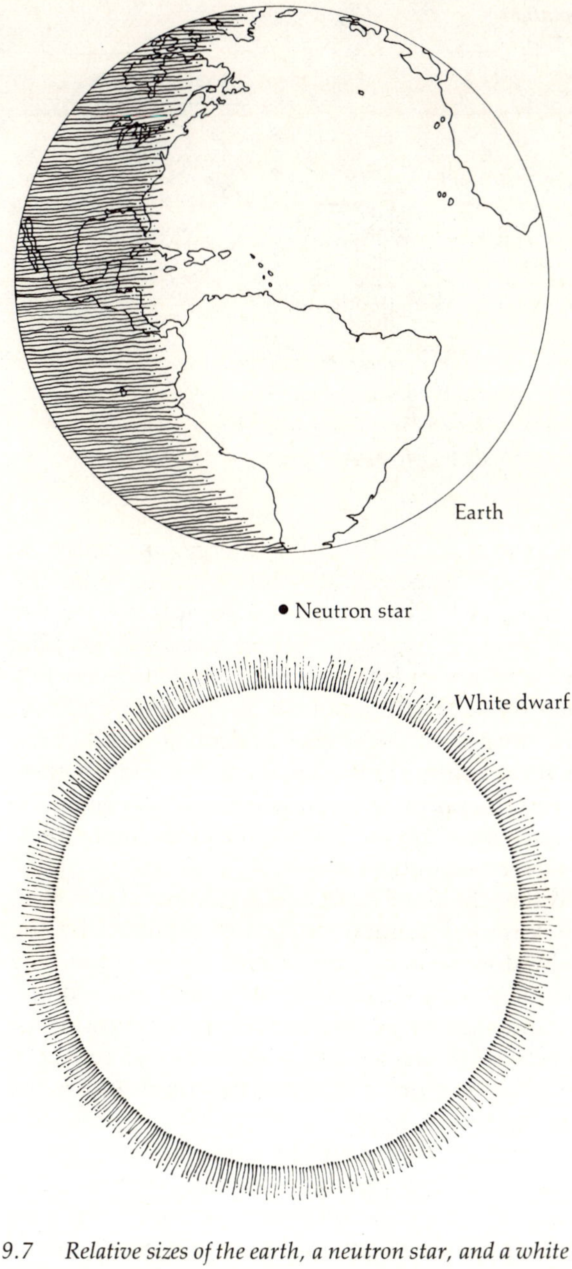

9.7 Relative sizes of the earth, a neutron star, and a white dwarf: A typical white dwarf has about the same radius as the earth, approximately 6000 kilometers, but its mass is comparable to that of the sun. Neutron stars are equally massive but are thought to have a radius of only 10 to 15 kilometers. To ensure visibility the dot representing the neutron star was made several times too large; the period at the end of this sentence would still be twice too large.

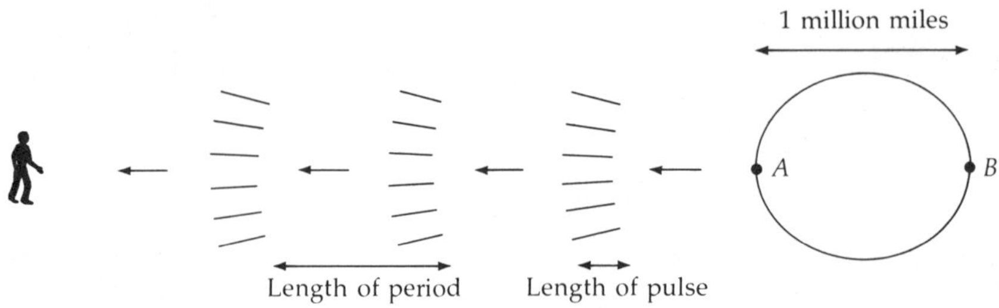

9.8 An "instantaneous" pulse of energy from an object one million miles across cannot be observed to be shorter than five seconds in duration because it would take five seconds longer for radiation from B to reach the observer than it would for radiation from A.

been found to pulse in x-ray energy as well as radio energy, and only one of these two flashes out in light energy as well. It is the pulsar located in the Crab nebula. After its radio pulses were detected, its light pulses were also discovered, flashing on and off at intervals of 1/30th of a second, keeping the same pace as the radio and x-ray pulses (see Fig. 9.9). However, this was a far shorter period than that of the previously discovered pulsars.

When the neutron star was first hypothesized, calculations indicated that such a star would spin very rapidly. This was due to the conservation of its angular momentum[1] when it contracted from the size of a normal star to its present size. Such a star would be but a few miles in diameter and would turn once on its axis every second.

It was soon established that the newly discovered radio pulsars were indeed neutron stars. The extremely regular pulsation is caused by the star's rapid rotation. Like a lighthouse beam, we receive a pulse of energy every time a particular "hotspot" on the surface whirls around in our direction. However, each burst of intense energy lasts but a few milliseconds, or about 1/30th of the period. Finding the pulsar within the nebulous region of the Crab nebula led to the theory that such neutron stars or pulsars are the result of supernovae explosions.

Extensive observations revealed that all of these pulsars had gradually decreasing pulse rates. This further supported the theory that the pulsing neutron star would decrease its rate of rotation as it lost energy through radiation. The pulse rate has therefore been used as a rough age determiner. The extremely short period of the Crab pulsar is an indication of its

---

[1] *Angular momentum* is loosely defined as the quantity of motion of a spinning body. The spinning ice skater conserves angular momentum by rapidly bringing the arms in, thus spinning faster.

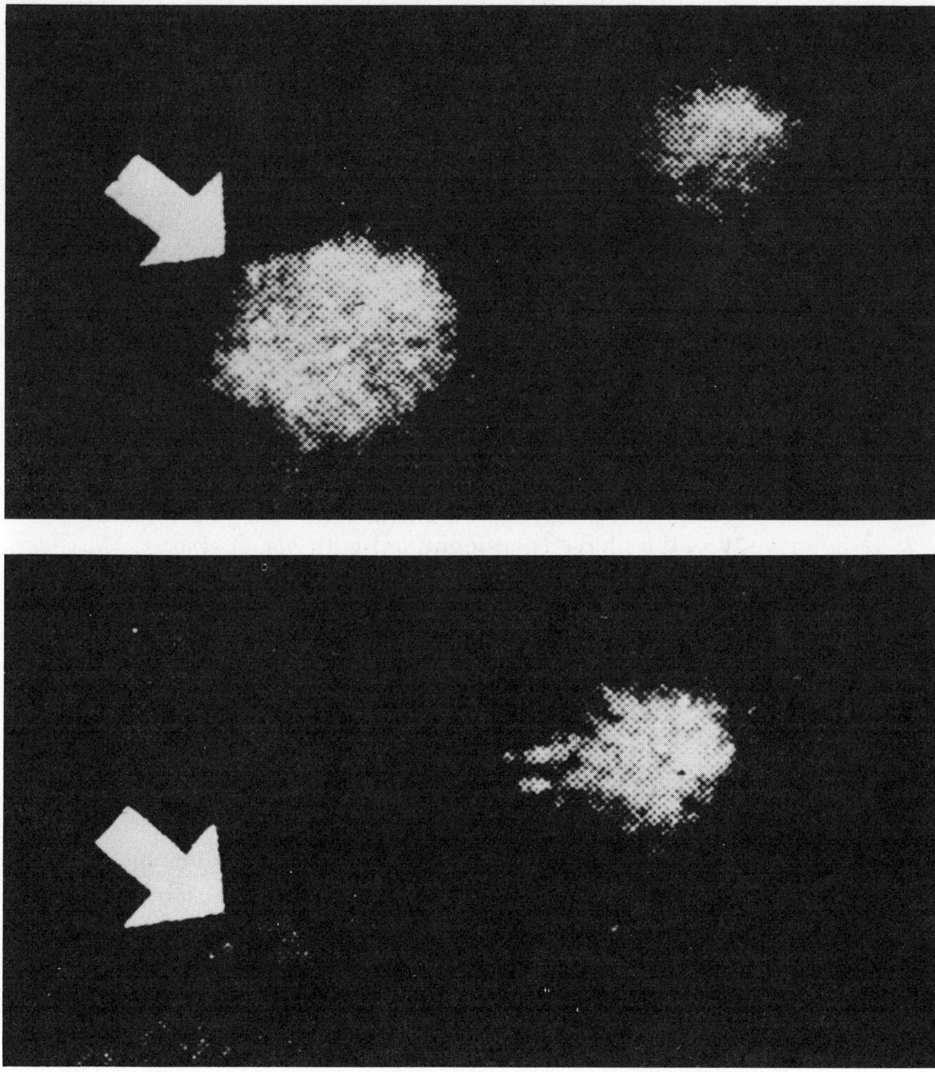

9.9 Photograph of a pulsar previously thought to be an ordinary star. However, a rapid scanning technique showed that the object actually flashes on (top) and off continuously in rapid succession. The object in the upper right of both photographs is an ordinary star (photograph from the Lick Observatory).

youth. The pulsar with the second shortest known period is found near the center of the large and fairly old supernova remnant in the constellation Vela. This pulsar is the only other one known to pulse in both radio and x-ray radiation. In 1976, it was finally identified with a very faint but hot star. Again these findings suggest that visible and x-ray radiation is associated with young pulsars. Actually the visible evidence of supernovae remnants lasts only a few thousand years after the explosion. The radio emission of the remnant may last a hundred thousand years, but the pulsar itself may send out radio emissions for millions of years.

In the early 1970's, with the use of satellites carrying x-ray telescopes, several sources of fast and regularly pulsed x-rays were discovered. Although similar to the radio pulsars in their rapidity, precision, and intensity of their radiation, none of them has any connection with the radio pulsars. Unlike most of the radio pulsars, they are all found to be members of binary systems whose companions can be visually identified. Finally in 1974, a radio pulsar was also found to be in a binary system, but it has no known x-ray radiation. Other than the Crab and Vela pulsars, none of the radio or x-ray pulsars have been identified with visual objects. However, research in this area is intense; current scientific journals should be consulted for more recent findings.

One of the more curious phenomena observed in the pulse rate of some neutron stars was their "glitches" or sudden speed-ups. Even though the overall pulse rate of each pulsar was found to be slowing up very gradually, every now and then, quite unpredictably, a very definite increase in the rate was noted. These were irregular both in timing and in amount. It is believed that the glitches are caused by "starquakes" in the supposedly solid crust of the neutron star.

The known pulsars or neutron stars are estimated to be between 100 and 10,000 light years from the sun. The Crab pulsar is about 6500 light years away. Assuming that our part of the galaxy is not unusual in this respect, there must be thousands of such objects throughout our galaxy and, of course, countless more in other galaxies throughout the universe.

These very small, dense stars represent a second evolutionary path that a star may follow in its "old age." The star is originally more massive than the type that contracts into a white dwarf, and its contraction is therefore more drastic. Because its mass is so great and so very concentrated and because it is spinning rapidly, it is believed to have a very strong magnetic field, far exceeding that of any other known object. Whereas the sun's magnetic fields may be as great as 4000 gauss,[2] and white dwarfs may have magnetic fields ranging from ten million to one hundred million gauss, a

[2] *Gauss* is the measure of the strength of a magnetic field. The earth's magnetic field, which activates our compasses, is only about one gauss.

neutron star should have a magnetic field in the billions of gauss.

The theory of general relativity and the curvature of space must be considered when we are dealing with matter in such unusual states. According to general relativity, matter as densely packed as that in a neutron star will curve the space around it to a much greater extent than would normal matter. Hence, the path of a passing light ray which follows the geodesic of the curvature of space, will also be bent more than when it passes normal matter in the universe (see p. 65). When light from a star passes near our sun, this deflection is no more than a few ten-thousandths of a degree. For a white dwarf it would be about 1/60th of a degree (one minute of arc). A light ray passing a neutron star, however, would be deflected by as much as 30° as seen by an observer far away.

Black Holes

We have seen that a white dwarf can have a mass no more than about 1.4 solar masses. A neutron star is similarly limited, to perhaps two to three solar masses. But some stars on the main sequence have far more mass than that. When their central core has finally used up its nuclear fuel, first of hydrogen, then helium, carbon, and on up the list of elements, it, too, must collapse, for it can no longer sustain the pressure necessary to hold up the overlying layers. And because it is so very massive, the resulting gravitational collapse is that much greater also. Electron pressure halted the white dwarf's contraction at about the size of the earth; nuclear forces between the neutrons stopped the neutron star at about 15 or 20 kilometers in diameter. But with a more massive star, the gravitational force, which is normally the weakest force of all, cannot be withstood. No known physical force can prevent the star from getting smaller and smaller. Theoretically, it will eventually crush itself out of existence!

As this massive star collapses and grows smaller, it becomes more and more dense. Consequently, the intensity of its gravitational field increases.[3] Instead of bending a ray of light 30°, as in the case of the neutron star, the light is literally bent back upon itself. It can never leave the confines of the star. This is true of all radiation and of all matter in the

---

[3] The strength of a gravitational field of an object is proportional not only to its mass but also to the inverse square of its radius, i.e., the smaller it is, the greater will be its gravitational field strength. Therefore, although two stars may have the same mass, the smaller star will have the stronger gravitational field.

vicinity of the star. In order to escape this gravitational field, velocities exceeding that of light would be necessary, and according to the theory of relativity, that is impossible.

The star becomes a *black hole,* so called because it is invisible. No matter or radiation can escape from it. We cannot observe it as we would other stars or celestial objects. Although we have as yet no definite proof that black holes exist, this third evolutionary path for aging stars was theoretically calculated to be possible in 1939 by Robert Oppenheimer, an American nuclear physicist. However, until the discovery of pulsars in the late 1960's and the subsequent identification of pulsars with the previously-theoretical neutron star, few astronomers or physicists took these theoretical possibilities very seriously.

Then in 1971, Uhuru,[4] the first satellite devoted entirely to x-ray radiation research, began sending back data on many x-ray sources which had been totally unknown before. Earth-bound astronomers had not been able to detect such x-ray radiation because of our protective atmosphere, through which very little such energy can pass.

One of the very strong x-ray sources detected by Uhuru was designated Cygnus X-1 and has become of particular interest in black hole studies. Comparisons between the observational data and the theoretical models of black holes strongly support the belief that this x-ray source is, in fact, a black hole.[5]

Cygnus X-1 is a binary star system about 7500 light years from the earth. It consists of a very hot blue supergiant star and an invisible x-ray source, about 10 million miles apart. From spectroscopic observations, it is known that glowing gases are continuously passing from the larger star toward the invisible companion. As the gases approach the invisible star, their velocities are greatly accelerated and they violently collide with gaseous matter already in orbit around the star. Such collisions heat the gases to multimillion-degree temperatures, enough to generate intense x-rays. This x-ray radiation, unlike the regularly pulsed bursts of the neutron star, erratically fluctuates in intensity. It has been compared to the sporadic discharges from a string of exploding firecrackers. Such wildly fluctuating x-ray radiation is believed to be emanating from the chaos of superheated matter as it plunges into the abyss of a black hole (see Fig. 9-10).

From the observations of this binary star, the mass of the supergiant

---

[4] Uhuru was launched in December 1970. Its name means "freedom" in Swahili.

[5] The name of this x-ray source is derived from the constellation in which it is found. It was the first one found in that constellation. Not all x-ray sources are designated in this manner and eventually an internationally agreed-upon convention will have to be adopted.

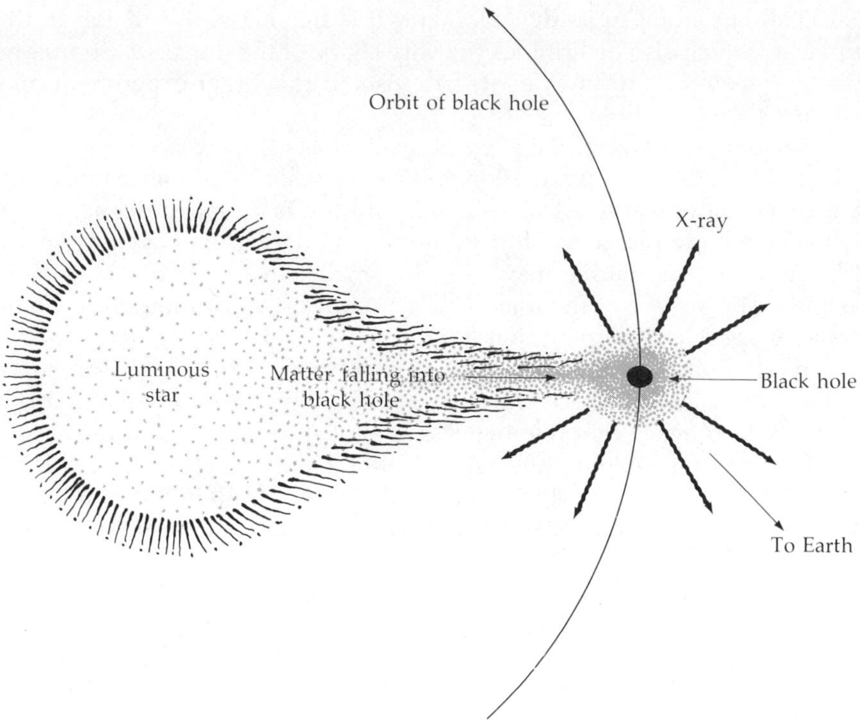

9.10 Model of a binary system in which one member is a black hole.

in the Cygnus X-1 system is found to be about 30 times as massive as the sun, whereas the x-ray source is about 10 times as massive. These massive bodies, each of which dwarf the sun, swing about each other in a little more than just 5½ days. Before the advent of x-ray astronomy, Cygnus X-1 was considered to be but one of many binary star systems with only one visible component. The unique and bizarre character of its invisible companion was revealed by its x-ray radiation.

Cygnus X-1 is not the only binary suspected of having a black hole as one of its members, but it is thus far the most likely candidate. Because such a star is otherwise invisible, detection is possible only by seeking the effects it might have on other objects nearby, such as its companion.

Thus we find three different possible evolutionary paths that a star may take in its last stages of existence. The path which will be taken is determined by its mass. Throughout its entire existence, the mass of the star is all-important in its development. It is this mass, this matter in the universe, which also determines the very shape of the universe. Strange as this may seem and difficult as it is to visualize, another experiment may perhaps clarify it a little.

Imagine a very large flat piece of foam rubber. If we place a ping-pong ball on its surface, the indentation will be very minor—if noticeable at all. A heavier, more massive ball will sink into the foam much deeper. Note that such a more massive ball need not be any larger than the ping-pong ball. A solid steel marble may even be smaller in radius but heavier in weight.[6] The surface of the foam rubber mattress can be thought of as the shape of space (or the path that a light ray will take through this space). The balls represent accumulations of matter in this space. The more massive the material object, the more space around it is curved or bent. If the mass is so heavy that the mattress cannot support it, it will push right through, making a gaping hole. Any matter coming near the hole will fall into it forever. Any light shining upon the hole will go into it and not be reflected back. This is the effect of a black hole on the curvature of space.

[6] *Weight* is the amount of gravitational attraction two objects have for each other. It depends upon the *masses* of the objects and their distances from each other. We measure this distance from each body's center of gravity. Your weight is determined by your mass, the earth's mass, and your distance to the center of the earth. If you go upstairs in a building, you will weigh slightly less.

Discussion Questions

1. It was suggested in this chapter that the next time you eat a steak, you might offer your compliments to a burned-out star. Why is this suggestion a valid one? Trace the steps by which this could be so.

2. Trace the steps by which the "guest star" of 1054 A.D. became associated with the 1967 discovery of "pulsars." Why are these believed to be neutron stars.

3. The mass of a star is an extremely important factor in stellar evolution. What part does it play in this life cycle?

4. What kind of star does the mass-radius relationship apply to? State this relationship.

5. What are the differences between a nova and a supernova?

6. Describe the three "paths to oblivion" that a star may follow. How are they similar? Dissimilar? How is each type of star detected?

10 *Between the Stars*

Interstellar space was once thought to be completely filled with an ethereal fluid or *aether*. How else could light travel from the sun and stars to us? If light is a wave motion, it obviously must have something to wave "in"—or so it was implicitly believed. The idea that light waves could move through "nothing," through no propagating medium as is required of sound waves, was unimaginable. This hypothetical aether was supposed to offer no resistance to any celestial objects moving through it. In fact, it proved to be completely undetectable at all, as demonstrated, for example, by the Michelson-Morley experiment.[1] There was no way of proving its existence and it was ultimately found to be an unnecessary hypothesis.

However, the space between the stars and even between the galaxies is not completely empty. It is not a vacuum, although it is still a far better vacuum than can be produced in laboratories here on earth. The rarified mixture of gas atoms and dust particles permeate, in varying densities, the entire universe. In Chapter 8 we saw how clouds of this gas are detected by the various energies they emit. They are also frequently detected by their blocking the light of background stars (Fig. 10.1). These are called *dark nebulae.* A good example of a dark nebula is the Horsehead nebula (see Fig. 8.8). Even if the starlight is not completely obscured, the dust particles will dim and redden it, much as the earth's atmosphere does to the sun at sunset. Most of this dimming is caused by the scattering of starlight in directions other than its original path. This diffuses the light, and if the cloud is dense enough and the nearby star bright enough, the cloud is actually illuminated by the starlight. The result is called a *reflection nebula* (see Fig. 10.2).

---

[1] The Michelson-Morley experiment was the most famous of the many experiments attempting to measure the absolute speed of the earth through the "aether," and therefore through space. As was discussed on p. 60, the speed of the earth can only be determined relative to some other object. Only the speed of light is an absolute quantity.

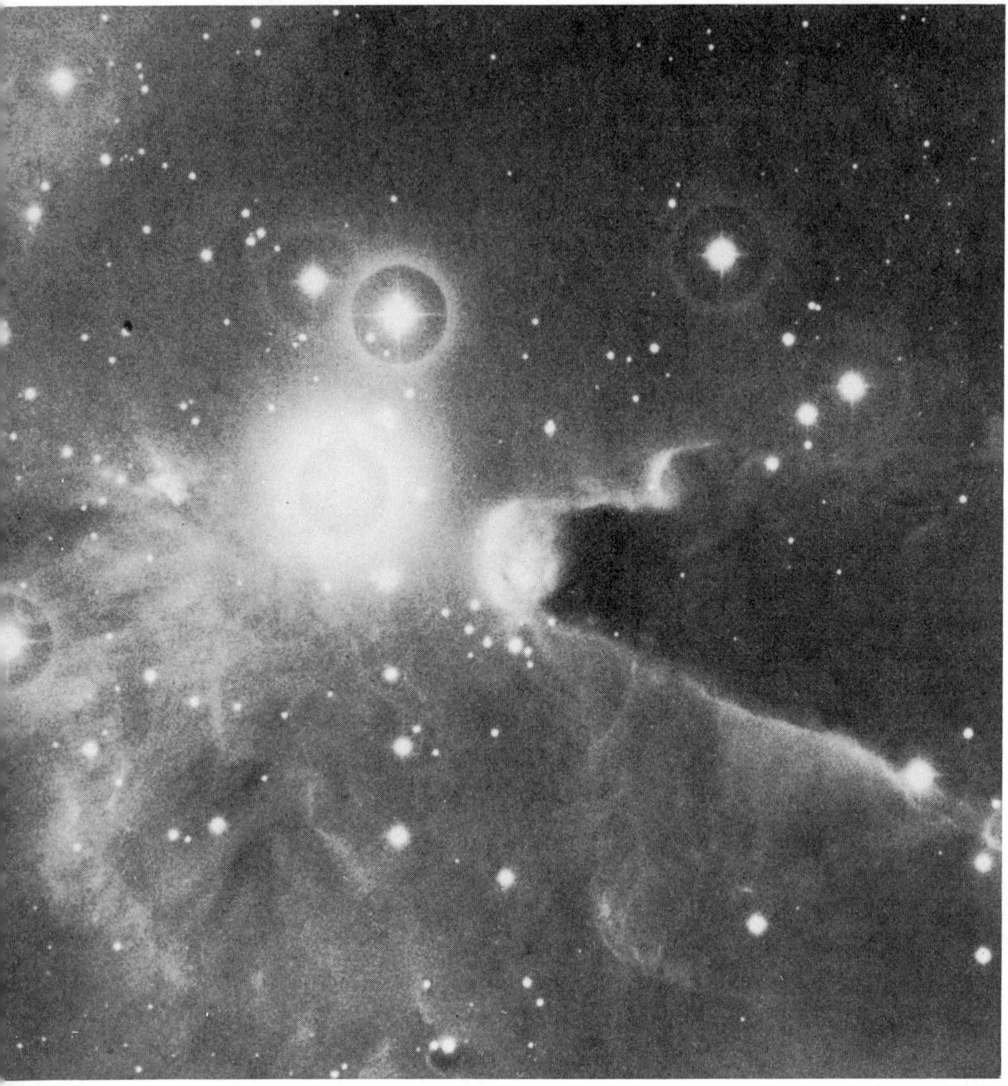

10.1 The Cone nebula in the constellation Monoceros. This nebulosity is situated in the outer region of a young open cluster of stars (photograph from the Hale Observatories).

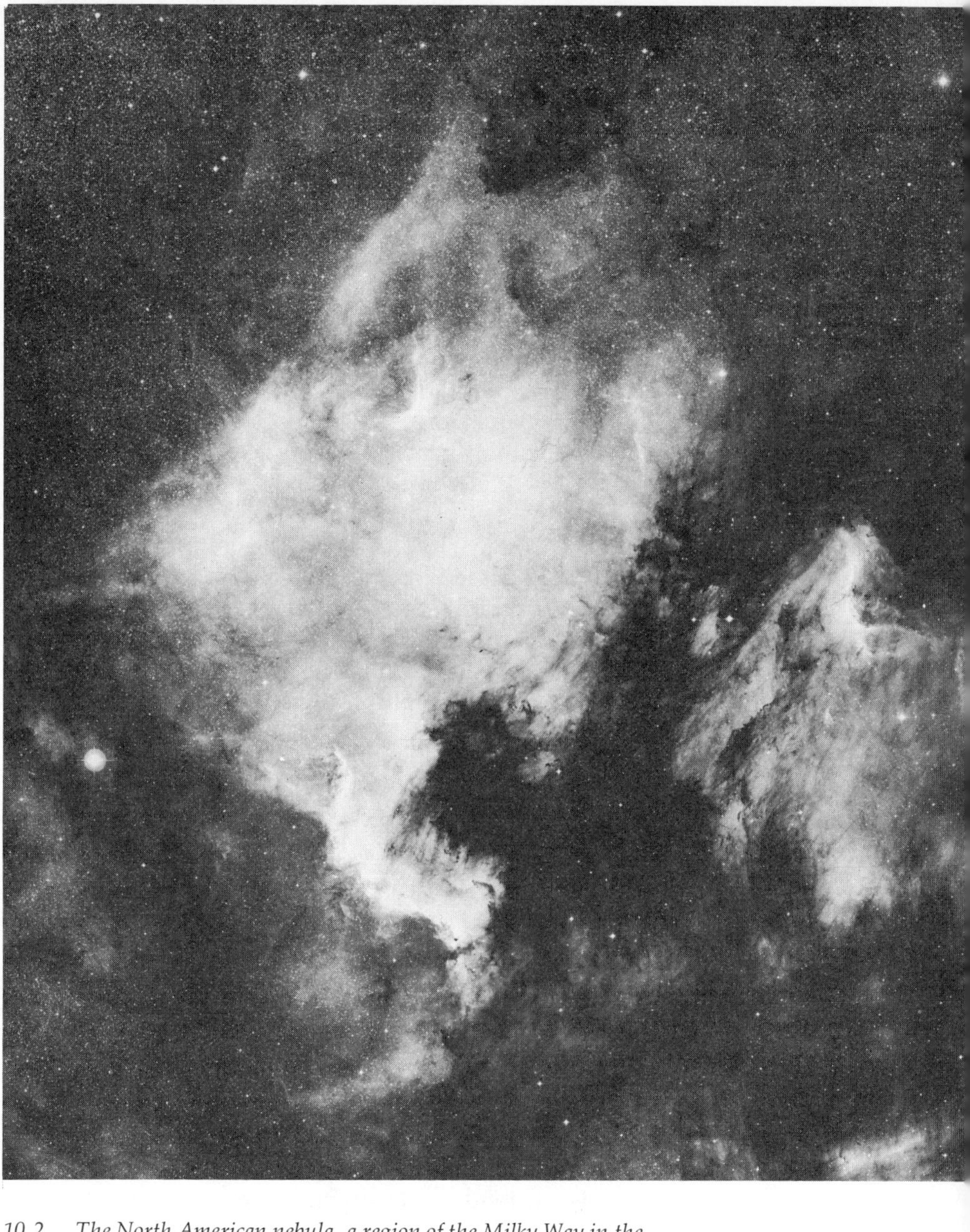

10.2 The North American nebula, a region of the Milky Way in the constellation Cygnus, photographed with the 48-inch Schmidt telescope. This nebula derives its appearance from the dust clouds in the area (photograph from the Hale Observatories).

Because blue light is scattered more than red,[2] the reflection nebula itself appears bluer than the star illuminating it. Since most bright stars in the plane of the Milky Way (where most of the dust is) are blue stars, reflection nebulae are usually very blue. A good example of such a nebula is found around the stars of the Pleiades (see Fig. 8.2).

Although the gas atoms in interstellar space are about 100 times more abundant in mass than the dust particles, they have less effect on starlight, since they are almost transparent. The difference is comparable to the visibility through dust-laden polluted air and through dry, crystal-clear air. However, interstellar gas is visible if it is near a hot star. It then fluoresces (see p. 153), making a very spectacular and colorful nebula called an *emission nebula* (Fig. 10.3).

When the gas is not close enough to a hot star to fluoresce, the atoms stay cold and neutral (that is, un-ionized). The gas in the cloud is invisible photographically but can be detected by the absorption lines it produces on spectra taken of stars that lie behind the cloud.

The stellar and cloud absorption lines can be distinguished primarily by the differences in their Doppler shifts. It would be quite coincidental if a mass of gas and a particular star had identical radial velocities, unless they were associated with one another in space. If they were associated, then the starlight would have other detectable effects upon the cloud, such as fluorescence. Stellar absorption lines can also be distinguished because a cloud's absorption lines are from neutral atoms in the lowest energy states, whereas a star's lines originate from hot gases. Further, the appearance of the lines themselves have characteristics which can be identified with neutral or ionized atoms.

Cool gas atoms can also be detected by their radio emissions, for example, the 21-centimeter line of hydrogen. In fact, most of the radio energy received from the Milky Way comes from interstellar gas. Hydrogen is by far the most abundant element, although other elements and molecules have been identified from their radio emission and absorption lines (see Fig. 10.4). These lines are produced in a manner similar to the way the visual lines are produced, except that lower-energy, much longer wavelengths are involved.

The 21-centimeter line of hydrogen was predicted in 1944, but equipment sensitive enough to detect it was not available until 1951. By mapping these radio emissions, the flat spiral shape of our galaxy was determined. Neutral hydrogen is almost exclusively found in the spiral arms in an extremely flat layer extending throughout the plane of the galaxy (Fig. 10.5).

---

[2] This is why the sun appears red when it sets; its blue light has been scattered out through the atmospehre.

10.3 The Trifid nebula (M22) in the constellation Sagittarius as photographed with the 200-inch telescope. The Trifid nebula is an example of an emission nebula (photograph from the Hale Observatories).

Diatomic (two-atom)			
	Inorganic		*Organic*
H_2	hydrogen (UV)	CH	methyladyne radical
OH	hydroxyl radical	CH^+	methyladyne ion (optical)
SiO	silicon monoxide	CN	cyanogen radical
NS	nitrogen sulfide	CO	carbon monoxide
SO	sulfur monoxide	CS	carbon monosulfide
Sis	silicon sulfide		

Triatomic (three-atom)			
	Inorganic		*Organic*
H_2O	water	CCH	ethynyl radical
N_2H^+	dinitrogen monohydride	HCN	hydrogen cyanide
H_2S	hydrogen sulfide	HNC	hydrogen isocyanide
SO_2	sulfur dioxide	HCO^+	formyl ion (X-ogen)
		HCO	formyl radical
		OCS	carbonyl sulfide

4-atomic			
	Inorganic		*Organic*
NH_3	ammonia	H_2CO	formaldehyde
		HNCO	isocyanic acid
		H_2CS	thioformaldehyde
		C_2H_2	acetylene
		C_3N	cyanoethnyl

5-atomic			
	Inorganic		*Organic*
		H_2CNH	methanimine
		H_2NCN	cyanamide
		HCOOH	formic acid
		HC_3N	cyanoacetylene

Interstellar Molecules

One of the more surprising developments in recent astronomical research has been the discovery of relatively complex molecules in space. Their existence was not considered possible a decade ago. In the vacuum of space, the formation of many-atomed molecules seemed unlikely, and if they had formed, it was believed that they would soon be disrupted by the ul-

6-atomic			
	Inorganic		*Organic*
		CH_3OH	methyl alcohol
		CH_3CN	methyl cyanide
		$HCONH_2$	formamide
7-atomic			
	Inorganic		*Organic*
		CH_3NH_2	methylamine
		CH_3C_2H	methylacetylene
		$HCOCH_3$	acetaldehyde
		H_2CCHCN	vinyl cyanide
		HC_5N	cyanodiacetylene
8-atomic			
	Inorganic		*Organic*
		$HCOOCH_3$	methyl formate
9-atomic			
	Inorganic		*Organic*
		$(CH_3)_2O$	dimethyl ether
		CH_3CH_2OH	ethyl alcohol

10.4 *Some of the interstellar molecules that have been discovered.*

traviolet light from the stars. It is now believed that the dust particles in the interstellar clouds act to shelter such molecules from destruction.

Very simple two-atomed molecules were known to exist in interstellar space since the early 1940's. These were detected by characteristic spectral lines superimposed on stellar spectra. The more complex molecules were not discovered until recently, because their characteristic spectral lines are in the microwave radio region of the electromagnetic spectrum, and were undetectable until the advent of sophisticated techniques in radio astronomy. Most molecules in the cold vacuum of outer space emit their detectable radiation in the microwave region because they are in their lower energy states. The radiation from these molecules arises largely when the

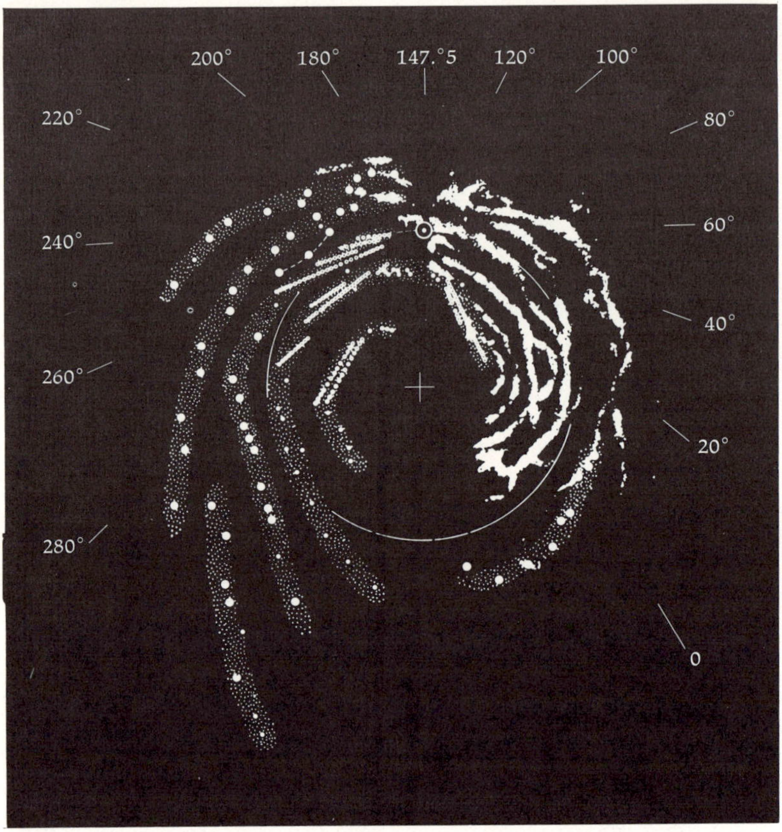

10.5 The spiral structure of the Milky Way galaxy. This composite diagram was made from the radio emissions received from neutral hydrogen in the galactic plane (photograph from C.S.I.R.O., Australia).

molecule as a whole changes from one rotation rate to another, much in the same manner that the neutral hydrogen atom radiates energy. These transitions are known to involve only minute energy changes, which means low frequencies and long wavelengths. That is why we must search in the radio part of the spectrum to identify these molecules in space.

The characteristic radiation "fingerprints" of the molecules are known from laboratory studies. When observations with radio telescopes reveal the same radio spectral patterns, the existence of that molecule in space is demonstrated. However, some molecules may exist in the unfamiliar conditions of space but cannot be reproduced on earth. Other molecules are not sufficiently common on earth to have been given a common name and are therefore known only by their chemical formula (Fig. 10.5).

In 1976, the 40th interstellar molecule was detected near the galactic center. This was cyanodiacetylene (HC_5N) which is capable of combining with two water molecules to form glycine, the simplest amino acid. None of the amino acids themselves have been detected as yet, but this may be because astrochemists have not yet determined the wavelength at which they radiate.

Most of the interstellar molecules exist in areas of space where the dust particles are the most dense. As we noted above, this may be a result of the protection the dust gives the particles against starlight, especially the ultraviolet radiation. The two areas of the sky in which most of these molecules have been detected are the Orion nebula and the clouds in Sagittarius that lie toward the center of our galaxy. Most contain carbon and are therefore organic molecules, i.e., members of the large class of compounds on which all life on earth is based. The existence of these relatively complex molecules may indicate that still more complex molecules, such as the amino acids necessary to life processes, may also exist. This does not mean that there is life in the depths of space but it does add greatly to the belief, which is fairly well accepted among astronomers, that our earth is not the only place in the universe where life may exist. The search has not ended.

Cosmic Rays

Although some of the interstellar matter is probably part of the original cloud or nebula from which it is believed our galaxy formed, much of it comes from the material spewed out by erupting stars such as supernovae in the later stages of their evolution. Such an ejection of matter sends electrons and the nuclei of many kinds of atoms hurtling through space at velocities approaching the speed of light. Most of the nuclei are those of hydrogen. These are the protons. The second most numerous are the alpha particles or nuclei of helium atoms. A very small percentage of the particles are the nuclei of heavier elements.

These very high energy particles are called *cosmic rays* because they were originally thought to be photons of energy rather than minute bits of matter. They permeate all of space and interact with the weak but extensive magnetic fields which also permeate outer space. This interaction causes the cosmic rays to spiral around the fields, and thus they appear to come from all directions.

When these cosmic rays hit the earth's magnetic field, those with the least energy are deflected. Only at the magnetic poles[3] are they able to interact with the atmosphere. The heavier, higher energy particles break through this magnetic barrier and interact with the atoms and molecules of the atmosphere. These atmospheric atoms and molecules are broken up, causing great showers of free electrons and nuclei to rain down upon the earth. With such a continuous abundance of electrons and ionized atoms, the air around us is constantly electrically charged. This fact was known a century ago through relatively simple experiments with electricity, but the cause was not discovered until many years later.

Most of these *secondary particles* (those created by the interaction of the cosmic rays and our atmosphere) are charged particles called *mesons*. From studies in laboratories, it has been found that their masses are between that of the proton and electron, although their charges equal that of the electron. The most common type of meson (at sea level) is the *mu meson* (see p. 63).

A mu meson's lifetime is only about two microseconds (i.e., two millionths of a second), changing into either an electron if it is negatively charged or a positron[4] if positively charged. Since it originally possessed over 200 times more mass than an electron, some of the extra mass is converted into energy and the rest becomes *neutrinos*, which are strange particles that have energy but no mass. Neutrinos travel at the speed of light and pass through just about all matter with practically no interaction. They easily pass through the body of a star and of course through the earth itself. Their detection is extremely difficult and one of the current searches for them involves the determination of their rate of production in the sun.

We are not physically aware of cosmic rays or the secondary particles that they produce, although they are constantly hitting the earth and even passing through our bodies, contributing to the general radiation that we receive. It is estimated that 10^{18} (a billion billion) particles enter the earth's atmosphere each second. By interacting with the atmosphere, they are responsible for the formation of the isotope carbon-14 which has been so useful in the carbon dating of historically ancient sites.

---

[3] The earth can be thought of as a simple bar magnet, but the poles of this magnet do not coincide with the geographic poles. They are separated by many hundreds of kilometers. Furthermore, the magnetic poles wander about over long periods of time.

[4] A positron is a particle with a mass equal to an electron but with a positive charge.

Discussion Questions

1. What is the significance of the discovery of relatively complex molecules in space? Which molecules are considered of particular importance?

2. How are molecules detected in space?

3. Define:

    a. dark nebula
    b. reflection nebula
    c. emission nebula

4. What are cosmic rays? How do they effect the earth? How do they effect us?

5. Why was the concept of an interstellar "aether" felt to be so important?

6. From Figure 10.4, what can be concluded about the shape and structure of the Milky Way galaxy?

11 *The Universe of Galaxies*

When we took an overall view of the universe in earlier chapters, a detailed study of the galaxies, and especially our own galaxy, was necessarily omitted. Now that we have studied the main components of the galaxies, the stars, we turn our attention once again to these building blocks of the universe.

Until the twentieth century, astronomers did not realize that the galaxies were in fact remote stellar systems far outside of our own Milky Way galaxy. Even today, the exact distances to these systems are still being studied. As we have seen, the uncertainty in cosmic distances strongly affects cosmological considerations.

Use of the Cepheid variable period-luminosity relationship (page 26) is still the most important method of distance determination for nearby galaxies. However, these relatively rare stars have only been useful to determine the distances of less than three dozen nearby galaxies. For galaxies at greater distances, bright supergiants, globular clusters, or supernova outbursts can be used by comparing their apparent magnitudes with what we believe to be their absolute magnitudes, which of course have been determined from observations of similar objects in our own galaxy or in neighboring galaxies. As the distance increases, these bright objects within the galaxies become less identifiable and their use as distance determiners becomes more unreliable.

If the galaxies were all of about the same intrinsic luminosity, their apparent magnitudes would enable us to determine their distances with considerable accuracy. However, galaxies have an enormous range of luminosities, so this is not possible. In large clusters, the very brightest galaxies can sometimes be used as guides to determine the distance to the entire cluster, thereby giving the approximate distances of all the galaxies in that cluster, but the results have been very uncertain when this method is used.

The *Hubble velocity-distance relationship* for galaxies (page 48), based on their Doppler shifts, gives us the distance directly. But it must be remembered that this method of distance determination had first to be calibrated—that is, to get the Hubble constant, the distances to a number of galaxies had to be independently determined beforehand.

11.1 Spiral galaxy in the constellation Ursa Major (M101). Note how clearly the illuminated gas clouds and bright blue-white stars outline the spiral arms (photograph from the Hale Observatories).

Although galaxies are not uniformly distributed throughout space but rather are grouped into clusters, the clusters themselves are found uniformly in all directions as far away as the most powerful telescopes can reach. The images of individual galaxies become smaller and less distinct at these tremendous distances and they are distinguishable from faint stars in our own galaxy only because of their slightly fuzzy appearance (see chap. 2, Figs. 2.14–2.17).

Types of Galaxies

Galaxies, like other objects in the universe, vary in size, shape, mass, and luminosity. No two are exactly alike, although the vast majority are either spiral or elliptical in shape. A small minority is classified as irregular galaxies.

Spiral Galaxies

A spiral galaxy, as its name implies, resembles a giant pinwheel (Fig. 11.1). It is much wider than it is thick and rotates about its center or *nucleus*, taking millions of years for one rotation. From radial velocity measurements, it has been determined that these spiral galaxies rotate like the Milky Way

11.2 *The Andromeda galaxy with two satellite galaxies (photograph from the Hale Observatories).*

galaxy, which is a large spiral galaxy. The more central regions, other than the nucleus itself, rotate faster than the outer portions.

The arms of the spiral galaxies vary in the tightness with which they are wound around the nucleus. The Milky Way and the Andromeda galaxies (Fig. 11.2) are typical large spirals with tightly wound arms, whereas the Whirlpool galaxy (Fig. 11.3) shows a more open spiral struc-

11.3 The Whirlpool galaxy in the constellation Canes Venatici. This galaxy is over 30 million light years away from us (photograph from the Hale Observatories).

11.4 Barred spiral galaxy in the constellation Eridanus (photograph from the Hale Observatories).

ture. About one-fourth of these spirals have "bars" running through their nuclei, with the spiral arms beginning at the ends of the bars. These are therefore called *barred spirals* (see Fig. 11.4).

The spiral arms consist of concentrations of stars, dust, and gas. In some, the dust lanes are quite evident, as for example when a galaxy is viewed edge-on (Fig. 11.5). Large amounts of gas in the arms are evident when the gas is illuminated by the bright stars associated with it, creating emission nebulae. It is these bright stars and emission nebulae that make the arms stand out so spectacularly. Here is where new stars are believed to have been formed—and are still being formed—from the abundant gas and dust. Open star clusters are found in these arms, whereas globular clusters are found surrounding the galactic nucleus, much like we found in the structure of the Milky Way (see p. 149).

The nuclei of the spirals can be either so extensive as to dominate the appearance of the galaxy, or small and compact in relation to the arms. In the nuclei we find the older stars and very little dust and gas. These nuclei, even that of the Milky Way, are not well understood, but apparently are the centers of intense and often explosive activity.

Spiral galaxies range in size from about 20,000 to more than 100,000 light years in diameters, and have absolute magnitudes of between −16

11.5 Spiral galaxy in the constellation Virgo. Because this galaxy is seen edge-on, the dust lanes along the central plane are very obvious (photograph from the Hale Observatories).

and −21. Their masses range from about one billion to 400 billion solar masses. The Milky Way, nearly 100,000 light years in diameter, is one of the larger ones and is estimated to contain more than 100 billion (10^{11}) stars.

However, the mass of the Milky Way, as calculated by various methods, is found to be considerably greater than the mass of its stars alone. In fact, since most of the stars in the Milky Way have masses less than that of the sun, their total mass would account for only about one-third of the total calculated mass of the galaxy. Some of this discrepancy is accounted for by the presence of dust and gas in the spiral arms. Undoubtedly, the remainder of the discrepancy is contributed to by white dwarfs, neutron stars, and possibly black holes. As we have already seen in Chapter 8, these relatively nonluminous components of the total mass represent both the beginnings and endings of stellar evolution.

Although differing in size, other spirals are basically the same as the Milky Way. If our sun were a member of another spiral system, we would very probably not detect much difference in the general makeup of the galaxy. Our telescopes would reveal globular clusters, gaseous nebulae, a variety of stars, and other stellar configurations characteristic of our own Milky Way galaxy. Of course, the constellations would be completely different.

Elliptical Galaxies

An elliptical galaxy looks like a spiral galaxy shorn of its arms (Fig. 11.6). They range from nearly spherical to highly flattened ellipsoidal shapes. Like the nuclei of the spirals, they have very little dust or gas and consist entirely of older stars. In the larger ellipticals, many globular clusters are found spherically distributed around the center.

In any given large volume of space, there are far more elliptical galaxies than spirals, even though the spiral galaxies represent more than two-thirds of the 1000 brightest galaxies in the sky. This is because ellipticals are generally less luminous than spirals and therefore cannot be seen at as great distances. We could create a "ladder of luminosity" for galaxies as we did for stars (see p. 133).

The stars in both spiral and elliptical galaxies are concentrated toward their centers and gradually thin out into space, making a precise determi-

11.6 An elliptical galaxy in the constellation Virgo (M87). This galaxy is also a radio source (photograph from the Hale Observatories).

nation of galaxy size almost impossible. However, ellipticals have a much greater range of size as well as of mass and luminosity. As can be seen in Figure 11.7, there is a great diversity in their shapes.

The dwarf ellipticals—which are the most common type—can be as small as 5000 light years in diameter. Their luminosity is only one million times that of the sun. Galaxies smaller than this approach the size of globular clusters, and if a globular cluster were indeed found alone in space, it would be defined as a dwarf elliptical galaxy.

At the other extreme, the giant ellipticals have absolute magnitudes of more than −23, which makes them 100 billion times more luminous than our sun. They have more than ten times the luminosity of the Andromeda galaxy, which is considered a very bright spiral galaxy. Such giants are several hundred thousand light years in diameter and have 100 times the mass of the Milky Way. They are frequently strong radio sources as well.

11.7 *An elliptical galaxy (NGC 205), one of the satellite galaxies of the Andromeda galaxy shown in Figure 11.2. It is also a member of the Local Group (photograph from the Hale Observatories).*

These giant elliptical galaxies stand out so prominently in a cluster that they can sometimes afford a rough indication of the cluster's distance from us (see Fig. 11.8).

11.8 A very faint cluster of galaxies containing a giant elliptical galaxy. Note how it stands out far more brilliantly than the rest. On the celestial sphere, the area represented by this photograph is less than one square degree of arc. Because this is a negative print, the background sky is light and the stars are dark (photograph from the Palomar Sky Survey: Hale Observatories).

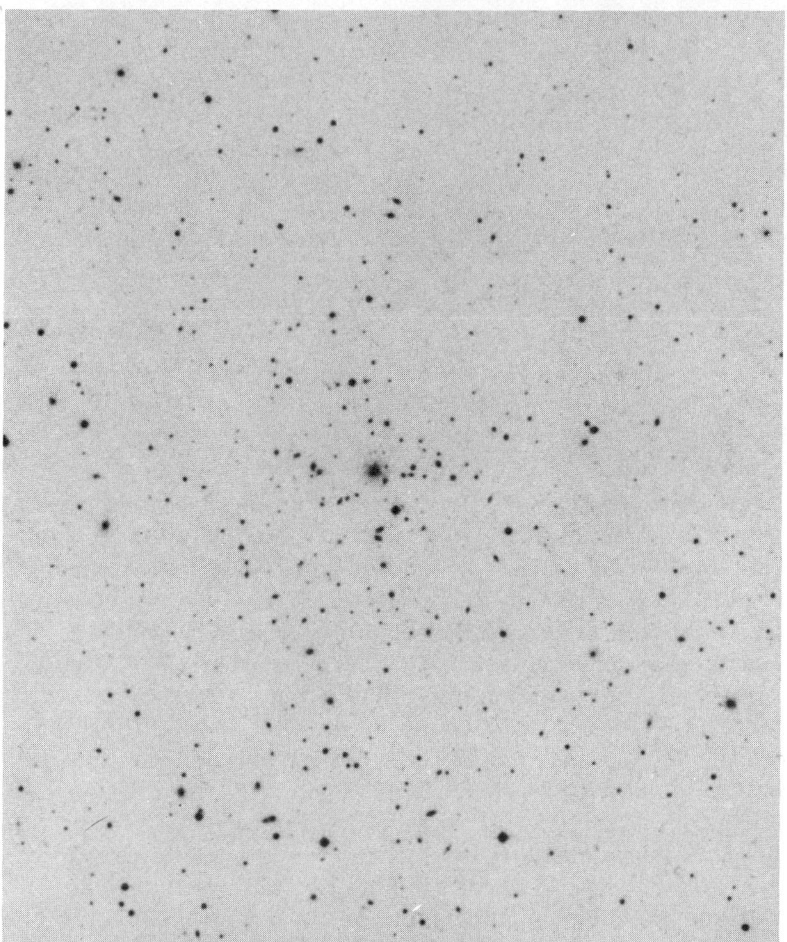

Look at the photograph of the cluster of galaxies in the constellation Hercules. With a small magnifying glass, examine each image and try to determine which are galaxies and which are stars in our own galaxy. Can you classify the galaxies? Are there any tightly wound spirals? Barred spirals? What shapes do the ellipticals take?

Cluster of galaxies in the constellation Hercules photographed with the 200-inch telescope. The stars are in our own galaxy (photograph from the Hale Observatories).

The least common type of galaxy is the irregular type. As the name implies, these are nonsymmetrical in appearance, having a rather chaotic-looking structure (Fig. 11.9). The Large and Small Magellanic clouds are the most famous of this type of galaxy (see Fig. 2.10). The cause of such irregularity is not well understood in all cases but can be caused by the gravitational effects of another nearby galaxy. When two galaxies actually "collide," there is little likelihood of collision between their individual stars, because of the great distances between the stars. It would be like two large swarms of gnats approaching each other, but the gnats in each swarm, if proportioned to the size of the stars and the distances between them in the galaxies, would be separated by several miles, and therefore the swarms could easily pass through each other without a single collision. However, in colliding galaxies, even though no stars would collide with another star, the gravitational attraction of one galaxy on the other can be

11.9 An interacting pair of peculiar spiral galaxies (NGC 4038-9) in the constellation Corvus; also a source of radio noise (photograph from the Hale Observatories).

great enough to distort their respective shapes. Very odd-looking galactic structures, which are often accompanied by strong radiation in the radio region, are usually produced (see Fig. 11.9).

Attempts have been made to use these classifications of galaxies in determining galactic evolution (Fig. 11.10). Do galaxies start as ellipticals and become spirals or the other way around? Or does a galaxy's structure endure throughout its entire existence? These are some of the lines of investigation currently being explored. Much more data is necessary before definitive answers can be made.

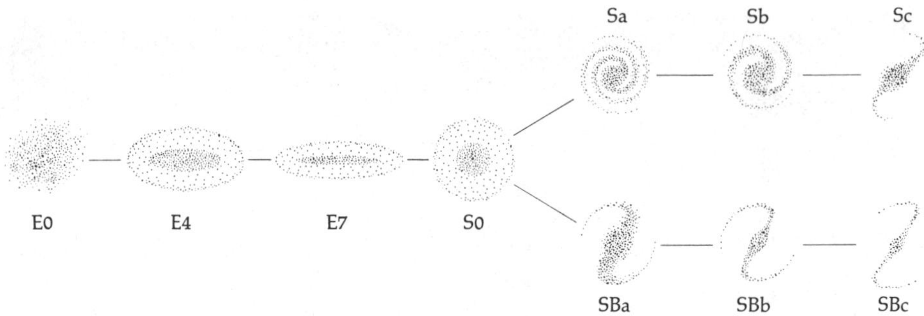

11.10 An early attempt by Edwin Hubble to classify galaxies. The diagram indicates the possible relationship between the various types. E = elliptical; S = spiral; SB = spiral barred.

Clusters of Galaxies

In Chapter 4, we saw how the Hubble relationship of recessional velocities to actual distance was used to determine the distances of galaxies and clusters of galaxies. This relationship can be used reliably for the distant clusters of galaxies because the overall speed of recession of the cluster as a whole is far greater than the motions of individual galaxies within the cluster (Fig. 11.11). On the other hand, members of our cluster of galaxies, the Local Group, are so close to us that their Hubble expansion motion is minute compared to the motions of the galaxies relative to each other. The Milky Way, for example, is moving at a speed of 100 to 150 kilometers per second with respect to the center of mass of the Local Group. This point is located approximately halfway between our galaxy and the Andromeda galaxy. The Andromeda galaxy is actually approaching us at 260 kilometers per second while the Large Magellanic cloud is receding at 276 kilometers per second. These velocities are far greater than any universal expansion motion and the latter motion is therefore undetectable at such close quarters.

The Local Group is some three million light years in diameter. It is believed to be a part of a much larger *supercluster* of galaxies, in which the Virgo cluster, which is the nearest large cluster of galaxies to us, is the major cluster. This local supercluster is estimated to be between 150 and 300 million light years in diameter and to have a mass of some 10^{15} solar masses, or about 10,000 times the mass of the Milky Way. Credence to the

11.11 Stephan's quintet: a small cluster of distant galaxies (photograph from the Lick Observatory).

local supercluster concept is strengthened by the fact that the concentration of clusters of galaxies seem to diminish as we approach 70 million light years' distance and does not increase again until much greater distances are reached. Other statistical measurements also appear to verify this hypothesis.

The clusters of galaxies, both in our local supercluster and those farther out, are classified according to their spherical symmetry. A *regular* cluster is spherical in shape with most of its members located towards its center—in a sense making it resemble a globular star cluster. Its galaxies are practically all ellipticals or very tightly wound spirals. It may also be a powerful x-ray source.

The *irregular* clusters, such as our Local Group, are sometimes called "open clusters" because of their resemblance to the open star clusters. They may have far fewer members than the regular clusters but are composed of all types of galaxies. Some irregulars such as the Virgo cluster have over 1000 members.

It is tempting to carry the comparison of the star clusters and galactic clusters even further and say that the regular clusters must be much older than the irregular clusters, much in the same way that the globular star clusters are older than the open clusters. This may eventually be proven to be true, but until such evidence is found, all we can say is that there is a parallel similarity.

A large number of these clusters of galaxies have been catalogued, including many of the far distant ones. The most remote clusters are some four billion light years away. Such a survey explores a volume of space 500,000 billion (5×10^{14}) times greater than the Milky Way galaxy. (Let us remember that only 50 or 60 years ago, the Milky Way itself was thought to be the entire universe!)

Radio Galaxies

Radio astronomy, which got its impetus from radar development during World War II, rapidly gained in importance as thousands of discrete radio sources were detected. These sources were distributed more or less uniformly across the sky and therefore were presumably not associated with the Milky Way. If they had been so associated, they would have shown a strong concentration on the galactic plane. However, many of these sources at that time could not be associated with any visual objects. In part, this was due to the inability of early radio telescopes to pinpoint the locations of the sources. Rapid improvements of radio astronomy techniques ultimately led to a greater location precision than is possible even with optical telescopes (see Chap. 12).

In 1951, the first identification of a radio source with an extra-galactic object was made. It was the strongest known source from outside our galaxy and was found to emanate from a peculiar pair of galaxies in the constellation of Cygnus. This was the first *radio galaxy* discovered; there are now well over 100 known. This first one was called "Cygnus A" because it was the strongest radio source found in the constellation Cygnus.

The radio energy from Cygnus A at first appeared to be associated with a pair of closely interacting galaxies. However, with the rapid improvement of radio techniques, it came as a surprise that the radio energy actually came from two independent regions a million light years on either side of the visual galaxies (Fig. 11.12).

From its red shift, Cygnus A was found to be about one billion light years away. Using this distance determination and the measurement of the radio energy that we receive from this source, astronomers soon realized that the radio energy is many times greater than the light energy of the associated galaxies.

How such a tremendous source of radio energy could be generated is not known. It appears to be associated with a type of violent, explosive activity. What is particularly amazing is that the radio energy emanates from the two lobes on either side of the visual galaxies, not from the galaxies themselves. If these tremendous lobes originally came from the visual galaxies of Cygnus A itself, they presumably took at least a million years to reach their present distance. They would have had to travel at the speed of light to be so far from the visual galaxies in one million years, yet they are still emitting an enormous amount of energy. Thus Cygnus A presents a twofold problem: how did these sources become separated to such great distances from the visual region, and how have they been able to maintain such a high level of energy over such a long period of time?

Other such *peculiar radio galaxies* emit as much or more radio energy as they do light energy, although visually they may look like ordinary large elliptical galaxies. Because of the tremendously elongated lobes of radio-emitting matter that stretch out in opposite directions from the visual galaxy, these radio galaxies are by far the largest "objects" known to astronomers. The largest one that has been detected (called 3C236) stretches over 16 million light years of space. After millions of years, it is still emitting energy and expanding outward.

Such peculiar galaxies have violent internal activity which causes them to eject matter without any known external influence. We detect this violence through their radio emissions as well as through other nonvisual electromagnetic radiation. Most of the ejected matter comes from the nucleus and may cause this central region to emit many times the energy of a

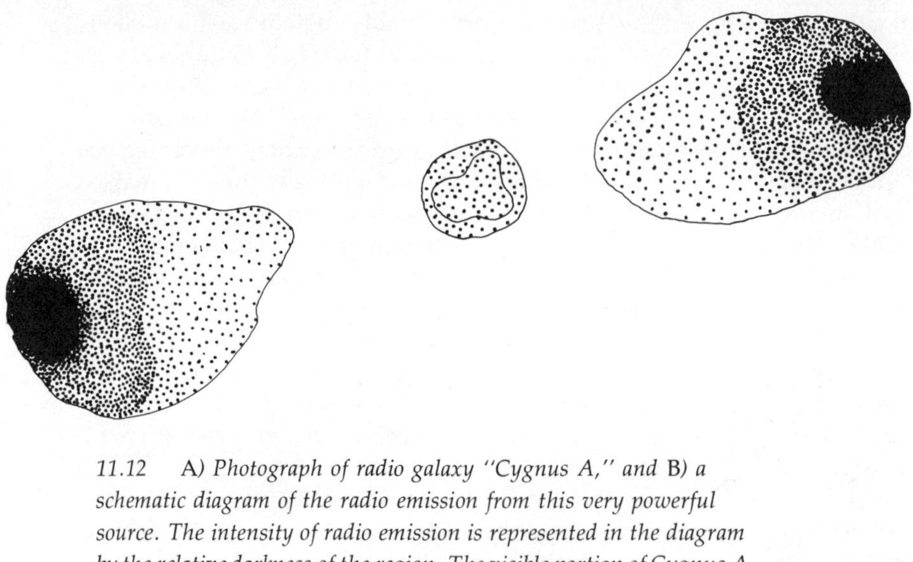

11.12 A) *Photograph of radio galaxy "Cygnus A," and* B) *a schematic diagram of the radio emission from this very powerful source. The intensity of radio emission is represented in the diagram by the relative darkness of the region. The visible portion of Cygnus A (center object in Fig. 11.12A) is within the central circular area in the diagram (photograph from the Hale Observatories).*

normal galaxy. The ejected matter may be traveling at thousands of kilometers per second away from its origin (Fig. 11.13).

Not all radio galaxies are peculiar, however. The "normal" ones are like our Milky Way, which emits a continuous radio energy much lower in intensity than its light energy. This relatively weak radio energy can be partially explained by the interaction of cosmic rays and interstellar matter. The electrons that are thus ejected emit energy as they spiral around the lines of force of the magnetic fields in the galaxy.

11.13 A peculiar galaxy in Centaurus (NGC 5128). It is believed that this galaxy is actually being torn apart by some tremendously violent explosion, which has also caused it to be one of the most powerful radio sources in the southern sky (photograph from the Hale Observatories).

Quasars

In the 1960's, during the search for optical objects that correspond to radio sources, astronomers found several that looked like ordinary stars rather than galaxies. In fact, these objects were listed as stars in older catalogues. Only when one of the star-like objects happened to be occulted (covered) by the moon was it shown to be the actual radio source: both the light and radio radiation stopped simultaneously when cut off by the moon. Because it looked like a star, it was at first called a *quasi-stellar object.* This name was later shortened and popularized to *quasar.*

Astronomers know that if our sun was even at the distance of the nearby stars, we would not be able to detect its radio emission. Therefore, astronomers realized that this star-like object must be rather unusual to produce so great a radio emission. When its spectrum was obtained, it was found to have a very large red shift. This placed it much farther away than any of the stars in the Milky Way, and indicated that visually as well as in radio radiation it was relatively a very bright object.

When the total energy of this first quasar was computed, it was found to be 100 times greater than the total energy of all the stars in the Milky Way galaxy. Yet its size, as inferred from its rapid brightness variations, was found to be quite small, perhaps one light year in diameter. An average galaxy has a diameter measured in thousands of light years.

Over 300 quasars have now been located, either by looking for optical objects where radio sources are detected or by studying star-like objects that are extremely blue in color (Fig. 11.14). The ones that have been found thus far all have very large red shifts, which indicate that they are receding from us at extremely high velocities. Some of these red shifts correspond to recessional velocities of 91% the speed of light! That is about 273 thousand kilometers per second or 170,000 miles per second. If the Hubble relationship is correct, these velocities imply distances of 16 to 18 billion light years, a distance which approaches the theoretical limits of the visible universe. If it were possible for an object to reach a recessional velocity of 186,000 miles per second, the speed of light, its radiation would no longer be detectable.

The luminosity of quasars is greater than the brightest peculiar radio galaxy or the brightest elliptical galaxy. Further, they all exhibit short irregular variations in their energy output. It is from this that their size is estimated.

In the mid-1960's, other star-like objects that were as blue as the quasars and had similarly large red shifts were detected, but these were not radio sources, or if they were, their radio energy was too weak to be detected. They seem to be far more numerous than the radio-emitting quasars, which leads to the idea that the latter represent one evolutionary stage before becoming radio-quiet, which may be a much longer stage.

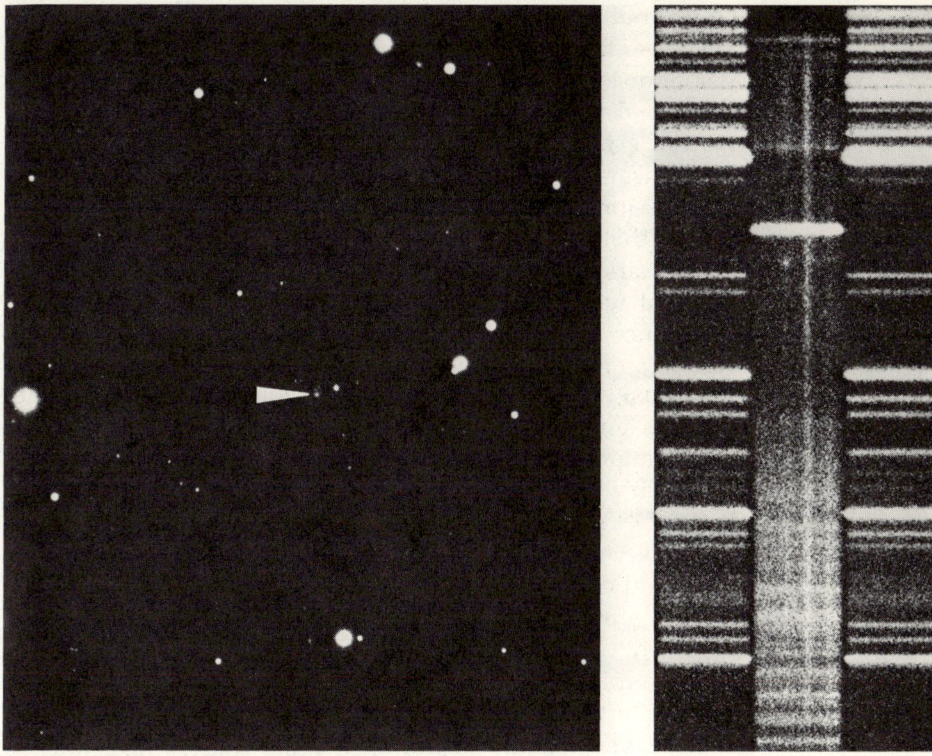

11.14 Quasar 3C-295 in Boötes. This is one of the most distant objects known. The spectrum on the right shows a red shift that indicates a recessional velocity of 36% of the speed of light (photograph from the Hale Observatories).

There has been a heated debate as to whether the quasars were really at such great distances. Various hypotheses have been proposed to explain their rapid recessional velocities, although none have been very convincing thus far. To counter these arguments, it has been found that those quasars which are believed to be the farthest away also appear the faintest.

Another piece of evidence that places the quasars at extreme distances comes from the detection of other objects at such distances. Some quasars appear to be associated with a cluster of galaxies, and all their red shifts are the same. These are the nearer quasars, since for those farther away, we would not be able to see any galaxies. At present, researchers are trying to find (and some believe they have found) a few very distant objects such as galaxies or intergalactic clouds that lie in front of a quasar. Since there would be no doubt that an intervening object's smaller red shift is an indication of distance, the quasar's larger red shift would be more acceptable as an accurate distance measurement.

There are several types of peculiar galaxies whose velocity-distance relationship is well established but which appear to have some of the characteristics of quasars, such as their strong ultraviolet and blue radiation and their radio emissions. Many of these also exhibit large red shifts. Astronomers suspect that these may represent various stages of galactic evolution, with quasars at the beginning stage. Just as astronomers identified many different types of stars and were able to determine in which stage of their existence they were and thereby formulate the basic tracks of stellar evolution, so may the same eventually be done for galaxies.

However, it must be remembered that the major determination of the distances of quasars and other very distant objects is based upon the velocity-distance relationship, which, as we have seen, has had to be frequently readjusted in the past to fit new observational data, and which may still need relatively minor changes in the future. It is based on distance determinations of objects much closer to us than the quasars, using parallax measurements and comparisons between apparent and absolute magnitude. We do not know what the maximum absolute magnitude for a quasar or giant galaxy can be and therefore we cannot use their apparent magnitude for determining their distances, except as a rough estimate. The best guess is that the farthest ones are between five and fifteen billion light years away. There still are many unidentified radio sources that may be even farther away, too far to be detectable by our most powerful optical telescopes.

Other Energy Sources

The visible universe—that which can be recorded on photographic plates—may be but a small part of the cosmos we will eventually be able to explore. Radio astronomy opened up an avenue of research that has proven unexpectedly fruitful. Similarly, space astronomy, by allowing astronomers to get above the atmosphere, has allowed observations of the universe in all regions of the electromagnetic spectrum.

For example, sharp, sudden x-ray bursts of energy have been detected in various parts of the sky. Some seem to be coming from the direction of the galactic center. These are usually point sources, as differentiated from sources that come from extended bodies such as nebulae. Most of these x-ray point sources have not been identified with visible objects, although it is thought that they come from neutron stars or black holes.

There are also some x-ray sources that fade after a few months of brilliance. For this reason they are called *x-ray novae.* All of these x-ray sources are brighter than the very diffuse x-ray radiation background that is coursing throughout all space, coming equally from all directions.

Even more unusual pulses have been received by *gamma ray telescopes.* There does not seem to be any connection between the x-ray and gamma ray types of radiation flares. The gamma ray bursts, of which there have been about a dozen recorded each year since they were discovered, are so strong that if they were in visible radiation, the cavemen would have been well aware of them. The energy output, if the source is assumed to be as close as the nearest stars, is 10^{34} ergs per second. This is considerably greater than the sun's energy. If the source is farther away than the nearby stars, its energy must be even more intense.

Yet because these gamma ray bursts are found to have rapid fluctuations, we can infer that their source is very small, perhaps one-tenth the diameter of the earth. This would mean that they are radiating some 10 million times more energy per square centimeter of surface area than the sun. Thus far no visible object has been found associated with any of these x-ray or gamma ray bursts. They remain at present one of the great mysteries of the cosmos, which are challenging and will continue to challenge astronomers for years to come.

Discussion Questions

1. Make a chart of the various types of visible galaxies and list their characteristics: size, structure, composition, etc.

2. Define: Radio galaxy
   Quasar
   Supercluster of galaxies

3. What are the two kinds of clusters of galaxies? How do they differ?

4. What are some of the lines of evidence that point to the fact that the quasars are very distant objects?

5. The sun is moving 150 miles per second relative to the center of the galaxy. How many AUs is that per second? How many mph?

6. How many cosmic years old is the sun? (Assume an age of about 6 billion earth years.)

7. The earth revolves around the sun in an elliptical orbit at 30 km/sec, while accompanying the sun around the galaxy at the speed of 240 km/sec. To orbit the sun takes one year, but to orbit the galaxy takes 200 million years (a cosmic year). How much of the sun's orbit around the galaxy is covered in one earth year? In terms of degrees of a circle, what does this represent in curvature? Do you think that, over the course of the year, this deviation from a straight line is more or less than the deviation of the center line of a typical straight highway over the course of a mile? What, then, would be the path of the earth as seen from outer space?

12 *The Tools of Astronomy*

Our concept of the universe, which is closely related to the way we regard ourselves, has been almost entirely a function of our instruments. Without the great radio and optical telescopes of the world, our knowledge of the architecture and dynamics of the universe could not have been attained. Yet for virtually all of the several million years that we have been on this planet, we have had only the naked eye with which to observe this universe. Until very recently, mankind had no idea of the existence of fainter stars, invisible to the naked eye. They did not know of the Milky Way galaxy nor of the billions of other galaxies in our universe. These could not be known until the invention of the telescope.

The Optical Telescope

In the early years of the seventeenth century a Dutch spectacle-maker's apprentice, so the story goes, arranged two lenses so as to produce the first telescope. Hans Lippershey, the spectacle-maker, saw the commercial value of such an instrument and made known his discovery. In 1609, on hearing of this new instrument, the famous Italian scientist Galileo Galilei constructed his own crude telescope and was the first to turn it toward the sky and record what he saw. The quality of such an instrument might compare with an inexpensive child's telescope today. Its magnification power was only three-fold. Even the best one he built had only a 30-power magnification (see Fig. 12.1). Yet with these telescopes Galileo made more important discoveries in a shorter time than anyone has done since. It is interesting to speculate on how history might have differed if the Egyptians or Greeks, who used lenses as "burning glasses," had discovered the principle of the telescope.

Galileo discovered myriads of previously unknown stars. He found moons orbiting Jupiter, craters on the moon, and spots on the sun. He saw Venus go through phases like the moon and that Saturn had protrusions on either side of its disk (these were Saturn's rings, but Galileo's telescope

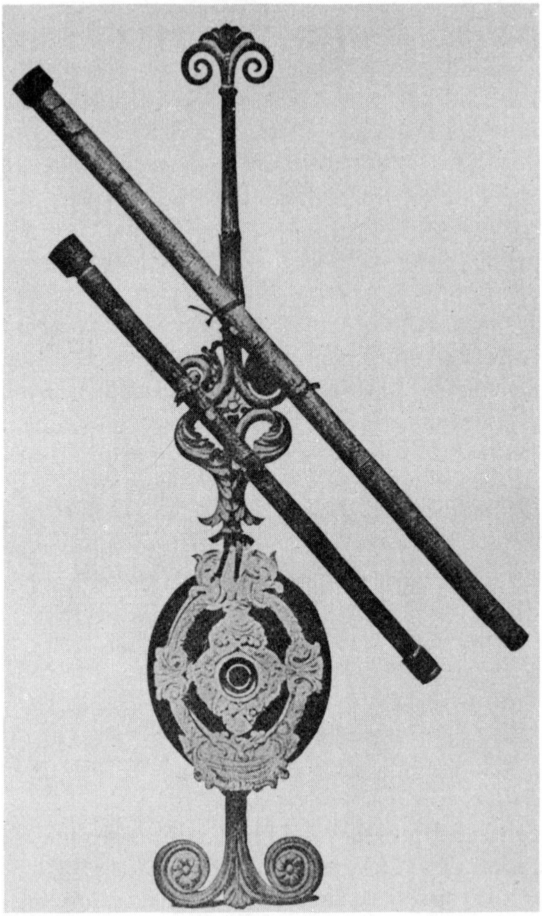

12.1 Galileo's telescopes, now in the Museum of the History of Science in Florence, Italy (photograph from the Yerkes Observatory).

was not powerful enough to show them in their true form). These revelations were very disturbing to the scientific and religious community of that time because they challenged widely accepted beliefs. Eventually, however, the truth became inescapable and man's knowledge of the universe moved to a new and higher level.

The optical telescope is still the most widely used instrument in astronomy. No matter what other type of radiation is detected, astronomers attempt to identify these sources with some visual object. In fact, until such an identification is made, the radiation data is usually referred to as a "source" rather than an "object."

Basically, a telescope does three things better than the naked eye.

First, it can resolve or separate objects, or the small details of a single object. The better a telescope's *resolution,* the more detail we can distinguish. We speak of a telescope "resolving" minute details such as craters or other features on the moon. A telescope with a higher resolution can separate stars that are very close to each other on the celestial sphere.

If more than one eyepiece is available with your telescope, compare the size of the field each shows. Which gives the greater field? Is this the more powerful magnification? Under what circumstances might a less powerful magnification be desirable?

Second, if an *eyepiece* is used, a telescope can *magnify* objects. However, even with the most powerful telescope, the stars themselves will not be truly magnified, even though they will look much brighter. More stars and more distant stars will be seen than with the naked eye, but any apparent magnification of their physical size is primarily the distorting effect of our own atmosphere. The planets, moon, and sun will appear larger through a telescope because they are close enough to us to appear as disks, but the stars are so very far away that they remain essentially points of light.

However, through the telescope each tiny area of the sky will appear enlarged, so that the space between the stars appears greater. The greater the magnification of the telescope, the smaller the area of sky that is seen at any one time. For example, the 200-inch telescope takes in such a small portion of the sky in each photograph that even if it could be used every night, it would take about 1000 years to photograph the entire sky.

Most important, the telescope is a *light-gathering instrument.* That is why we can see stars through a telescope that are too faint to be seen with the naked eye. The larger the diameter of the lens or mirror, the greater the telescope's light-gathering ability (as well as its power to resolve details), and the fainter the stars it will detect. However, when they first look through a telescope, many people are disappointed. After seeing magnificent star-studded photographs showing the wonders of the universe, they expect to see such sights with just the aid of the telescope. They do not realize that these spectacular pictures are the results of long photographic exposures, sometimes requiring many hours (Fig. 12.2).

Photographic films and plates are capable of accumulating the radiation reaching them. This is not so with the human eye, which dissipates such light relatively quickly to make ready for new stimuli. This characteristic is essential in daily life, but for deep space viewing it presents an

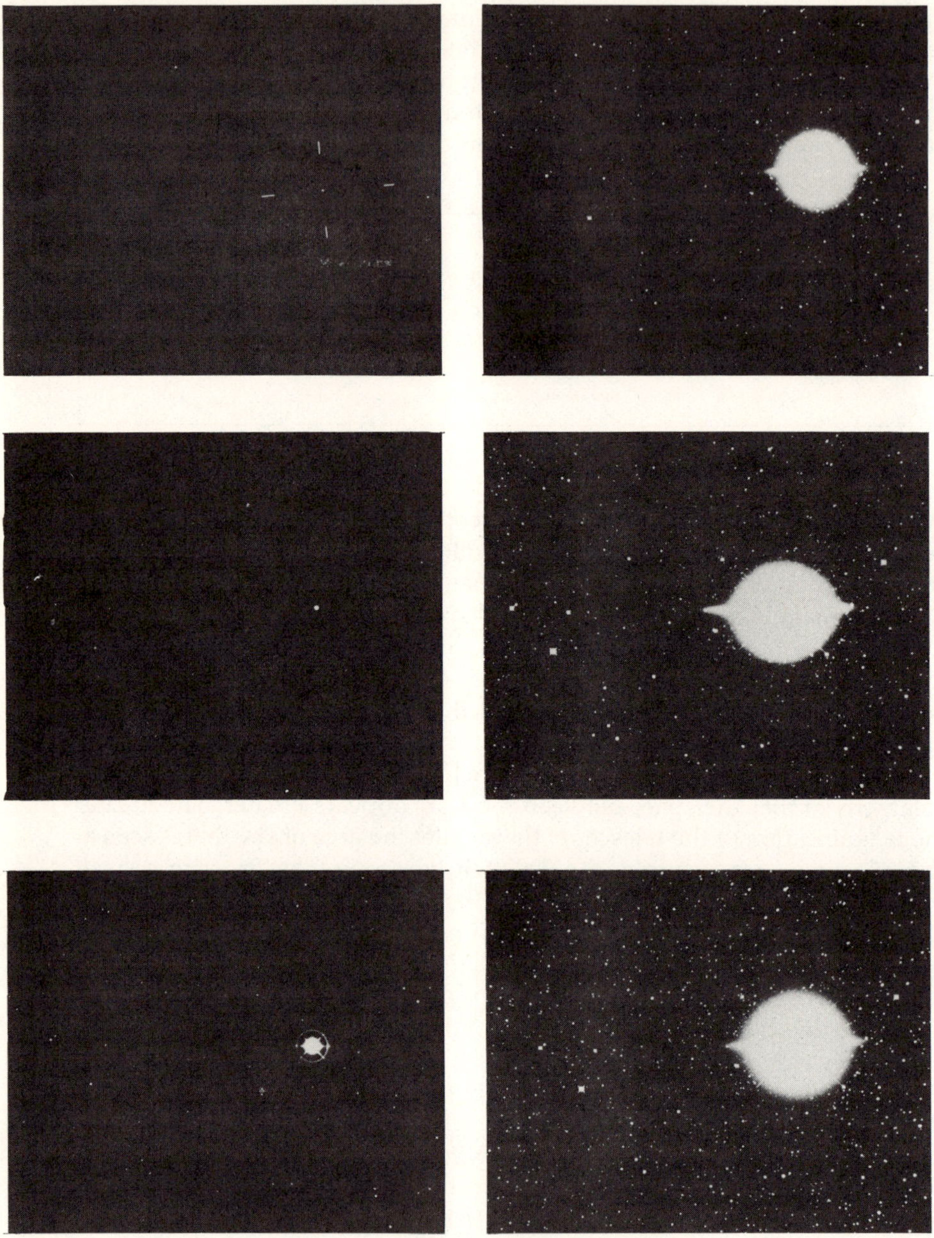

12.2 Series of photographs of the same star field using longer exposure times (photographs from the Hale Observatories).

obstacle. However, when the moon and planets are viewed through the eyepiece of a telescope, their brightness allows the eye to take advantage of the instants of "good seeing" (moments of nonturbulent atmosphere) to see fine details that photographs cannot record.

Using a camera held very steady on a tripod, take several time exposures of the same area of the sky. Vary the exposure times from 5 minutes to 30 minutes to one hour or more. Compare your results.

There are two basic types of optical telescopes. The type that Galileo used is called a *refractor* or *refracting telescope,* because it uses a lens which refracts or bends the incoming light to form an image of a distant object (Fig. 12.3). To use such a telescope visually, one looks through an eyepiece

12.3 *Diagram of a refracting telescope. Light is gathered through the* objective lens, *which determines the size of the telescope. This first lens bends the rays and carries them to the* eyepiece, *where they are focused by a magnifying lens or eyepiece.*

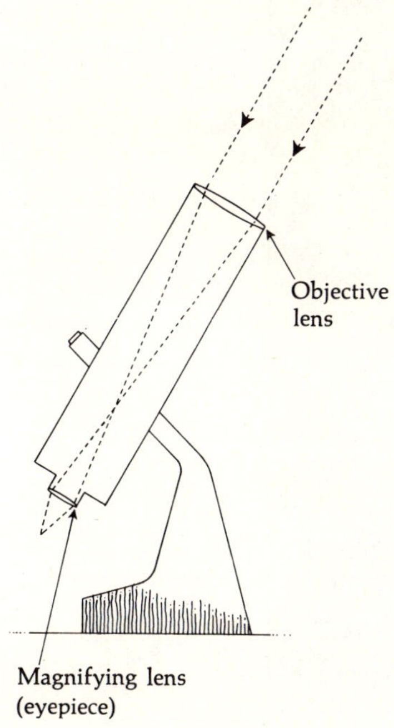

(which is basically a very good hand magnifier) to magnify the image. When we speak of the *size* of a refracting telescope, we refer to the diameter of the primary lens. The largest refracting telescope has a 40-inch lens and is at Yerkes Observatory in Wisconsin (Fig. 12.4). The construction of refracting telescopes larger than 40 inches is not practical, since a larger lens would have to be made thicker—too thick to allow enough undistorted light through to be useful in astronomy.

Even with the early refractors of the seventeenth century (Fig. 12.5), problems arose due to the rainbow effect created by the lens itself. To lessen this *chromatic* or *color distortion,* very long telescopes were built,

12.4 The 40-inch refracting telescope at Yerkes Observatory in Williams Bay, Wisconsin (photograph from the Yerkes Observatory).

some as long as 200 feet. Since these were difficult to use, Sir Isaac Newton, about 1670, invented a different type of telescope, the *reflector,* which uses a mirror instead of a lens to collect light (Fig. 12.6). Although it was necessary to curve the mirror in just such a shape (parabolic) so that it, too,

12.5 *Seventeenth and eighteenth century European telescopes. Because early telescopes had a serious problem with chromatic aberration (rings of color around sighted objects), they were made with very long focal lengths, using lenses with very slight curvatures. These telescopes are optically very similar to those of Galileo (courtesy Adler Planetarium, Chicago).*

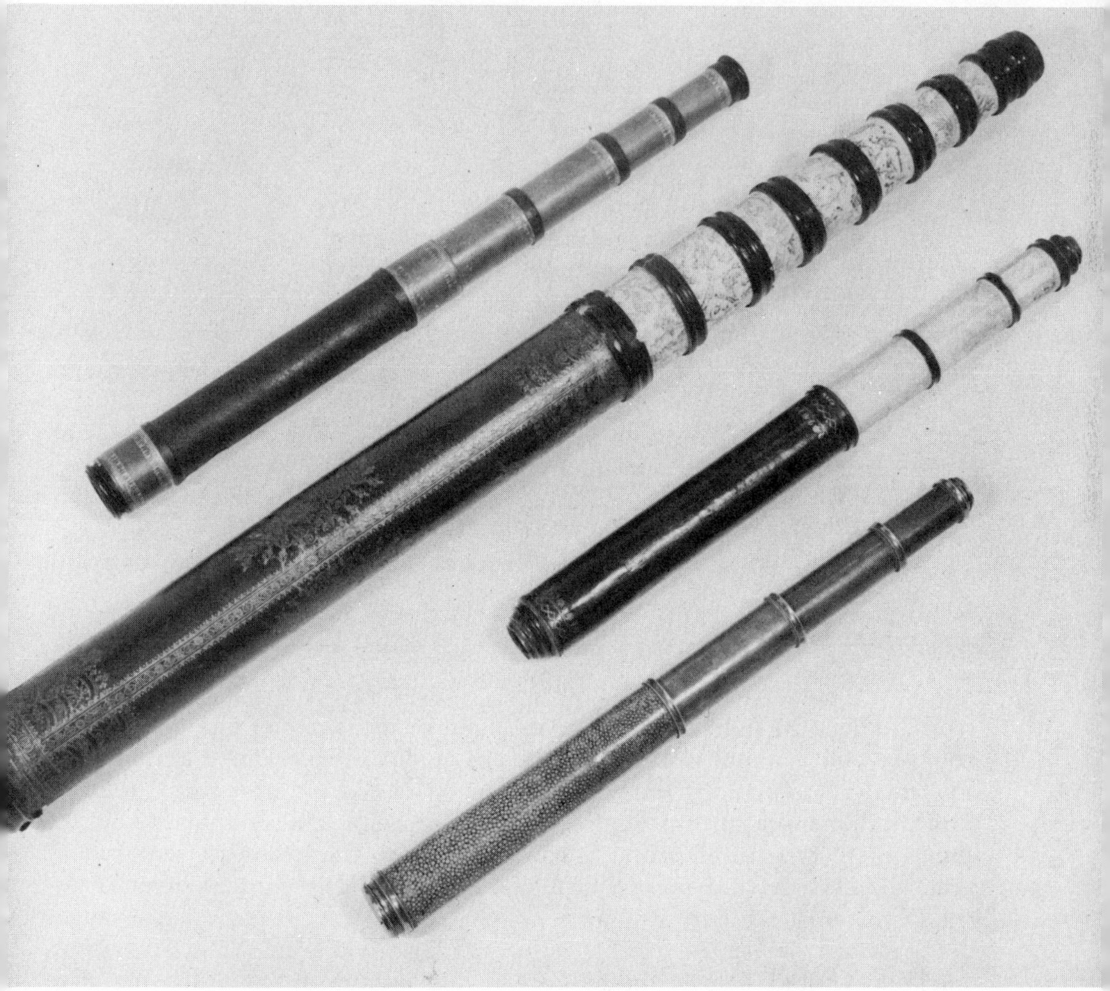

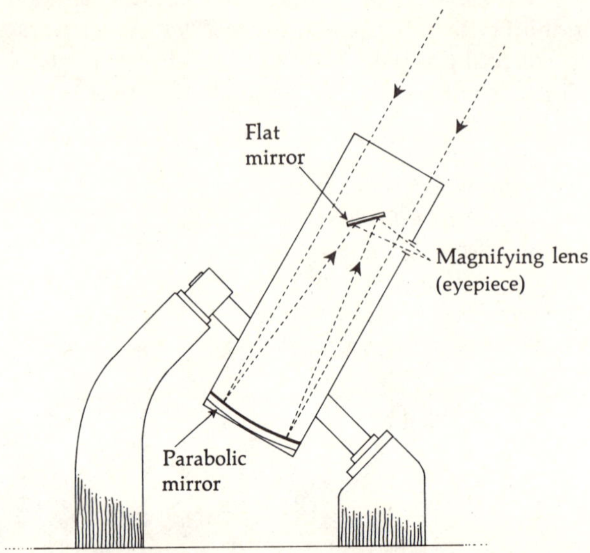

12.6 Diagram of a reflecting telescope (Newtonian). Light is gathered in the long tube at the bottom of which is a parabolic mirror. The size of this mirror is the size of the telescope. The mirror reflects the light back to a focal point near the top of the tube, where a small flat mirror deflects the light out through a hole in the side of the tube. There an eyepiece magnifies the image.

If a telescope is available, determine what kind it is. Draw a diagram similar to Figure 12.3 showing how the light is gathered and focused. What size is the telescope?

---

would reflect the light back to a given point or focus, larger instruments could be built without worrying about lens quality or color distortion. Before modern technology, however, adequate telescope mirrors were too difficult to manufacture successfully. Therefore, refractors continued to be the primary type of telescope in astronomy until the twentieth century (Fig. 12.7). When we speak of the *size* of the reflecting telescope, we refer to the diameter of the mirror.

Practically all large optical telescopes today are reflectors. The largest one in the United States is the 200-inch Hale telescope on Mount Palomar in California (Fig. 12.8). The largest optical telescope in the world is the

12.7 William Huggins' 8-inch refractor telescope, fitted with a spectroscope to obtain star spectra. In the 1860's Huggins studied the spectra of many bright stars and identified many of the spectral lines (courtesy Trustees of the British Museum).

six-meter (237-inch) Russian telescope located on Mt. Pastukhov in the Caucasus. In recent years other large telescopes have been constructed in the southern hemisphere, for example in Chile and in Australia, to enable astronomers to explore the southern skies as thoroughly as they have the northern ones.

12.8a The 200-inch Hale telescope on Mount Palomar pointing toward the zenith (photograph from the Hale Observatories).

Larger telescopes could be built, but the limitations imposed by atmospheric turbulence and by the light of the night sky as well as by man-made atmospheric pollution make such giant earth-bound instruments impractical. Much optical telescope development today is concentrated on instruments that can be used above the earth's atmosphere, where they are

---

Take a reading magnifying glass and a hand magnifier and hold them out with the reading glass in front of the hand magnifier. Experiment until you are able to magnify a distant object. Why is the image inverted? About how much magnification did you achieve? Draw a diagram showing the paths of the light ray.

12.8b The 200-inch Hale telescope, showing the observer sitting in the prime focus cage. Note the reflecting surface of the telescope at the far end facing him (photograph from the Hale Observatories.)

much more efficient. By the early 1980's, scientists expect to install the Large Space Telescope (LST) in the "Space Shuttle," a reusable spacecraft the size of a modern jetliner (Fig. 12.9). The Space Shuttle will be able to carry both men and equipment in orbit around the earth. The projected LST will be 95 inches or 240 centimeters in diameter, but because it will be used above the distorting atmosphere of the earth it will have a resolving power ten times greater than the most powerful telescopes on earth and will be able to detect objects 100 times as faint.

A question frequently asked is how "far" into space we can actually see from here on earth. Even without a telescope we can see objects that are hundreds of trillions of miles away. The nearest star (other than the sun) is 4⅓rd light years away, which is about 25 trillion miles or 42 million kilometers. The Andromeda galaxy, the only galaxy visible to the

12-9 The Shuttle Orbiter 101 "Enterprise" riding "piggy-back" atop the NASA 747 aircraft during an unmanned approach and landing test. Later in this flight the two craft were separated so that the shuttle could be tested for low speed performance and aircraft-type landing. In actual use, the shuttle will be put into orbit from launching pads.

naked eye in the northern hemisphere, is approximately two million light years away, which is about ten million trillion miles. The naked eye can see objects with an apparent magnitude of +6 if viewing conditions are excellent. The most powerful telescopes, with the aid of photography, can record images 10 million times fainter.

Ideally, if an object was bright enough and if there was no interfering matter in the line of sight, there would be no limit to how far one could see, even with the naked eye. However, even with the most powerful telescopes, using the most sophisticated photographic equipment, there is the limit imposed by our own atmosphere. On the darkest nights the atmosphere has its own very slight glow. It scatters light, whether from the stars or from city streets, and creates a faint glow which obscures the very faintest stars. Charged particles from the sun and outer space cause the atoms in the upper atmosphere to fluoresce, adding to this faint glow. Stars and galaxies fainter than the limit set by this atmospheric glow are unobservable from the surface of the earth. To even approach this limit, photographic techniques must be employed, since the eye is not capable of a "time exposure."

However, as valuable as the photographic process has been in the recording of distant celestial objects, even more sensitive methods have recently become available to aid the astronomer. Photoelectronic devices, such as those used in television, have markedly increased our ability to record distant celestial objects in much shorter exposure times. The increase in sensitivity has in some cases been as much as 100 times. These *image intensifiers* do not increase our ability to "see" farther into space, since the atmospheric glow is still a limiting factor, but they cut drastically the exposure time needed to record distant objects. Short exposure times are especially important in studying rapid changes in celestial objects (Fig. 12.10).

Such modern instruments have changed many of the observational astronomer's methods. No longer must one always sit in a freezing dome, adjusting a telescope with hand-held controls while viewing the star field through a very cold eyepiece. Instead, many observations can be done in a warm room, seated at a computer console with a television screen above (Fig. 12.11). The astronomer instructs the computer to operate the telescope, televise the viewing field, and photograph the resulting data. Through the computer, whole programs of study can be set up. Once the data have been recorded, other computers are used to analyze them. Mathematical computations that once took many hours can now be completed in a matter of minutes or seconds with high-speed computers.

Seeing

Even with all these improvements, however, the observational astronomer is still beset with difficulties, many of which arise from the earth's atmosphere. Besides the limiting factor of the night sky glow, the turbulence of the air also creates problems. Close to the ground, thermal disturbances distort the incoming light waves and cause a blurry, fuzzy image. Instabilities much higher in the atmosphere, in the region of the jet stream, cause the brightness of the image to vary or "twinkle." We say that the "seeing" is good when the stars appear as sharp, steady points of light. Twinkling is more obvious near the horizon than overhead because the starlight must pass through more atmosphere.

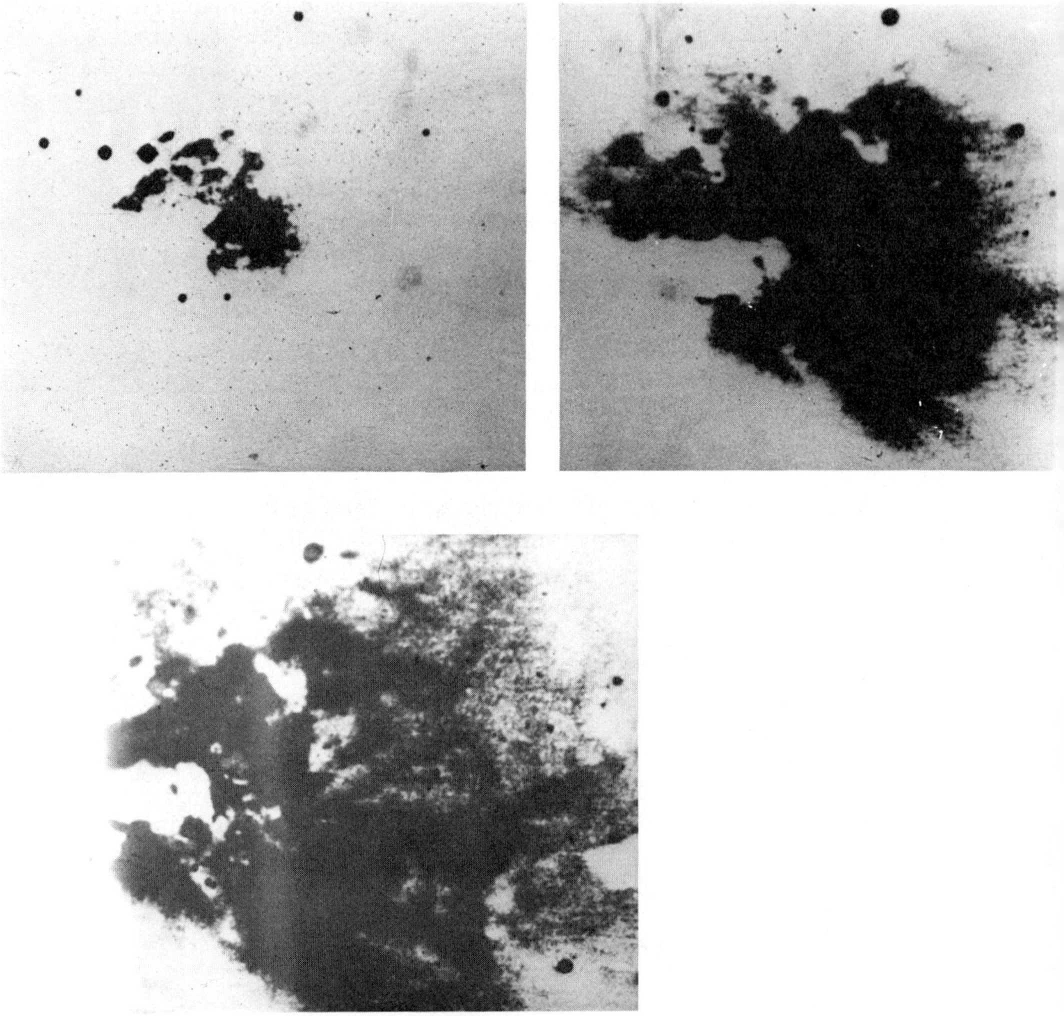

12.10 The Orion nebula, as recorded by the image orthicon system at the Northwestern University Corralitos Observatory in New Mexico. The three images (negative reproductions) were produced by exposures of 1/20, 3/8, and 3/4 seconds respectively. Compare these with Figure 2.9, which took many minutes of exposure time with a much larger telescope. (Photographs from Dearborn Observatory, Northwestern University.)

12.11 The image orthicon system at Corralitos Observatory, New Mexico. Justus Dunlap, resident director, is at the control panel (photograph from Dearborn Observatory, Northwestern University.)

To see how much distortion there is, try this simple experiment. Cross your eyes and look at a bright star near the horizon.[1] The two images of the star that you will see will not flicker in the same way at the

---

[1] If you have trouble crossing your eyes, the same effect can be achieved by holding your finger at arm's length, and while looking at it, being conscious of the star twinkling behind it. If this is done properly, you should see two images of the same star with two different twinkling effects.

same time. They may even appear to be two different colors at the same moment. Actually, different rays of light from the distant star are entering each eye. The rays that enter the right eye have had a slightly different journey through the atmosphere than those entering the left. The eddies of turbulent air which cause most of this disturbance can be very small, so that even rays separated by a few inches may be received differently.

In general, planets do not twinkle unless they, too, are very near the horizon. Whereas a star, because of its great distance, is effectively a point of light, a planet, which is much closer, is viewed as a whole area made up of many points of light. These points tend to cancel each other's flickerings, and the overall image becomes a relatively smooth one.

It is largely because of the problem of limited range of receivable radiation and the problem of atmospheric turbulence that astronomers are so desirous of having observatories above the earth's atmosphere. Even placing a telescope on a high mountain or on an elevated structure (Fig. 12.12) helps eliminate some of the heavier and thicker parts of the atmosphere so that twinkling is partially reduced. Also observatory domes are purposely not heated because warm air rising upward tends to cause even greater distortion. This is easily seen in the air rising from a hot stove or a car whose engine is still very warm.

The Advent of Radio Astronomy

As we have seen, an entirely new look at the universe was made possible by the development of radio astronomy. Although it is a relatively new field (less than 50 years old), radio astronomy has in some respects actually surpassed optical astronomy in surveying the universe.

As with so many other discoveries in science, the beginnings of radio astronomy were accidential. In 1931, Karl Jansky, an engineer for the Bell Telephone Laboratory, was trying to better understand the many causes of static interference in transoceanic communications which used shortwave radio. After allowance was made for known causes of static (thunderstorms, manmade electrical interferences, etc.), there was still an unexplained radio hiss in the background. Using his highly directional radio antenna, he determined that the hissing noise had a preferential direction in space. Jansky did not know it, but he had actually built the world's first radio telescope.

He found that the unknown source of the static crossed the meridian four minutes earlier each day. This immediately ruled out the sun, which crosses the meridian about noon daily (see chap. 15). It was later determined that the unknown source of the static came from the direction of the center of the Milky Way galaxy.

12.12 Lindheimer Astronomical Research Center (LARC), Northwestern University, Evanston, Illinois (photograph from Dearborn Observatory, Northwestern University.)

Jansky described his findings in an electrical engineering journal but no one paid much attention to them at the time. Then in 1937, Grote Reber, a radio "ham" and amateur astronomer from Wheaton, Illinois, recognized the astronomical significance of Jansky's discovery. In his backyard he built a directional radio antenna that could be pointed to all directions in the sky. Using this crude radio telescope, Reber made the first radio maps of the heavens and submitted his results to the *Astrophysical Journal.*

Again, no one gave the subject much attention, this time because astronomers believed that even if radio waves came from space, they would be of limited value to us. They thought that adequate resolution could not be obtained from such long waves, since in order to pinpoint exact locations in the sky, the longer the wavelength of the radiation involved, the larger the telescope must be. For example, the 21-cm radio wavelength of hydrogen is 350,000 times longer than the average light wave. To have the same resolution as a 6-inch optical telescope, a radio telescope receiving this wavelength would have to be 33 miles in diameter!

Obviously, such large receivers were impossible. However, after World War II many huge radio telescopes were constructed and proved to be of great value (Fig. 12.13). Using the technology developed in radar research, particularly precision amplifiers and recording devices, scientists in Australia, Holland, and England were the first to start the search for celestial radio sources.

It should be pointed out here that although we commonly think of the

12.13 The "dish" of this radio telescope is 64 meters (210 feet) in diameter. It is fully steerable and has a resolution power to within one second of arc. It is located in Parkes, Australia (photograph from CSIRO, Division of Radio Physics, Australia).

radio as something to be listened to, radio waves themselves do not make any sound. It is the apparatus within our radio sets which converts the radio energy into the sound waves that we hear. Radio energy, like light energy, is part of the electromagnetic spectrum. Sound, on the other hand, is caused by the physical vibration of matter and cannot travel through empty space.

Like its optical counterpart, a typical radio telescope employs a parabolic surface, but one which is usually made of wire mesh instead of a glass mirror (see Fig. 12.14). The radio telescope "sees" by recording the intensity of the radio radiation it receives. The parabolic surface or *dish* is usually mounted on a base so that it can be turned to any direction in the sky. Reber's first radio telescope had a dish 31 feet in diameter. The radio waves that the dish gathers are focused onto an antenna, creating an electric current. This current is fed into a receiver and amplified. The resulting "radio" signals are recorded. The receiver can be tuned much like a normal radio, so that radio astronomers can study a specific wavelength of radio radiation in the same way that you can tune in to a favorite station.

12.14 *Schematic diagram of a radio telescope. The parabolic "dish" reflects the radio waves to a focus at the antenna, which picks up these radio signals and feeds them into an amplifier. This in turn feeds the amplified signals into a recording device which records the intensity of the energy received.*

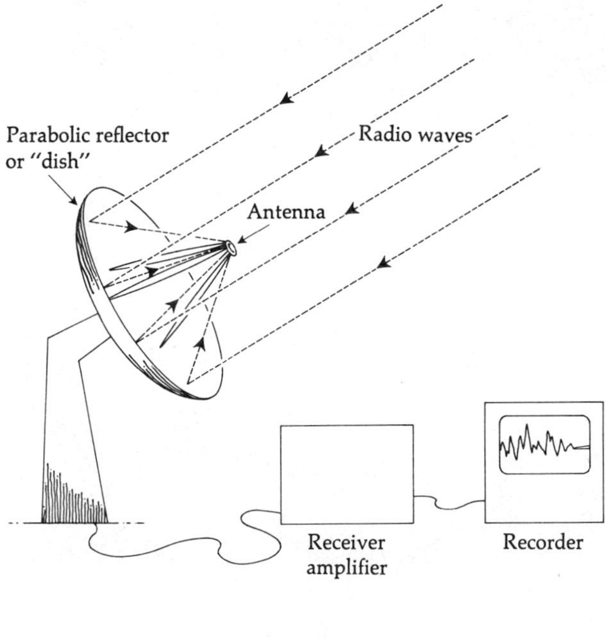

A radio telescope is measured by the diameter of its dish. This is usually in meters or feet rather than in centimeters or inches. One of the largest movable radio telescopes is the 300-foot National Radio Astronomy Observatory at Greenbank, West Virginia (Fig. 12.15). Any dish much larger than that would be difficult to support and to move. However, since a radio telescope is basically a huge radio antenna, it can take many and varied forms, just as can more conventional antennas. Some are literally carved out of the ground and lined with wire mesh. These dishes are unmovable and are therefore more limited than the steerable radio telescopes. However, because they are larger, their resolving power is greater. The largest dish of this type is the Arecibo Observatory built into the hills of Puerto Rico. It measures 1000 feet or 305 meters across. One favorite measure of its

12.15 The 300-foot Transit Radio Telescope at Green Bank, West Virginia. This instrument weighs 500 tons, of which 460 tons are movable (photograph from the National Radio Astronomy Observatory).

12.16 The National Astronomy and Ionosphere Center, the 1000-foot radio telescope located in Arecibo, Puerto Rico. The small valley has been contoured to allow the construction of the parabolic mesh dish. Above the dish hangs the antenna supported by three towers. (National Radio Astronomy Observatory)

size is that it has the capacity to hold the entire annual beer consumption for the entire world (Fig. 12.16).

Other radio telescopes are arrays of wire antennas stretched for great distances across the ground. These may be several kilometers in size. The Mills Cross Radio Telescope in Australia is a good example of this type (Fig. 12.17).

Because radio waves are so long, even these huge instruments have very poor resolving power. The one at Arecibo has a resolving power less

12.17 The Molonglo Radio Telescope ("Mills Cross") with arms one mile long was the first such large "cross antenna" arrangement to be constructed (CSIRO, Division of Radio Physics, Australia).

than that of the human eye. However, this difficulty has been largely overcome with the use of *interferometers* (Fig. 12.18). These are composed of two or more radio telescopes which are positioned over a great area. Radio waves received by these separate instruments are combined electronically to produce a "radio image" of the object being studied. The resolution or detail attained is as great as it would be if a telescope the size of the entire array had been used.

In the late 1970's and 1980's such an array will be completed in New Mexico. Called the VLA (Very Large Array) Radio Telescope, it will eventually have 27 dish-shaped radio telescopes, each 85 feet across, mounted on railroad tracks in a Y-shaped pattern, each arm of which will be 13 miles long (Fig. 12.19). The farther apart on the tracks the dishes are moved, the better the resolution. This resolution will be the equivalent of a single radio telescope 17 miles across (if one could ever be constructed). Astronomers

12.18 An interferometer: Caltech's twin 90-foot (27 m) radio telescopes in Owens Valley, California (photograph by Dr. Alan T. Moffet, ORVO, Caltech).

estimate that its radio "pictures" should be able to match or exceed those of any earth-based optical telescope.

Since better resolution could be achieved with a still longer base line, the ultimate step for the earth-bound astronomer is to have one dish in, say, the United States, and the other in Australia. This has been done using computers instead of telescopes on railroad tracks. The data from these telescopes, which are separated by continental distances, is recorded on magnetic tape and combined by a simultaneous "playback" in a computer. This radio telescope is called the VLBI (Very Long Baseline Interferometer). It actually has a far greater resolution than any optical telescope.

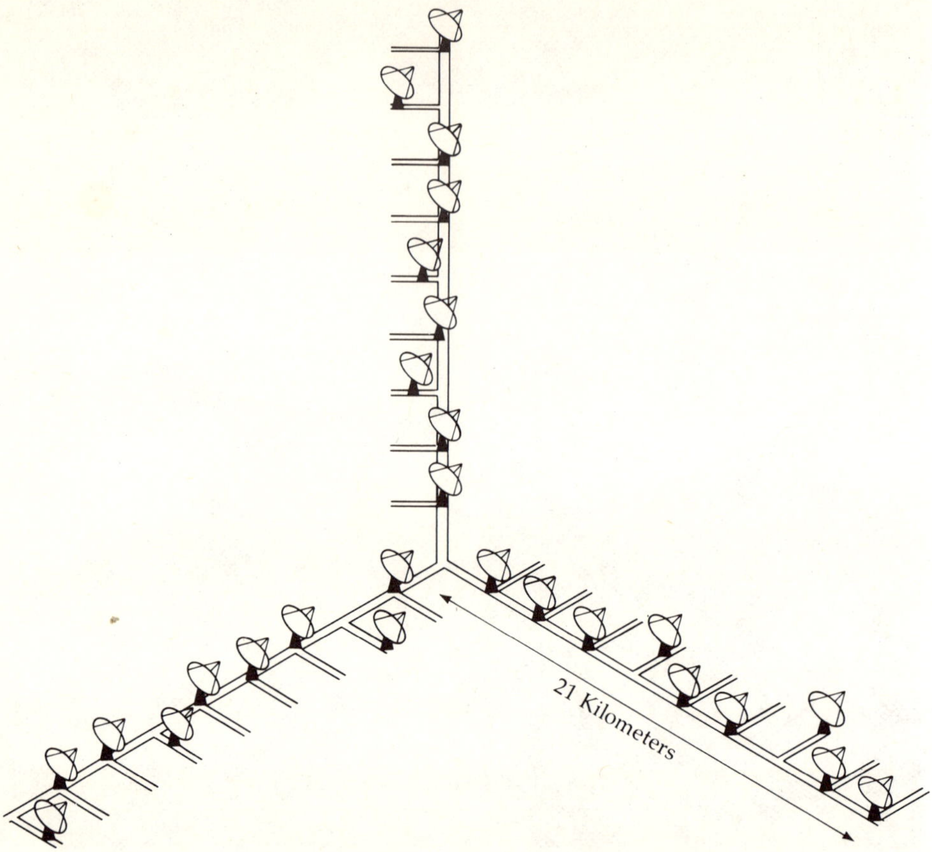

12.19 A schematic drawing of the VLA (Very Large Array). Each arm is 21 kilometers (13 miles) of standard-gauge railroad tracks and has nine spur tracks on which any one of the 27 200-ton antennas can be positioned for observing. This allows complete freedom in arranging various antennas as well as eliminating the need for roads. The completed installation will be equal in resolution to that of an antenna 27 kilometers (17 miles) in diameter.

As bigger and better radio telescopes were built, the search for celestial radio sources accelerated. Finding them was not difficult; identifying them with a visual object in the sky became the main problem. In addition to the source that Jansky had found, the sun and later Jupiter were also quickly discovered to emit radio waves (although not of the same wavelength that Jansky had been studying). The hundreds of radio sources detected were numbered and catalogued. The third edition of a catalogue prepared in Cambridge, England is frequently cited. Examples of entries in

Plate A Flash spectrum of the solar chromosphere. During a solar eclipse, in the brief seconds after the moon's disk has covered the sun's photosphere but before totality is reached, radiation from the solar chromosphere can be recorded. The emission spectrum indicates that this layer of the sun's atmosphere consists of hot gases that are emitting light (photograph from the Dearborn Observatory).

Plate B The stars of the Hyades open cluster as seen with the aid of a prism. The light from each star is spread out into its component colors. Spectra obtained in this manner are called "objective prism spectra" because the prism is placed in front of the objective lens or mirror of the telescope (photograph from the Department of Astronomy, University of Michigan).

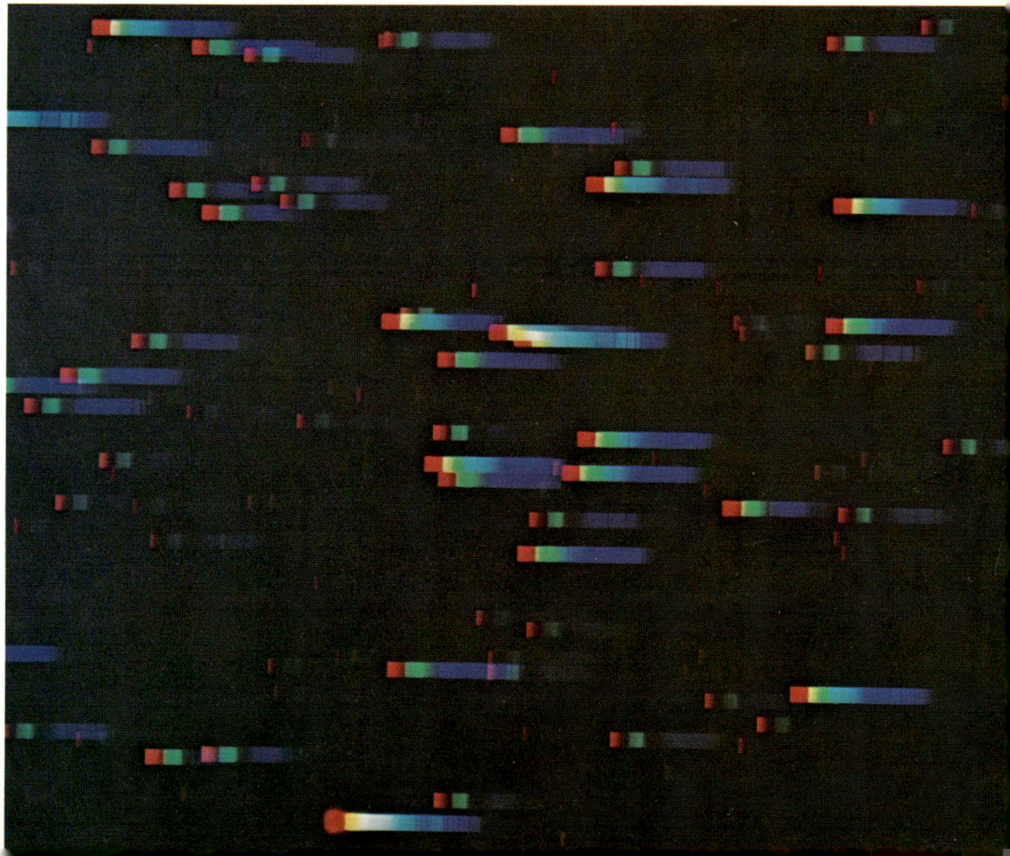

Plate C *Martian landscape as photographed by Viking 1 after it landed on the surface in July 1976. The cable on the spacecraft is bright orange in color, indicating that the intense colors of the Martian surface, as seen in the photograph, are, in fact, real. Orange-red surface materials form a thin veneer over darker bedrock, while the sky presents a reddish cast, probably due to scattering and reflection from reddish sediment suspended in the lower atmosphere. (JPL-NASA).*

Plate D On the moon. An Apollo 15 astronaut is seated in the lunar rover; at left is the lunar module. About 3 miles away in the distance is St. George crater on the slope of Hadley Delta (photograph from NASA).

Plate E Moon rock: a photomicrograph of a thin section of one of the Apollo 17 lunar samples.

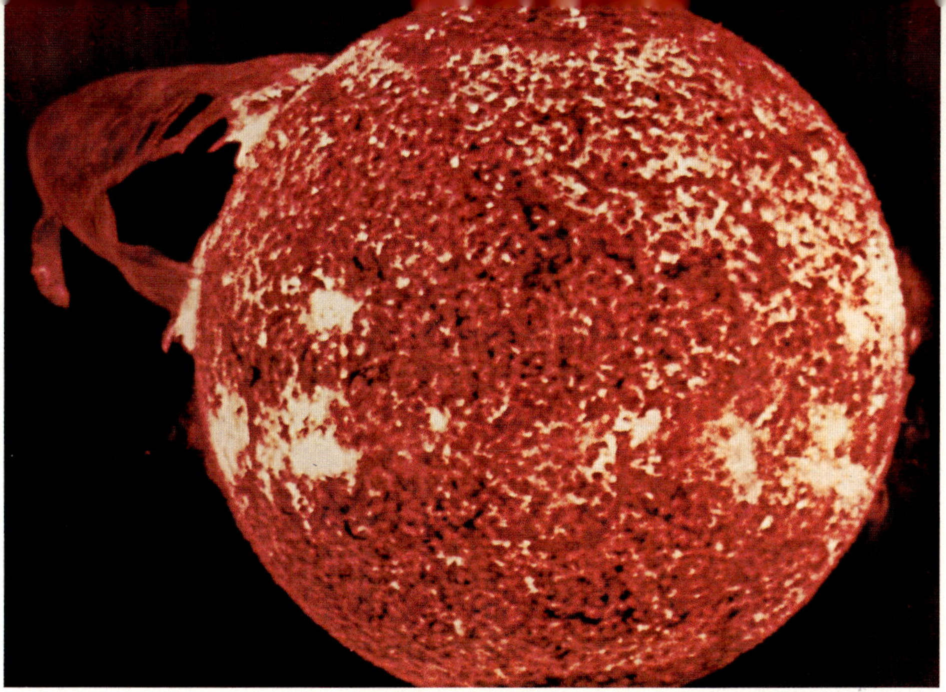

Plate H (above) *The sun, photographed by an extreme ultraviolet spectroheliograph instrument during the third Skylab mission, shows one of the most spectacular solar flares ever recorded, spanning more than 588,000 kilometers (365,000 miles) across the solar surface. The flare gives the distinct impression of a twisted sheet of gas in the process of unwinding itself. (NASA)*

Plate I (right) *The planet Jupiter as photographed by Pioneer 11 in 1974. The dark spot is the shadow of Io, the nearest of the Galilean moon, as it crosses in front of the planet, creating a solar eclipse in that area. (NASA)*

Plate F (opposite, top) *"Rosette" nebula in the constellation Monoceros (photograph from the Hale Observatories).*

Plate G (opposite, bottom) *"Dumbbell" nebula in the constellation Vulpecula (photograph from the Hale Observatories).*

Plate J *The diamond ring effect photographed during the total solar eclipse, July 20, 1963. The last rays of sunlight are seen shining through the mountain peaks on the moon just before totality is reached (photograph from the Dearborn Observatory).*

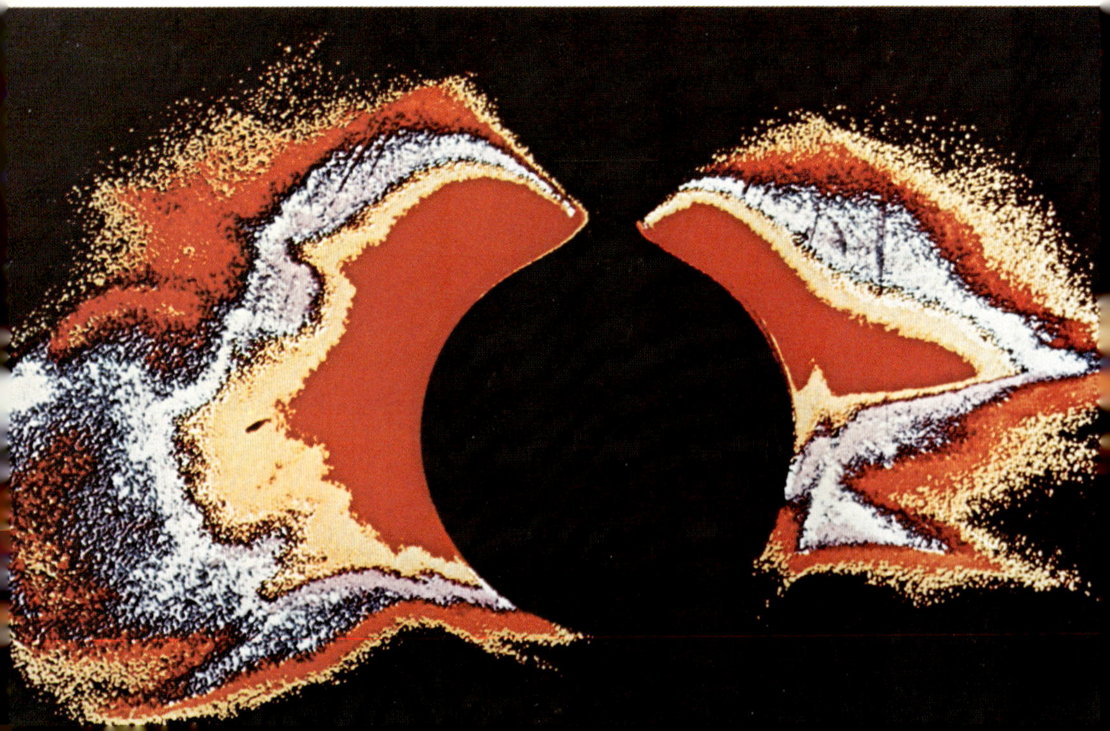

Plate K   Solar eclipse. Only when a totality is reached can the magnificent solar corona be seen. This photograph was taken July 20, 1963 at Moose River Crossing, Ontario, Canada by members of the Dearborn Observatory expedition (photograph from the Dearborn Observatory).

Plate L   (opposite) The solar corona: color coded to distinguish levels of brightness. Skylab astronauts used a coronagraph which masks the sun's disk and creates artificial eclipses. The original exposure for this image was made on August 10, 1973.

Plate M Space Colony: a segment is shown of a space colony during its final construction. The colony, 150 yards in internal diameter, is visible through the 100-foot strip windows encircling the wheel-like colony.

this catalogue are 3C273 or 3C48, which refer to the 273rd or 48th entry, respectively, in the Third Cambridge Catalogue.

Telescopes in Space

Both optical and radio telescopes can be located on the earth's surface because they are capable of receiving and recording electromagnetic radiation that can pass through our atmosphere. But in order to detect the radiation in other regions of the spectrum, instruments must be carried above the atmosphere. Balloons, rockets, and spacecraft have all been employed, opening up new fields of astronomy to be explored.

Balloons, for example, were found very useful because they are relatively inexpensive and stay aloft long enough for sufficient data to be recorded. Of course, they cannot go completely out of the earth's atmosphere if they are to be recovered, but the heights that they have achieved have been quite adequate, for example, in infrared astronomy.

Rockets, which can go much higher than balloons, have a very limited time before they parachute back to earth. Further, their expense keeps most private experimenters away. Still, the United States and other nations have individually and jointly worked on rocket probes to detect x-ray, gamma ray, and ultraviolet radiation from outer space.

However, satellites that remain in orbit around the earth are now the most important in such research. It was the Vela satellites, which were designed to monitor the world for nuclear bomb explosions, that detected the mysterious gamma ray bursts (see p. 219). The Uhuru x-ray satellite was the first to detect the x-ray sources that are thought by some to be associated with black holes. Other such satellites have also enormously increased our knowledge of these and other electromagnetic radiation sources.

Such a discussion of space probes should mention manned spacecraft, such as the Apollo missions to the moon (Fig. 12.20), Skylab, and the future Space Shuttle. Probes to the planets with Pioneer, Mariner, and Viking spacecraft have greatly extended our knowledge of the solar system. These discoveries will be discussed in Chapters 18 and 19.

For each type of radiation received from outer space, a specially designed "telescope" is necessary. Just as radio telescopes differ radically from optical ones, so do x-ray and gamma ray telescopes differ from each other, and from ultraviolet or infrared telescopes. The wavelength of each type of radiation largely determines the kind of instrument required—its size, its method of receiving the data, its means of focusing this data, and its method of recording it for future interpretation and analysis.

12.20 The Apollo 15 Command and Service Modules in lunar orbit as photographed from the Lunar Module. The Service Module's Scientific Instruments Module (SIM) bay, which holds 8 orbital science experiments, is clearly visible (NASA photograph).

Discussion Questions

1. What are the advantages of placing a telescope on the moon or in an orbiting space laboratory?

2. How might human history have been changed if the telescope had been invented in 500 B.C. instead of about 2000 years later?

3. If an astronomer is looking through a telescope of moderate size while a moth flutters a few inches in front of the instrument, what effect will this have on the astronomer's observations? Can he see the moth through the telescope?

4. What are the two major kinds of optical telescopes? Describe how each gathers and focuses the light it receives. What are the advantages and disadvantages of each?

5. Compare the components of the radio and optical telescopes. Which part gathers the radiation? Focuses the radiation? Records?

6. What advantages does a telescope have over observing with the naked eye? What kind of objects can a telescope magnify?

7. The pupil of the eye is about 1/5th inch in diameter. How much farther out can stars be seen with a 20-inch telescope?

13 *Evolution and Dynamics of the Solar System*

An observer viewing our sun from the vicinity of the next nearest star, Alpha Centauri, would see it as a bright star in the constellation of Cassiopeia. If the observer were able to make very careful observations and measurements of the sun's motion, he or she might be able to detect a very slight wobble caused by Jupiter's gravitational effect as it orbits the sun. But he would be unable to see Jupiter, or any of the other planets, including Earth. In fact, he would not be able to obtain any evidence at all of the existence of the earth, using the instruments available to us today.

Similarly, if other planetary systems do exist, their detection from the earth would be equally as difficult. Planets are relatively small bodies and shine only by the reflected light of their parent star. From distances such as those that exist between stars, a planet's dim light is lost in the glare of the star. Only by detecting the minute wobble of a star as it moves through space can we infer the presence of an orbiting planet. Even for a very massive planet orbiting a star in the solar neighborhood, this wobble would be barely within our measuring limits.

We know that stars have been and are currently being formed from the clouds in interstellar space. Most develop into binary or multi-stellar systems, but some are single stars like our sun. It is widely believed that many of these single stars can have planetary systems similar to our own, but thus far we know of only one—our solar system. Let us therefore concentrate our attention on this solar system and find out how it relates to the rest of the cosmos.

Origin of the Solar System

Most astronomers agree that about four and a half billion years ago a tremendous cloud of gas and dust in one of the spiral arms of the Milky Way galaxy began to contract under the gravitational attraction of its own mass. As it grew smaller, its rotational speed increased, due to the conservation of angular momentum (see p. 178). This rotation tended to flatten the cloud and it eventually formed into a spinning disk. Most of the mass accumulated at the center of the disk, and as it contracted the disk collected even more of the cloud's material. In time this body of matter became hot and dense enough for nuclear reactions to begin in its inner core. It became a star—the sun.

In the outer regions of this disk, *globules* of matter accumulated as the cloud grew smaller. Although nowhere as massive as the central body, they were nevertheless large enough to condense into spherical bodies and continue to attract other particles in their immediate neighborhoods, thus increasing their masses (Fig. 13.1). These protoplanets all preserved the original rotational direction of the primordial disk and began to orbit the sun, all rotating in the same direction and in about the same plane, and all but Venus spinning on their polar axis in the same direction as their orbital rotation. Seen from many millions of miles above the North Pole, this direction would be counterclockwise (Fig. 13.2).

Smaller accretions of matter formed into various-sized bodies that were not pulled into the larger planetary bodies, but eventually found themselves in orbit around the protoplanets rather than the sun. These became the natural satellites or *moons* that today orbit six of the nine planets. Again, most of them proceeded to orbit their respective planet in the original rotational direction of the disk, that is, counterclockwise. However, a few, for reasons that are still unknown today, established orbits in the opposite direction. We find that most of the moons in the solar system, regardless of the direction of their orbits, revolve close to the plane of their planet's equator rather than the plane of the solar sytem itself. Our moon, which revolves more closely to the plane of the ecliptic, is a notable exception.

Once the planets and moons had formed, a large quantity of gas and dust remained circling the sun. However, most young stars pass through a stage where they forcefully eject much of their accumulated matter before settling down on the main sequence. This "wind" was strong enough to have pushed out of the solar system most of the minute bits of matter that were not already incorporated into one of the planets or moons. The few small bodies still orbiting the sun after this initial "clean-up" of interplanetary space were constantly subjected to the gravitational effects of the various planets and moons, and to the pressure exerted by the light from the sun itself. Most of them collided with one of the planets or moons (photographs of the surfaces of Mercury, Mars, and our moon show a multitude of craters, which are ample evidence of this early bombardment) or were so perturbed in their orbital paths as to be ejected from the solar system completely.

The larger, more massive planets tended to form farther from the sun. Their greater mass enabled them to gravitationally hold most of their origi-

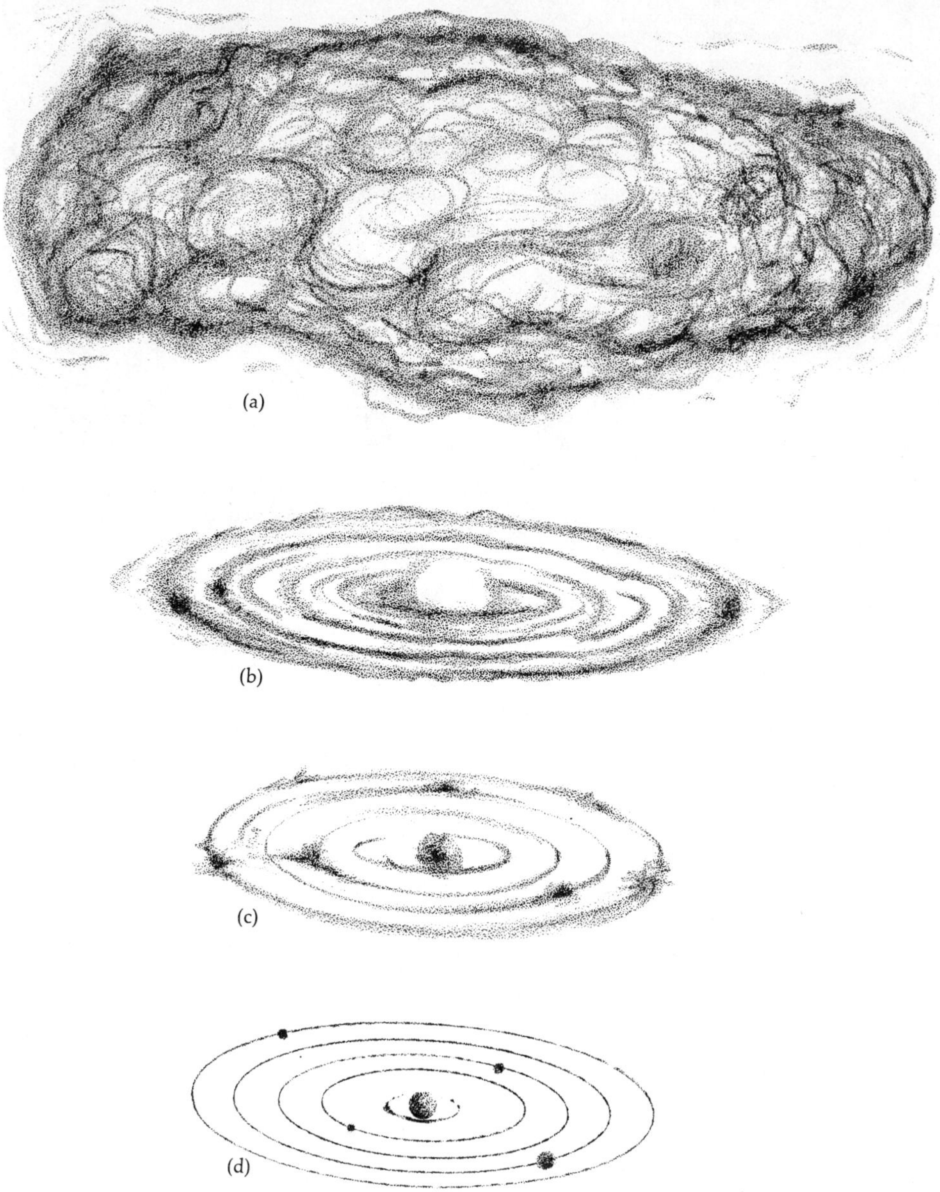

13.1 An artist's conception of how the solar system may have been formed: (a) A vast cloud of dust and gas begins to contract; (b) rotation of the cloud increases due to conservation of angular momentum; (c) most of the mass condenses into a central body while lesser accumulations condense and begin to revolve around this larger body; (d) gradually much of the original dust and gas becomes part of one of these bodies (or their moons).

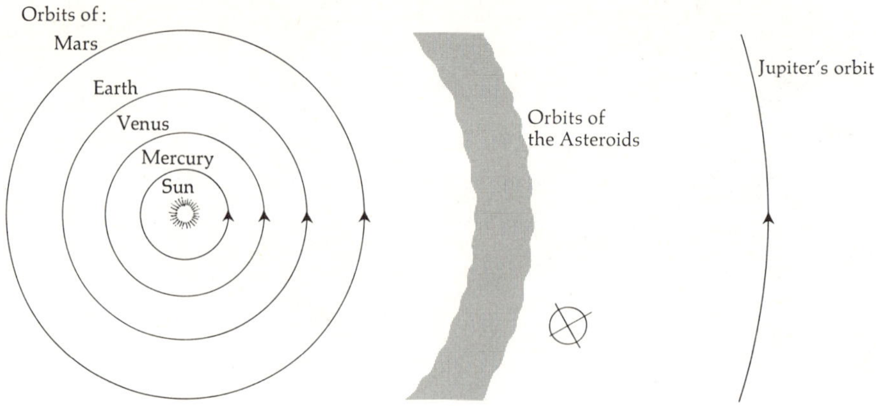

(a) The inner planets

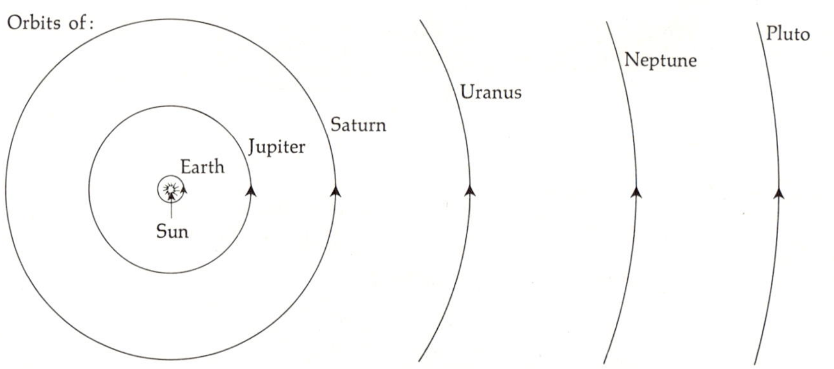

(b) The outer planets

13.2 *Schematic drawing of the planetary orbits of the solar system. In reality, the orbits are all ellipses of very low eccentricity. The planet with the most eccentric orbit is Pluto—one section of its orbit actually comes closer to the sun than does Neptune's orbit at that point. In diagram (a), Jupiter's orbit is shown so that its position can be related to the inner planets. Similarly, the innermost orbit shown in diagram (b) is the earth's.*

nal material, particularly hydrogen and helium. On the other hand, the inner planets like the earth were too small in size and mass to retain the lighter gases and therefore evolved into planets with a much higher percentage of the heavier elements—the minerals.

Until recently, theories of the solar system's origin and evolution were based on very few observational facts. The earth had long ago erased, through erosion, practically all evidence of its early history. Astronomers had noted the way the planets were spaced, and their orbital motions and rotations. They had also measured the angular momentum of the sun and the planets as well as their respective masses. But they knew more about the evolution of the stars than about the formation of our own solar system.

We have acquired a tremendous amount of information about the solar system in the last fifteen years. The study of meteorites, which are considered to be samples of the matter of the primordial solar sytem, tells us much about these early conditions. The data obtained from the various space probes in recent years have been invaluable. There are still many unanswered questions, but the currently planned space probes may well solve many of them. Of course, as has happened so often in the past, they may also raise many more problems and questions than they solve. Some of the projected space missions the United States has in the advanced planning stages are listed in Figure 13.3.

13.3 Projected United States space missions.

Name	*Launch Date*	*Destination*	*Date of Arrival*
Pioneer II	(has already flown by Jupiter in 1974)	Saturn	Sept. 1979
Voyager I (Originally designated as Mariner II or MJS-1)	August 1977	Jupiter Saturn Uranus* Neptune*	April 1979 June 1981 Jan. 1986 Sept. 1989
Voyager 2 (Originally designated as Mariner 12 or MJS-2)	Sept. 1977	Jupiter Saturn	Dec. 1978** Aug. 1980
Pioneer Venus A	May 1978	Venus	Dec. 1978
Pioneer Venus B (These are also referred to as Pioneer 12 and 13)	Aug. 1978	Venus	Dec. 1978
Space Shuttle	March 1979	First orbital flight around earth.	
Space Telescope	1983	To be flown on Space Shuttle in orbital flight	
Galileo	Jan. 1982	Jupiter and Galilean satellites	Jan. 1986

*These destinations are still very questionable.

**Note that although Voyager 1 was launched first, it will arrive at its destination *after* Voyager 2.

The Problem of Planetary Motion

Solar system dynamics describes and explains the motions and interactions of the system's many members. It is a part of the branch of astronomy called *celestial mechanics,* which had its beginning in the study of the motions of the planets. Like the theories of the solar system's origin and evolution, such explanations date back to prehistoric times. As with many scientific theories, they were based not only on the available information and observations of the time, but also on the social climate in which the scientist-philosopher lived.

Except for a very few early philosophers, the universally accepted belief in ancient times was that the earth was the stationary center of a universe which was placed there specifically for mankind. An outstanding exception to this was the famous Greek astronomer Aristarchus of Samos (240 B.C.). He boldly placed the sun at the center of the universe and made the earth one of the planets. For this and other "heresies" he was ostracized. How much different the course of history might have been had Aristarchus' concept been accepted! Mankind's view of itself and its place in the universe would have been profoundly changed. Unfortunately, the concept of the earth-centered universe prevailed. Until relatively modern times its main proponent was the famous astronomer Ptolemy (140 A.D.).

The classical problem for scholars and astrologers alike was the riddle of planetary motion. From careful observations astronomers before Ptolemy were well aware that the planets did not move at an even pace in their apparent paths through the stars. They sometimes stopped, went backwards (retrograded), and then continued forward again. To the astrologers, such erratic behavior held great meaning. Astrologers were not as much interested in the causal relationships as they were in the message that the gods were trying to convey through the behavior of the planets. These astrologers, and many theologians of the Middle Ages, believed that man could never understand the workings of the universe, and some even considered it sacrilegious to inquire into such matters. Nevertheless, attempts were made to formulate a mechanism which would explain the planets' erratic motions and permit a prediction of their future positions. Of course, such an explanation had to be based on a completely stationary earth at the center of the universe.

The cause of retrograde planetary motion is a very simple one, once the true relation of the sun and planets is known. The earth is traveling faster around the sun than, say Mars, and therefore will overtake and pass Mars at some point in its orbit, causing Mars to appear to be moving backwards for a short period of time (Fig. 13.4). This effect is much the same as one observes when in a fast train passing a slower train. For a moment one has the impression that the slower train is moving backwards.

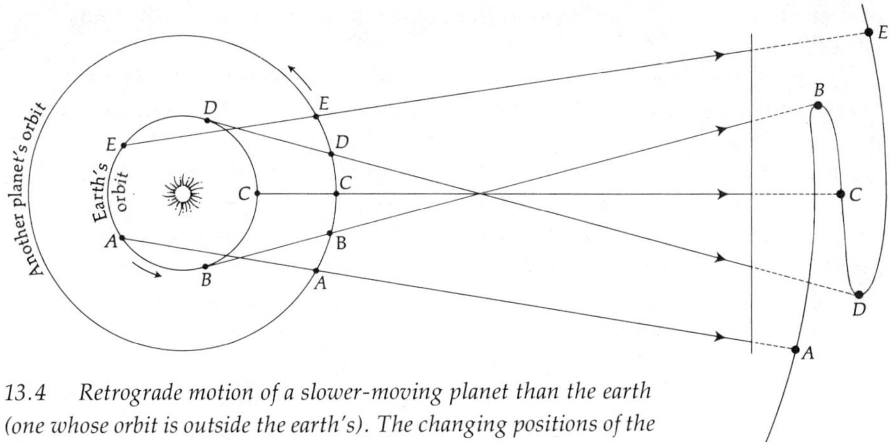

13.4 Retrograde motion of a slower-moving planet than the earth (one whose orbit is outside the earth's). The changing positions of the earth and planet relative to each other produce the apparent path of the planet on the celestial sphere against the background of distant stars.

Chart the observed path of one of the planets on the celestial sphere for several months. Mars, Jupiter, or Saturn are easiest to follow since they can be visible long after the sun has set, when the stars are brightest. Draw a diagram similar to Figure 13.4 showing the relative positions of the earth and the planet during these months.

However, in early societies, in which the earth was held to be stationary and centrally located, such an explanation was not credible. Another solution was devised, which developed for many centuries and was finally codified by Ptolemy in 140 A.D. In his scholarly work, the *Almagest,* he brought together many of the schemes of his day, resulting in the Ptolemaic system of planetary motion.

Ptolemy's attempt to explain and predict the motions of the planets was not done primarily to depict reality. That was not the main concern. Rather, the primary concern was to be able to interpret the astrological meaning involved in the motions and positions of the planets. Since the astrologers were not interested in discovering natural laws of the universe, the Ptolemaic system proved to be reasonably satisfactory for their purposes.

In the Middle Ages, the *Almagest* took on some of the aura of religious dogma and was rarely questioned. The Ptolemaic system took on the appearance of truth and became the basic text in astronomy for many centuries.

The Ptolemaic system placed the earth as a stationary object at the center of the universe, with the sun, planets, and stars revolving around it in the course of a day.[1] In order to explain retrograde motion, Ptolemy postulated that the planets moved uniformly on small circular orbits, the centers of which moved with uniform speed on a larger circular orbit around the earth. All the orbits were thought to be circular because the ancient Greeks considered the circle to be a noble and perfect form—the only one befitting the gods. The major orbits were called *deferents* and the smaller ones were called *epicycles.* (Fig. 13.5).

With the exception of Venus and Mercury, the line joining the center of the epicycle and the planet always had to be parallel to the line joining the earth and sun (note Mars in Fig. 13.5). Because Mercury and Venus were never seen far from the sun on the celestial sphere (for example, never seen at midnight), Ptolemy's system made the center of their epi-

13.5 *The Ptolemaic system, showing two of the five planets known to the ancient Greeks, in relation to the earth and sun.*

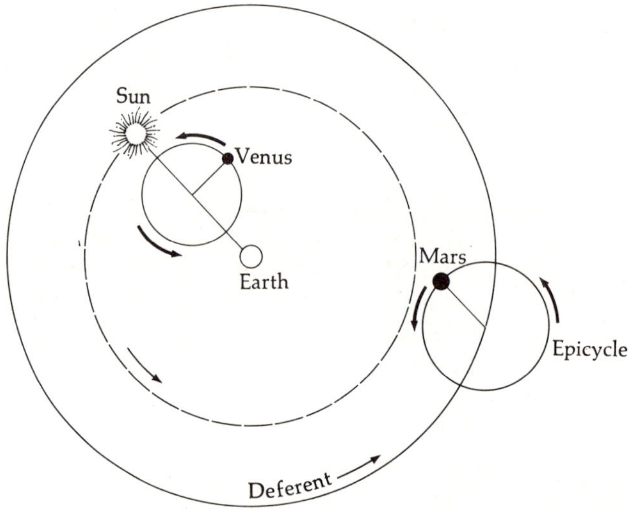

---

[1] To be precise, the earth was not located at the exact center of the sun's orbit but was somewhat displaced in order to account for the observed nonuniformity of the sun's annual motion among the stars. The point from which a hypothetical observer would have seen the sun move uniformly among the stars was called the *equant,* a mathematical fiction needed to "save the phenomenon."

cycles as lying *on* the line joining the earth and sun. Although these and many other devices helped to explain retrograde motion and to predict future positions of the planets, they were insufficient for accurate prediction (see Fig. 13.6). In an attempt to remedy the discrepancies between the theory and the actual observations, more epicycles had to be added, and eventuallv the whole scheme became hopelessly complicated.

Not until the sixteenth century did anyone acquire the observational precision to question the Ptolemaic system. Even then, the Polish astronomer Nicolaus Copernicus delayed the publication of his "heretical" book until just before his death in 1543. His major contribution was that he proclaimed the earth to be a planet and that it and its fellow planets all revolved around the sun. Because he still clung to the belief that the orbits of the planets had to be circles—the perfect curves—Copernicus was still forced to use epicycles to explain the observed planetary motions.

Not until the astronomers who came considerably after Copernicus showed that his heliocentric system was correct did his book create much attention and controversy. The Copernican theory declared not only that the earth was a moving planet, but that man was dethroned from his geocentric place. As with many new theories, this greatly disturbed the vested interests of the time, and created powerful opposition.

Three years after Copernicus' death, Tycho Brahe was born. As a rich Danish nobleman, he was able to fully pursue his interest in astronomy,

13.6 Demonstration of how the epicycle concept of the Ptolemaic system explained retrograde motion. This diagram is oversimplified, because while Mars moved on its epicycle, the center of the epicycle, of course, moved along the deferent.

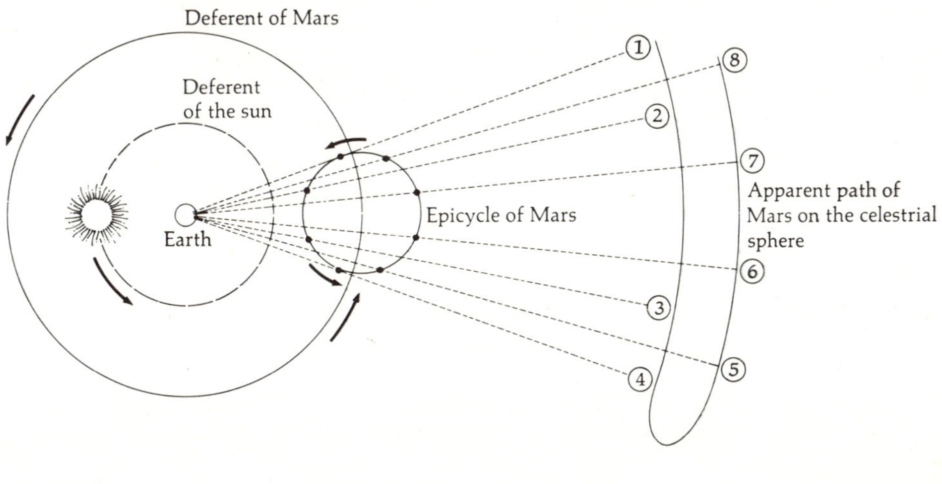

and he built a magnificent observatory called Uraniborg. Equipped with the finest instruments available, most of which he designed and constructed himself (the telescope had not yet been invented), Tycho made his scientific reputation as an excellent observer. Because he could not detect any stellar parallaxes, he rejected Copernicus' theory that the earth was in motion. Instead, he theorized that the earth was stationary with the sun revolving around it, but he adopted the idea that the planets revolve around the sun. In a sense, therefore, Tycho took both a step forward and a step backward in his attempt to solve the problem of planetary motion.

His fame rests on his outstanding abilities as an observer. He was particularly fascinated by the motions of Mars and concentrated his attention on this planet. Had it not been for his fortuitous association with Johannes Kepler, his careful observations might have gone for naught. After Tycho's death in 1601, Kepler took his valuable data and tried to plot the orbit of Mars. Kepler, who was essentially a mystic, believed that God was waiting for some human to decode His methods, which Kepler believed must be mathematically sound.

Starting with the heliocentric hypothesis, he found that circular orbits would not fit Tycho's observations. At times they came close to working—in one instance they were off only by eight minutes of arc—but Kepler had such faith in Tycho's measurements that he felt the "Master" could not have made such a large observational error. After many years of painstaking handwritten calculations and diagrams, Kepler finally discarded the concept of circular orbits. Instead, he found that he could plot an elliptical orbit for Mars using Tycho's observations. Again, this meant giving up a firmly entrenched preconceived notion of how things ought to be according to philosophical and religious beliefs. To give up the idea of circular orbits took great scientific courage, since it overthrew centuries of indoctrination.

One way of investigating the properties of an ellipse is to draw several of them, of different sizes and shapes. To do this, you will need paper, two thumbtacks, a piece of string, and a pencil. Tie the ends of the string together. Stick the thumbtacks into the paper about twelve centimeters (five inches) apart and loop the string around them. Holding the string taut with the pencil, move the pencil along the string. The resulting line will describe an ellipse.

Keeping the same length of string, move the tacks to different distances apart and draw other ellipses. How do these differ from the original one?

Now put the tacks back in their original positions and try varying the length of the string. Draw a series of ellipses in this manner. How do these differ from the original?

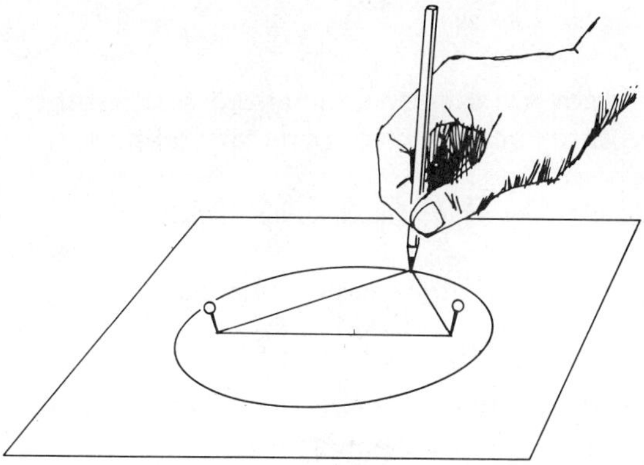

If both tacks are placed on the same spot (or just one tack is used), what shape is your ellipse?

The positions of the tacks in each ellipse are called the foci *of the ellipse (the singular of foci is* focus*). A circle results when both foci are in the same place. The widest part of the ellipse is called the* major axis *and half of this distance is the* semimajor axis *(below, left), which is used in computing statistics about orbits. What do we call the major and semimajor axis of a circle?*

The size of the ellipse depends upon the length of the major axis, whereas the shape of the ellipse depends upon how close the two foci are compared to the major axis. The ratio of these two distances or lengths (the distance between the foci and the length of the major axis) is called the eccentricity *of the ellipse (below, right). A circle has an eccentricity of zero, whereas a parabola has an eccentricity of 1.0. (which would be an ellipse with one focus at an infinite distance). Ellipses, then, have eccentricities between zero and one, depending upon their shapes. The further from a circular shape they are, the greater their eccentricity.*

One interesting characteristic of an ellipse is that from any point on the curve, the sum of the distances to the two foci is always the same. From your string drawings, can you see why this is so?

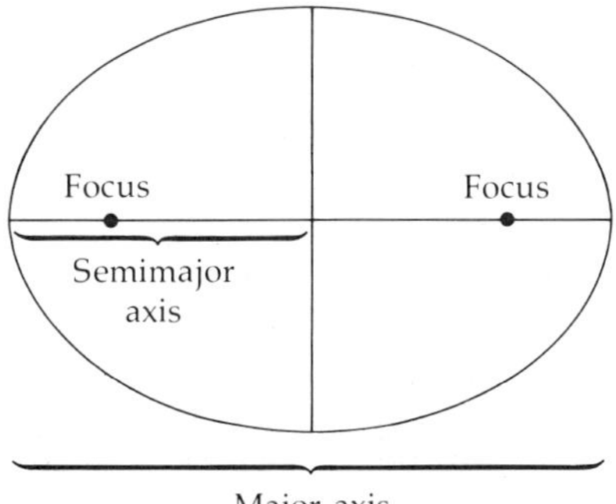

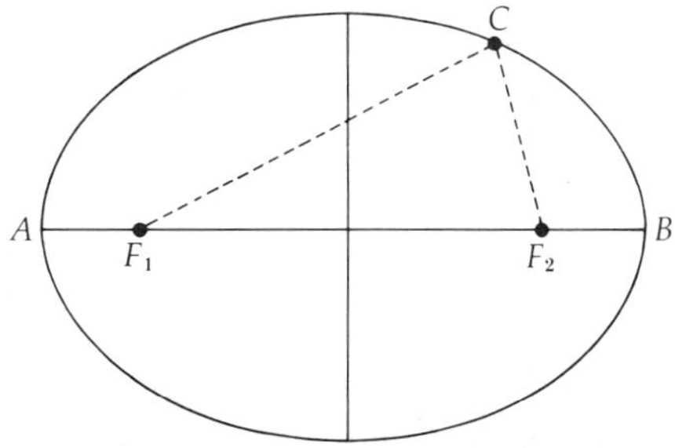

$$\frac{F_1 F_2}{AB} = \text{eccentricity}$$

From this revolutionary discovery, Kepler generalized to the other planets and set forth two of his three basic laws of planetary motion:

1. *Each planet moves around the sun in an elliptical orbit, with the sun at one focus of the ellipse.*
2. *The straight line joining a planet and the sun sweeps out equal areas in space in equal intervals of time.*

This second law, which is called the *Law of equal areas,* was also derived from Tycho's observations. It states that as the planet orbits the sun, the closer it gets, the faster it will move. In Figure 13.7, the hypothetical

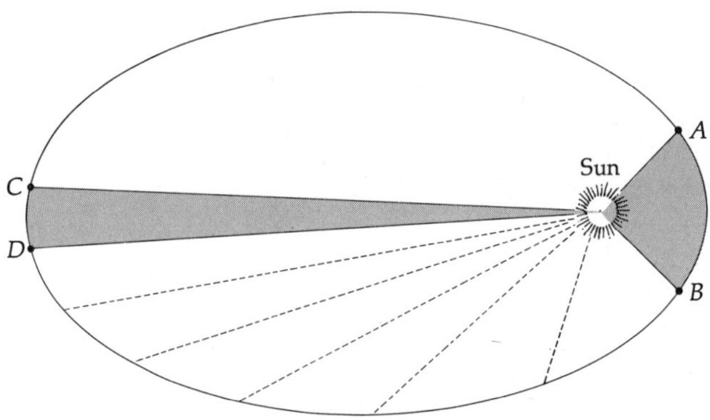

13.7 Kepler's Law of Equal Areas. The two shaded areas are equal. The planet moves faster when it is closer to the sun than when it is farther away, thus covering distances AB *and* CD *in equal times.*

planet takes the same time to go from *A* to *B* as it does from C to *D* because it moves faster in its orbit when closer to the sun. This finally explained what the ancients had already observed, namely that the sun appeared to move faster with respect to the stars at certain times of the year. (The sun's apparent motion is, of course, a consequence of the earth's motion around the sun.) We know today that Kepler's second law is a consequence of the principle of conservation of angular momentum (see p. 178).

With these two laws, Kepler then tried to discover the numerical relationship between the planets and their orbits. Because of his mystical beliefs in an orderly universe, he was certain that this relationship could not be haphazard but had to also follow some definite divine rule. It took him ten more years to finally discover and propound his *harmonic law,* the third law of planetary motions:

3. *The squares of the sidereal periods of the planets are in direct proportion to the cubes of the semimajor axes of their orbits.*

The *period* is the time it takes for the planet to orbit the sun; the *semimajor axis* is its average distance from the sun. If we use the earth's period of one year for the unit of time and the earth's semimajor axis (1 AU) for the unit of distance, this third law can be written as:

$$p^2 = d^3$$

where p is the period and d is the semimajor axis. Thus for Mars, whose period is 1.88 earth years and whose distance is 1.52 AU:

$$(1.88)^2 = (1.52)^3 = 3.5$$

If any two planets are compared, the formula must be written:

$$\frac{p_1^2}{p_2^2} = \frac{d_1^3}{d_2^3}$$

where the period and distance are measured in years and astronomical units.

In Chapter 6 we saw how this relationship between period and distance is used to determine the masses of binary stars. Kepler, of course, did not realize that the masses of the two objects involved were important parts of the equation because even the largest planet has only a minute fraction of the stellar mass.

While Kepler was struggling to compute the true planetary orbits, other post-Renaissance scientists were also overturning long-held beliefs. Galileo Galilei was turning his newly invented telescope to the sky as well as doing experiments in mechanics. As the first "physical thinker" in the modern sense of the term he introduced the basic scientific attitude that it is legitimate and not sacrilegious to question nature—that is, if you want to find out something about the natural world, you perform experiments; you do not rely upon the authority of an ancient sage.

Show that Kepler's third law works for two other planets besides Mars and the earth.

This led the way toward the development of a whole new attitude toward science and the world. After twelve centuries in which very few scientific discoveries were made, in less than one century the entire scientific world was revolutionized. This in turn eventually affected the philosophical thought of the day, the way in which man thought about himself and his place in the universe. But before this could occur, science needed someone to bring together in a unified theory the work of these post-Renaissance scientists. This was to come in the person of Sir Isaac Newton.

Modern Celestial Mechanics

Sir Isaac Newton has been called by some the greatest genius that ever lived. Certainly he deserves to be ranked among the top scientists of any age. In his *Principia,* published in 1687, he propounded basic physical laws, including his famous three laws of mtoion and the law of gravitation.

Kepler had stated empirical laws of planetary motion but had no idea of why these laws were true. Newton provided the underlying theory. This major contribution was based on his laws of motion and his law of universal gravitation. From these basically simple laws, scientists who followed Newton developed the complex subject of celestial mechanics.

Newton's main interest was the discovery of laws that governed all moving bodies, whether they were cannon balls, falling apples, planets, or moons. At one point in his studies, he was forced to invent his own mathematics—calculus—because no previous mathematical discipline was adequate to deal with the study of motion. Well aware of his indebtedness to Kepler, Galileo, and others, he said: "If I have seen farther than others, it is because I have stood on the shoulders of giants."

Newton's first law of motion states:

> 1. *A body remains at rest or in uniform motion in a straight line unless acted upon by a force.*

Attach a small object to the end of a piece of string. Holding the other end, whirl the object rapidly around your head. If the object broke loose while being whirled around, what would be its trajectory? What principle is demonstrated here?

It is this force that makes the body change its direction and speed. The force needed for a given change depends upon the mass of the object involved. Newton's second law defines this force:

2. *Force equals the mass multiplied by the acceleration*[2] *(F =* ma*).*

Thus, to change the direction of a butterfly requires much less force than to do the same thing to a stampeding buffalo.

Newton's third law of motion states:

3. *To every action there is an equal and opposite reaction.*

Blow up a balloon but do not seal its end. Release it. What physical principle is involved in the subsequent flight of the balloon.

The astronauts were very aware of this phenomenon when they moved about in their weightless capsules or left them for short "space walks." The slightest push in one direction sent them off in the opposite direction. When we attempt to climb out of a rowboat, our push to get on to the dock may very well send the boat out into the water again (see Fig. 13.8).

Newton's law of universal gravitation states:

The gravitational force between two objects is proportional to the product *of their masses divided by the* square *of the distance between them.*

This can be expressed by the formula:

$$F = G\,\frac{m_1 m_2}{d^2}$$

in which G is the universal constant of gravitation, m_1 and m_2 are the respective masses of the two objects, and d is the distance between them. This seemingly simple relation describes the gravitational attraction between *any* two material objects in the universe, for example a milk bottle in San Francisco and a bulldozer in Kansas City. Of course, the force in this

---

[2] In physics the term *acceleration* means to change either the speed or direction of motion (or both) of a body. In daily experience, we are more accustomed to thinking of acceleration merely as a change in speed alone. However, a car rounding a sharp curve at a constant speed does accelerate—we experience this acceleration as we are thrown toward one side.

13.8 Newton's Third Law: To every action there is an equal and opposite reaction.

instance is extremely small because the mass of the objects is tiny compared to the distance involved, but it does exist nevertheless.

Newton would not speculate as to the nature of this force of gravitation and even questioned its "reality." However, after the famous

---

Isaac Newton supposedly was led to the formulation of the Law of gravitation when an apple fell from the tree under which he was sitting. Compare the falling apple with the moon revolving around the earth. Are both in orbit? If we wanted to find the gravitational forces involved in these two cases, what would be the two masses in each case? What would the distance between the objects be?

---

mathematician Albert Einstein formulated his theory of general relativity in the twentieth century, the "force" became regarded as the effect of the "warping" of space-time by matter (see Chap. 4). In a sense, the force becomes a fiction but remains a useful concept to describe and predict the motions of objects in the universe. One can see a striking similarity to the Ptolemaic system in that both were schemes for prediction, but neither could be said to depict the "true" nature of things. The Newtonian theory was, of course, much more physically precise than the Ptolemaic theory.

Velocities of Rockets and Satellites

From many experiments, starting with Galileo's famous experiment (supposedly from the top of the leaning tower of Pisa), it is well known that a body here on earth, regardless of its mass, will fall from a "rest" position approximately 16 feet in one second. (The resistance of the atmosphere must be discounted.) It is also known that the earth curves about 16 feet every five miles.

On one of the Apollo missions one of the astronauts standing on the moon dropped a feather and a metal tool simultaneously. They both hit the lunar surface at the same time. What natural law did this experiment demonstrate? Why is it not usually possible to perform such a feat on the earth? Under what conditions could such an experiment be performed on earth?

Let us assume for the moment that the surface of the earth is circular and completely smooth, with no mountains, valleys, or atmosphere to contend with. If we could then shoot a rocket parallel to the ground so that it traveled at a velocity of five miles per second, it would never land but would remain in orbit just above the earth's surface (Fig. 13.9). Actually, it would be continually falling toward the earth but at the exact rate which the curvature of the earth causes the land to "move" away from it. In the same way, the moon in its orbit around the earth is continually "falling" toward the earth.

If we decrease the velocity of our rocket, it will eventually fall to the ground. Until it does, however, it is in orbit around the earth. This is equally true for a thrown baseball or for a bullet ejected from a gun. If in

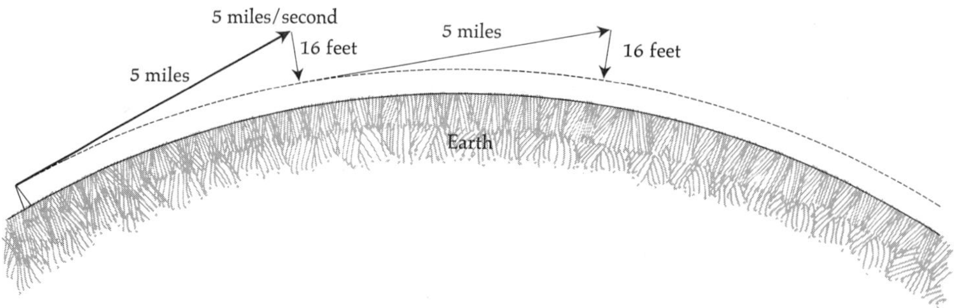

13.9 The earth curves 16 feet every 5 miles, and a free-falling object will fall 16 feet in a one-second fall. A ball projected at 5 miles per second parallel to the earth's surface will go into a circular orbit around the earth (dashed line). This assumes a circular and smooth earth with no atmosphere.

some way the mass of the earth could be instanteously compressed into a tiny ball at its present center just after the baseball was thrown so that the baseball would not hit any solid ground, it would continue in its orbit completely around the tiny earth (Fig. 13.10).

If we increased the velocity of the rocket to more than five miles per second, its orbit would become elongated or elliptical rather than circular,

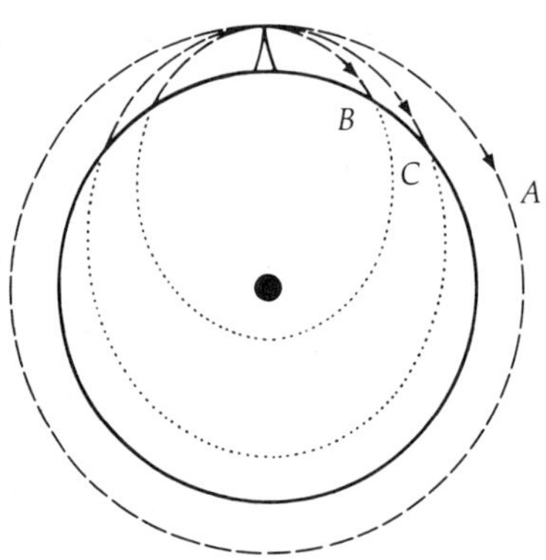

13.10 Orbits of satellites projected parallel to the earth's surface with different velocities. Satellite A *was projected at exactly 5 miles/second and has assumed a circular orbit. Satellites* B *and* C *were projected at less than 5 miles/second and therefore fall to earth before making a complete orbit. If the earth were suddenly contracted to a small ball after the launch of* B *and* C*, they would continue in elliptical orbits about this ball (dotted line).*

but it would continue to go around the earth. A still greater increase in the velocity would cause the orbit to become so elongated that eventually it would extend out beyond the moon itself (Fig. 13.11). Not until a velocity of seven miles per second was reached would the rocket's orbit change from elliptical to parabolic, thus allowing the rocket to escape the gravitational pull of the earth and go out into space, never to return. Seven miles per second is the *escape velocity* of the earth.

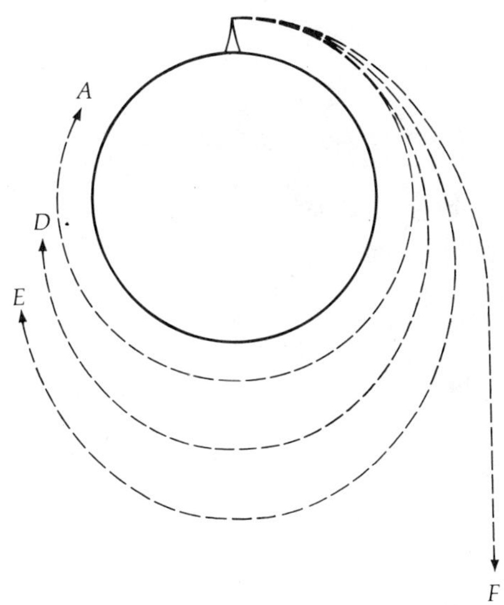

13.11 *Orbits of satellites projected parallel to the earth's surface with different velocities. Satellite* A *was projected at exactly 5 miles/second and has assumed a circular orbit. Satellites* D *and* E *were projected at more than 5 miles/second but less than 7 miles/second and are in elliptical orbits around the earth. Satellite* F *was projected at 7 miles/second and is therefore in a parabolic orbit. It has achieved escape velocity from the earth and therefore will not return.*

Take several conical drinking cups and with a very sharp knife or razor blade, slice through them at different angles, as shown in figure following. The curve created at the cut edge is called a conic section.

If the cut is made parallel to the cone's widest point (or base), the resulting conic section will be a circle *(a).*

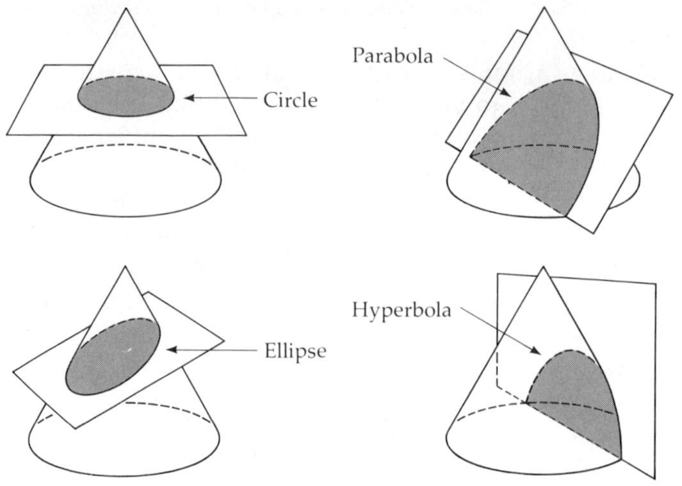

If the cut is made at an angle to the base but still cuts completely through the cone, the resulting curve or conic section will be an ellipse *(b).*

If the plane of the cutting knife is tilted even further, so that it becomes parallel to one side of the cone, the curve is called a parabola. *This curve, as can be seen in (c), never quite cuts all the way through the cone, and the curve of intersection is open at one end.*

If the plane of the cutting knife is tilted still further, another open curve results, called an hyperbola *(d). Obviously, once the plane passes the point at which it is perpendicular to the base of the cone, any further tilting will repeat the above configurations in reverse order.*

All four types of conic sections are important in celestial mechanics because the motion of every object in space is described by one of these curves. The most common is the ellipse, which is the shape of most orbits in space.

Although we derived the five mile/second rate in a simple geometrical way, this value is a consequence of a much more general formula, that describes all motions under gravity. It is one of the most simple, yet elegant, formulas in celestial mechanics:

$$V^2 = GM\left(\frac{2}{r} - \frac{1}{d}\right)$$

where V is the velocity of the object, G is the universal constant of gravitation, M is the *sum* of the masses of the two objects, r is the distance between their centers (in the above example the distance between the center of the earth and the rocket), and d is the semimajor axis of the orbit (the rocket's orbit about the earth as measured from the earth's center).

When (and only when) the rocket is traveling in a circular orbit, r and d are equal and the formula can be simplified to:

$$V^2 = \frac{GM}{r}$$

In this way we can compute the speed necessary for a rocket to either maintain a circular orbit or escape from the earth entirely in a parabolic orbit (in the latter case d is infinite and therefore $\frac{1}{d} = 0$). The speed necessary to attain a parabolic orbit is called the *escape velocity* of the earth for any object and is equal to 25,000 mph (11.2 km/sec).

If we wish to know the velocity of Jupiter in its orbit at any instant, we need only to insert the present distance of Jupiter from the sun *(r)*, at this instant, and the semimajor axis of Jupiter's orbit *(d)*. Because planetary orbits are all ellipses, these two values are not the same except at specific points in the planet's orbit. Similarly, if we wish to know the speed of an astronaut at any distance (r) from the center of the earth, with d being the semimajor axis of his orbit, this formula will immediately give us the information. These principles and formulas apply to any free-falling object, whether it is a rocket, a thrown baseball, a bullet, or the moon. Since the earth's atmosphere has such a great effect upon objects traveling through it, the real value of these formulas lies in astronomical research.

Note that although in the seventeenth century Newton was able to compute from this formula the orbit of any object moving solely under gravitation (including rockets and satellites), the technology of rocketry did not come until the twentieth century. This is an excellent example of how far technology has lagged behind scientific discovery. Only in recent years has the gap become markedly smaller.

The triumphs of Newton and his followers in the development of celestial mechanics led later scientists to assume that these same laws applied to the microscopic world as well as to the macroscopic world. Consequently, they thought that they could predict the exact position of any particle of matter at any given time in the same way that they could accurately predict the positions of the planets. This concept had deep repercussions on the whole body of philosophy, introducing the ideas of complete determinism and fatalism.

In the 1920's, the development of quantum mechanics showed how unpredictable the microscopic world could be. It was found impossible to know exactly both the position and speed of a subatomic particle. Only probable values of these quantities could be computed. This put an end to the concepts of mechanism and determinism in the scientific world, and the impact was gradually felt in the philosophical world as well.

Discussion Questions

1. Which of the following objects are members of the solar sytem?

a)	Planets	e)	Moons
b)	Sun	f)	Milky Way
c)	Quasars	g)	Constellations
d)	Polaris		

2. How was our solar sytem believed to have been formed?

3. What characteristics of our solar system do you believe would be generally true of other planetary systems, based on the processes that are believed to have formed it?

4. It takes about eight minutes for light to travel from the sun to earth. How long would it take for sunlight to reach Pluto? How long would radio waves take to go from the earth to Mars?

5. If civilization had arisen on the moon instead of on the earth, how would this have affected early cosmology?

6. How did the discovery of the laws of celestial mechanics influence the philosophical thought of the last few centuries? What new discoveries helped change this way of thinking?

7. What is "escape velocity"? Do Jupiter and the Earth have the same escape velocity? Explain.

8. Why is five miles/sec the critical velocity that a rocket must have in order to stay in an orbit around a smooth circular earth?

9. What facts must we know in order to compute the distance of an astronaut from the earth.

10. What is the "semimajor axis of a circle"?

11. What are Kepler's three laws of planetary motion? Use words, equations, or diagrams to explain.

12. What was the original purpose of the Ptolemaic theory?

13. Define "acceleration" as it is used in science. Which of the following are examples of acceleration?

    a) An airplane taking off
    b) The physicists in the moving towers
    c) A train rounding a curve
    d) A car speeding along the highway at 80 mph.

14. Show that Kepler's third law works for three other planets other than Mars.

15. From Kepler's laws and your study of binary stars in Chapter 6, which would have longer periods, visual or spectroscopic binaries? Why?

16. Explain how a stone dropped from a tall building and an astronaut going to the moon obey the same laws of celestial mechanics (assume there is no atmosphere).

17. Explain (with diagrams) the Ptolemaic theory. Use Mercury and Jupiter as illustrations. Show how the theory explained retrograde motion.

18. It was once proposed that a tenth planet might be revolving around the sun at a distance of about 64 AU. If such a planet existed, what would be its period in earth years?

14 The Sun—Star of the Solar System

From the beginning of history, mankind has worshipped the sun and its life-giving power. Early humans were aware that the sun was in some way responsible for much that happened on earth and although they did not understand how this causal relationship operated, they made the sun god one of the most powerful and awesome of their deities.

Although we no longer worship the sun or consider its power supernatural, there is much about this star that is still mysterious to us. It is 270,000 times closer to us than the next nearest star and yet we can only directly study its atmosphere and a very thin exterior shell of its body. What is within the sun we can know only through mathematical calculations. However, using our knowledge of the way gaseous matter behaves at various temperatures and densities, a realistic model of the inner sun can be developed (Fig. 14.1).

The sun has a diameter 109 times that of the earth. Over one million earths could fit into the volume of space it occupies. Its mass, which is over 330,000 times that of the earth, comprises 99% of the entire matter in the solar system. It is a typical main sequence star with a spectral class of G2.

Daily observation of the sun reveals an apparent motion of the sunspots across its face. This is due to the rotation of the sun about its axis. Galileo was the first to use sunspots to demonstrate the sun's rotation. The sun rotates faster at the equator than at the poles; at the equator the period of rotation is about 25 days, whereas near the poles it is more than 30 days. This differential rotation is possible because the sun is a gaseous body. Its average period of rotation is about 27 days.

As with all the stars on the main sequence, the sun's energy is created deep in its central core by the nuclear transformation of hydrogen into helium, its two main components. The calculated model indicates a central temperature of well over 10 million degrees and a density of about one hundred times that of water. The photons of energy created there make their way very slowly to the surface of the sun by constant absorptions and emissions in all directions. It may take a million years for such a distorted journey.

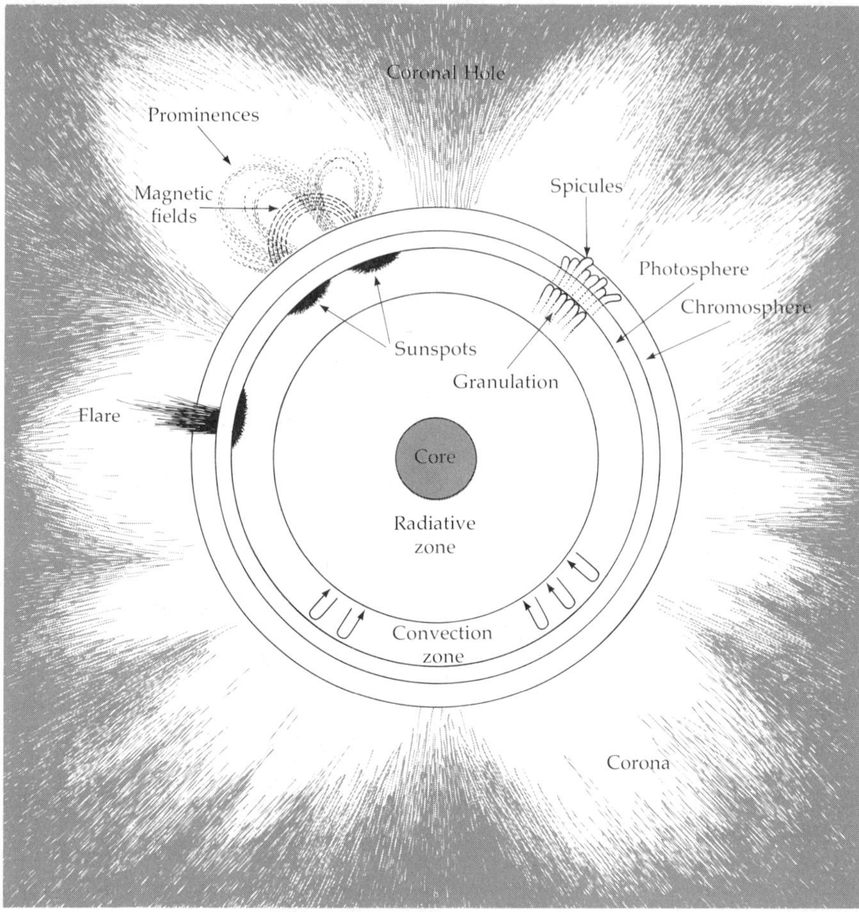

14.1 Schematic drawing of the sun. Exterior features are observable, whereas internal structure is hypothetical based on calculated models.

The Solar Atmosphere

However, once the photons reach the visible surface or *photosphere,* they can escape into space at the speed of light. We see and feel them as sunlight eight minutes later. Thus when we speak of the photosphere as the "surface" of the sun, we are referring to a 200 km thick shell of gas through which the solar radiation is finally released. Because of the ease with which radiation can pass through it, the photosphere can also be considered as the lowest solar atmospheric layer. It is from the photosphere that the sun

Focus sunlight through a telescope onto a piece of white paper so that a clear image of the solar disk is seen. The paper must be moved back and forth from the telescope until a clear image is formed. Draping a piece of black cloth over the paper and the end of the telescope will keep out extraneous light and give more contrast to the image. UNDER NO CIRCUMSTANCES SHOULD YOU LOOK DIRECTLY AT THE SUN EITHER WITH OR WITHOUT A TELESCOPE. *From a day-to-day viewing, the progression of the sunspots across the solar disk can be followed. Make a chart of the positions of the spots, and their number and duration. Estimate the speed of rotation of the sun from your data.*

Measure the size of the projected spots near the center of the disk and then calculate their actual size. (The sun's real diameter is 864,380 miles; therefore, a 1/16-inch diameter spot on a projected four-inch disk would represent a spot about 13,500 miles in diameter). How large a projected spot would be the size of the earth?

obtains its G2 spectral classification. Its temperature is 6000 K, the typical surface temperature of a G2 main sequence star.

To the naked eye, the photosphere appears smooth and gives the sun its distinct circular boundary. But under the magnification of a telescope, a mottled structure on its surface can be seen (Fig. 14.2). This is called *granulation* and is caused by columns of hotter gases rising from below the photosphere, cooling off, and sinking back down again. Each column or *granule* is about 500 to 1000 kilometers in diameter and may last only a few minutes on the solar surface.

Above the photosphere lies the *chromosphere,* a layer of gas two to three thousand kilometers thick. It is practically transparent to radiation and therefore does not seriously impede the flow of energy from the photosphere. Until this century, the chromosphere could only be observed during a solar eclipse because the brilliance of the photosphere completely obscures this much dimmer layer.

It is visible only for the few seconds during a total solar eclipse just after the moon has covered the photosphere but before it covers the chromosphere as well. It has usually been described as a thin reddish line around the edge of the eclipsing moon. However, it was not until photographs of it were obtained that its existence was definitely proven. Today special instruments are used, especially in spacecraft, to create the solar eclipse effect so that a more detailed study can be made.

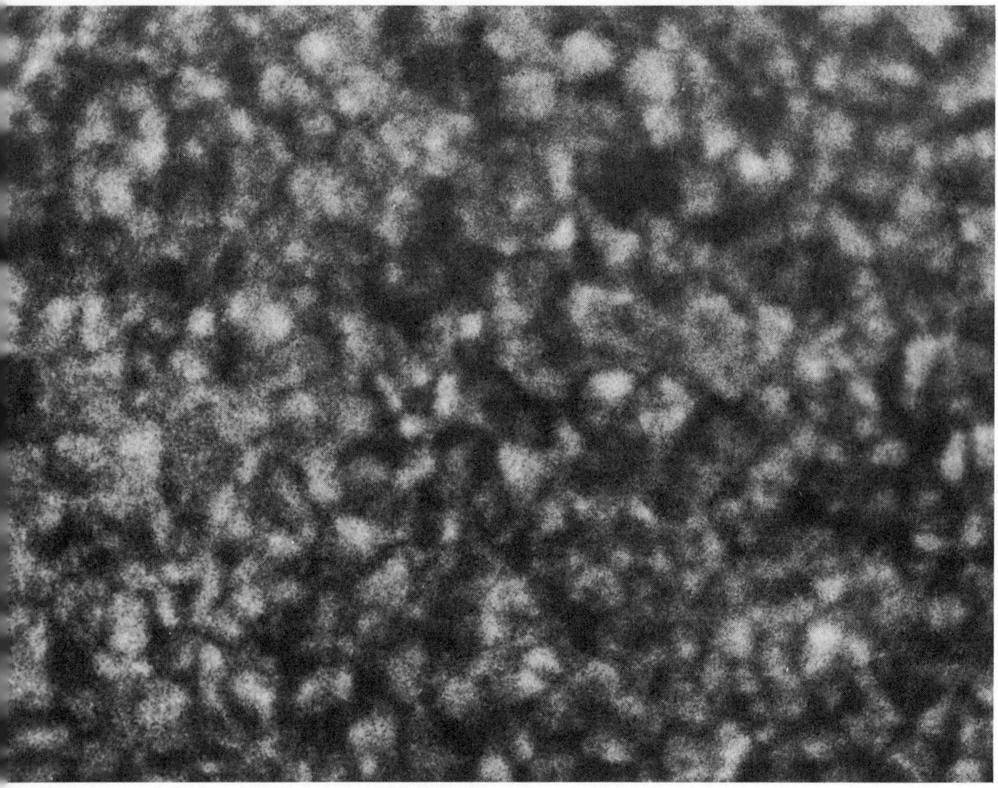

14.2 Solar granulation: a highly magnified section of the sun's surface (photograph from Hale Observatories).

Because of the short period of time in which the chromosphere is visible during an eclipse, spectrograms of it were called *flash spectra*. These showed an emission spectrum, indicating that the chromosphere is made up of hot gases that are emitting light. One of the strongest of these emission lines is that of *hydrogen alpha,* a bright red line. It is this emission which gives the chromosphere its reddish color and its name (see Color Plate A).

Constantly rising from the chromosphere are jets of matter called *spicules* which may be related to the *granules* further down. The spicules are believed to be important in transporting energy outward. Although they also last only a few minutes, they may rise over 10,000 kilometers above the photosphere, sinking back down again quite rapidly. Because they rise above the upper boundary of the chromosphere, they give this region a very jagged appearance (Fig. 14.3).

14.3 Spicules rising from the photosphere and piercing the chromosphere, making a jagged border. (Photograph from Hale Observations)

During a total solar eclipse, when the moon has finally covered both the photosphere and the chromosphere, the outermost layer of the sun's atmosphere is seen most spectacularly (Fig. 14.4). This is the *corona.* Its lower boundary gradually blends with the chromosphere and then it extends outward more than a million miles above the photosphere, its brightness comparable to about one-half that of the full moon. Under normal conditions, its light, like that of the chromosphere, is overpowered by the brilliance of the photosphere.

The density of the corona is extremely low, so that even at its densest region just above the chromosphere, it is only one ten-millionth the density of the earth's atmosphere at sea level. It quickly becomes even thinner, thinner even than what would be considered an excellent vacuum here on earth. The temperature of the corona (as measured from its spectral lines) rises to about a million degrees. This is a measure of the energy state of its atoms, i.e., the extent of their ionization. It is not a temperature in the

14.4 The solar corona as seen during a solar eclipse. Do not confuse this with the glow seen around the sun during the daytime, which is caused by the scattering of the sun's rays in our own atmosphere (photograph from Dearborn Observatory, Northwestern University).

ordinary sense. A typical thermometer registers the heat it absorbs from its surroundings. A thermometer placed in the corona, but shielded from the direct light of the sun, would register an extremely low temperature. The atoms there are so very highly ionized that most of the radiation occurs in the far ultraviolet region of the spectrum. It is from the corona that we also receive most of the x-ray radiation from the sun.

Beyond the visible portion of the corona, the sun's atmosphere, constantly becoming thinner, extends out to great depths of the solar system. The streaming of this highly rarified material, the *solar wind,* has been detected to distances well beyond the earth by instruments in spacecraft and on the moon. These charged atomic particles are continuously emitted from the sun. By the time they reach the earth, they are traveling about 400 km/sec.

When these particles reach the vicinity of the earth, they are prevented from penetrating to the terrestrial surface by the earth's magnetic field and its atmosphere. Although some of the more energetic particles may pass through the magnetic field and interact with the particles in the earth's upper atmosphere, most are either deflected toward the magnetic poles or trapped temporarily in the magnetic field. The interaction of the solar wind particles with the upper atmosphere near the earth's magnetic poles is the main cause of the auroras (Fig. 14.5).

Sunspots

When they are present, sunspots are the most conspicuous feature on the photosphere. Occasionally they are even big enough to be visible to the naked eye. Such spots were recorded as early as the fourth century B.C.

14.5 The Aurora Borealis: photo by Robert Colling (courtesy The Adler Planetarium and Mr. Colling).

However, when Galileo saw them through his telescope in the seventeenth century, such a discovery was very disturbing to some of the vested interests in the current thinking of his day. The idea that there could be "blemishes" on the handiwork of God was unthinkable and rejected by many. There were some who would not look through his telescope to see the spots, because "they could not be there."

However, Galileo was finally able to show that sunspots were actually on the surface of the sun itself. They were not clouds or shadows but were integral parts of the solar surface. In fact, by studying their movement around the sun, Galileo was able to determine the solar rotation rate.

For centuries after Galileo, it was believed that the spots were holes in the sun through which we could peer into the supposedly cooler interior. One famous astronomer even conjectured that the solar interior might be cool enough to support life. We now know that they are not holes but broad, shallow depressions with surfaces a few hundred kilometers below the solar surface.

Take a piece of cardboard and make a thin slit in it. Tilt it toward the sun so that the light makes a sharp image of the slit on another piece of paper held some distance behind the cardboard. Now take a prism and orient it so that the light rays passing through the slit also pass through the prism and are refracted onto the paper (the edge of the prism should be parallel to the slit). In what order are the colors of the spectrum thus produced? Why? Draw a diagram showing how the light is bent (refracted). What type of spectrum do you see on the paper? With more sophisticated equipment, what more could you expect to see in this spectrum?

In the twentieth century, astronomers found that the sunspots are caused by very strong localized magnetic fields on the solar surface. These fields have up to 1000 times the strength of nearby undisturbed areas, and are 6000 times stronger than the earth's magnetic field. At such strengths, an iron object would easily be pulled from one's grip. These very strong magnetic fields effectively block much of the radiation beneath them from emerging and therefore are much cooler than the surrounding areas. Their 1500° C cooler temperature makes them appear darker than the rest of the surface. Nevertheless they are hotter than the surfaces of many stars. If you could look at a sunspot alone, it would be blindingly bright.

Sunspots usually appear in groups. Frequently such groups have two major spots, designated the *leader* and *follower*. The leader is the one that crosses the center line of the sun's disk first as it is carried by the rotation of the sun. Such pairs of spots will have opposite magnetic polarity[1] so that magnetic lines of force rise from the leader and descend into the follower, forming a magnetic arch above the surface. When the leading spots in the northern hemisphere of the sun are positive, the leading spots in the southern hemisphere are negative.

As can be seen in Figure 14.6, the center of each spot, the *umbra*, is

14.6 Sunspots on the solar surface, February 2, 1971 (photograph from Sacramento Peak Observatory, AFCRL).

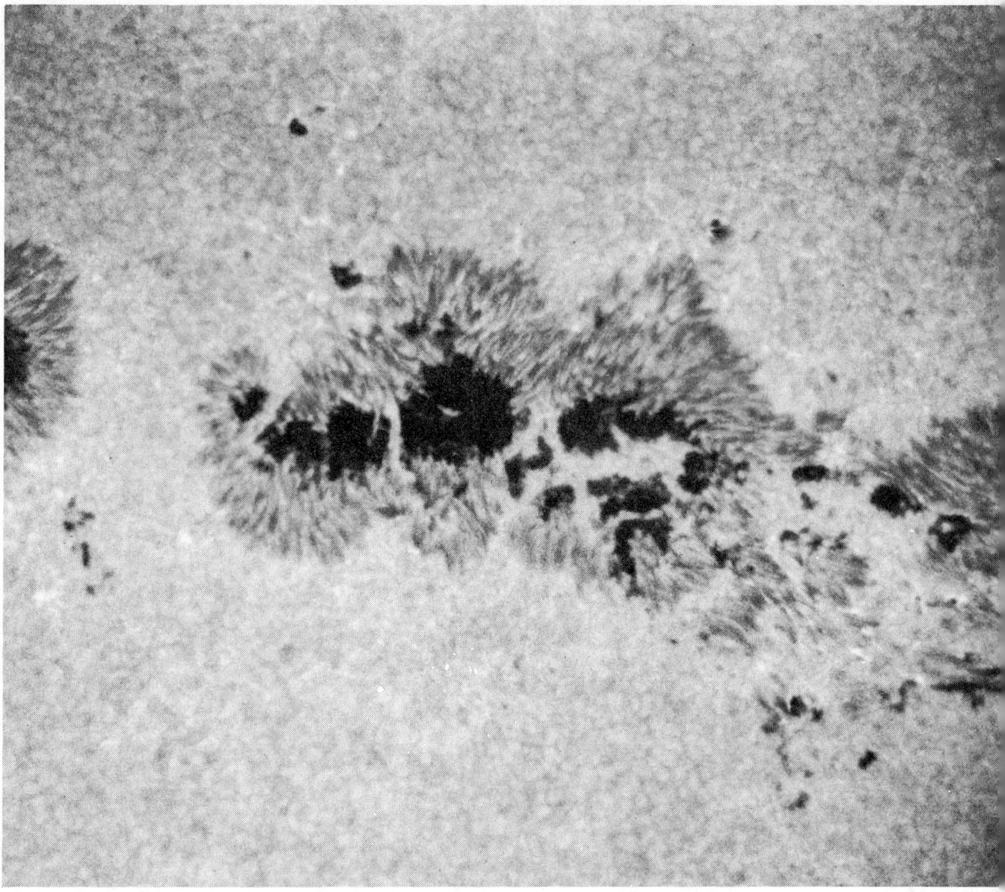

[1] Positive and negative poles just like an ordinary bar magnet.

very dark and solid-looking. This is where the magnetic field is the strongest. The umbra is surrounded by the *penumbra* in which thin strands can be observed. These are the lines of magnetic force made visible by the streams of matter caught up in the lines arching from one spot to another. There is evidence that the spots of opposite polarity are connected by a tube-like structure below the surface of the sun, the ends of which are visible as the pairs of spots on the surface. Thus the sunspots are only a manifestation of the changing pattern of the solar magnetic fields. It is these magnetic fields that are really basic to all solar activity.

The average sunspot is over 1000 miles in diameter, although some grow as large as the earth and a few have been found to be 100,000 miles in diameter (Fig. 14.7). Individual spots are short-lived, most of them disappearing within a day or so, but the average number of spots visible on the sun at any one time varies (at present) over a period of about 11 years. This is called the *sunspot cycle.* It starts with the appearance of small spots in the higher latitudes of both solar hemispheres. As the cycle progresses, new and larger groups of spots form closer and closer to the solar equator. The last spots of a cycle appear 5° to 10° from the equator. As these die out, small spots of the new cycle may already have started to appear in the higher latitudes (40° to 45°). When the locations of spots are plotted against latitude over the course of a sunspot cycle, the resulting diagram bears a close resemblance to butterfly wings (Fig. 14.8).

During a given sunspot cycle, if all the leaders in the northern hemisphere of the sun are positive, then in the following cycle they are negative. In the southern hemisphere, the opposite is true. This indicates that the true sunspot cycle is 22 years rather than 11 years long.

Much speculation has been made, particularly in the popular press, about the relation of sunspots to events on earth. Attempts have been made to correlate stock market activity and even political events with the sunspot cycle. Most such speculation is without foundation. However, there does appear to be a long-term weather cycle associated with the sunspot cycle. Total annual rainfall varies slightly in an 11-year cycle and in arid regions particularly this fluctuation can be a factor in the growth of plants. Such a cyclical variation may be reflected in the annual tree-ring growths (Fig. 14.9). Correlations have been attempted between rings of maximum thickness (corresponding to years of maximum rainfall) and the times of sunspot maxima.

Although the sunspot cycle seems fairly regular now, it has not always been so. In fact, shortly after their discovery by Galileo there was a period of 70 years in which the spots almost entirely disappeared. This period, from 1645 to 1715, is called the *Maunder Minimum* after the scientist who first called attention to it. It is perhaps because of the lack of spots during this period that it took over 200 years for the sunspot cycle to be rediscovered. In the twentieth century, until recently, astronomers were so

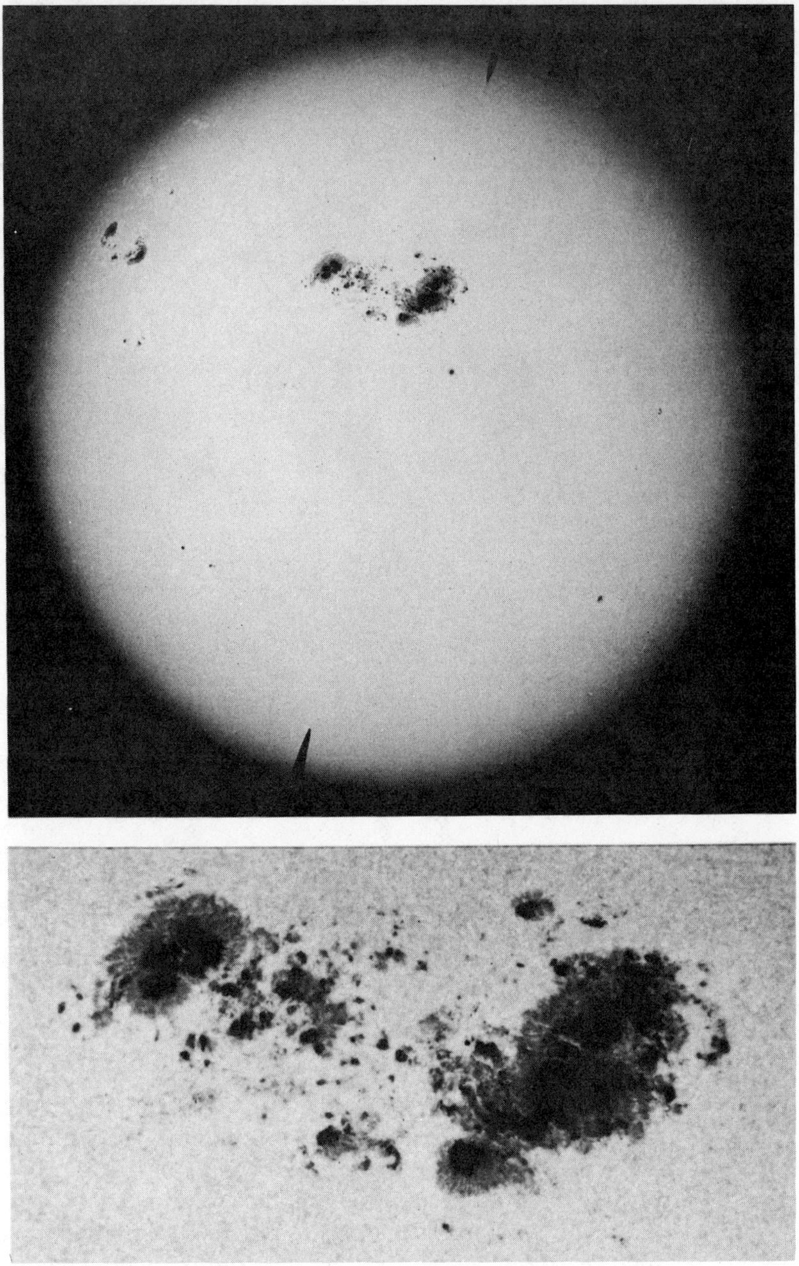

(a)

(b)

14.7 *(a) Solar disk showing an exceptionally large sunspot group (April 7, 1947). (b) An enlarged view of the sunspot group (photographs from the Hale Observatories).*

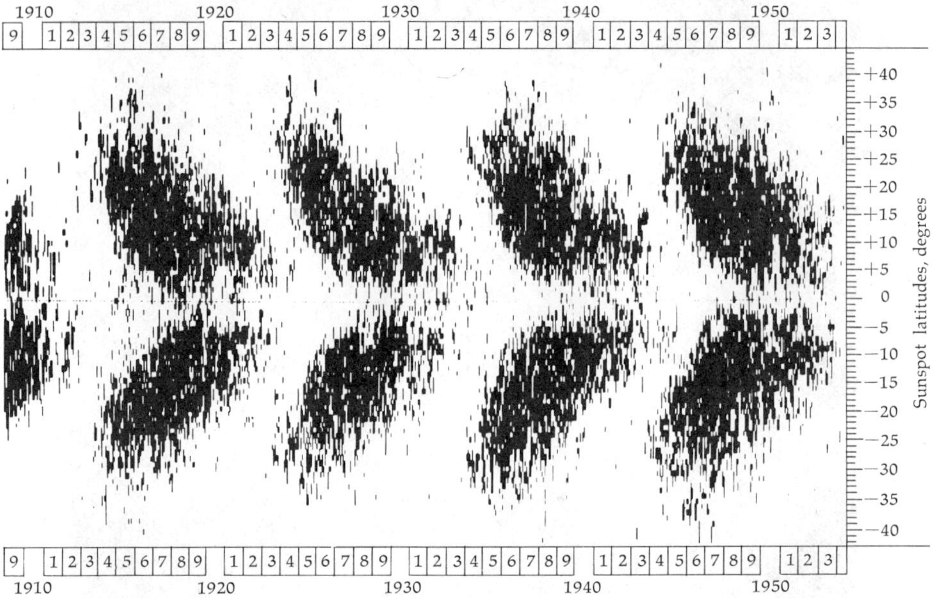

14.8 The butterfly diagram. *If sunspots are plotted each year by the latitude in which they appear, the "butterfly diagram" is formed. With each 11-year cycle, the sunspots start forming in higher latitudes and gradually form nearer and nearer to the solar equator. (Adapted from* Sunspot and Geomagnetic-Storm Data 1874–1954, *Royal Greenwich Observatory.)*

convinced of the regularity of the cycle that this 70-year minimum was ignored.

However, further evidence that solar activity was at a minimum during the Maunder Minimum is the reported lack of a visible solar corona during eclipses of this period. The corona derives much of its energy from the activity on the solar surface and that was evidently very minor at that time. For the same reason, very few if any displays of the aurora borealis were seen during this long period of sunspot minimum.

New findings now suggest that in the last 5000 years there may have been other such Maunder Minimums. More attention is now being paid to these long term changes in solar activity, especially since the dates of the Maunder Minimum closely coincide with what is called the "Little Ice Age" here on earth (1600–1850).

14.9 In this tree ring section from Prussia there appears to be a rough connection between tree-ring growth and the 11-year cycle.

Flares

In the solar atmosphere above the regions where sunspots are located, other major phenomena occur. They are much more common during sunspot maxima than during minima and are therefore believed to be closely related to this solar activity. Solar *flares,* for example, are caused by a sudden release of energy that had been blocked by the sunspots. The energy bursts forth in a nearby region in the form of a solar flare. The energy released from just one flare has been estimated to be enough to supply the earth's needs for more than 100,000 years!

Flares usually reach their peak in a very few minutes and then die away in about an hour. Around the time of sunspot maximum, there may be several small flares a day, with a major one every few weeks. Through special instruments we can see them as brilliant bursts of light, but much of their radiation is in the x-ray and ultraviolet regions of the spectrum. Although the flare covers only about one ten-thousandth of the solar surface, its energy can be equal to one-thousandth of the total solar energy. The matter spewed forth in such a violent eruption may reach 15,000 kilometers above the solar surface, traveling hundreds of kilometers per second before falling back.

About 50 hours after a flare occurs on the sun, the released energy reaches the earth. When that happens, short wave radio communications on earth suffer much static and may even be completely stopped. There may also be disturbances in major power line networks or even power outages at these installations if the flare is strong enough. Should there be any astronauts in space at that time, it is necessary that they be warned to return to the protection of their spacecraft where the increased radiation will effect them less. Aboard the Skylab craft an alarm system alerted the astronauts of the beginning of a solar flare so that they could study it more thoroughly. To venture outside the craft for a spacewalk would have been extremely dangerous in the period following such a flare because of the intense deadly radiation.

Prominences

Solar prominences, although far more spectacular than flares, are actually not as eruptive in nature. During solar eclipses they are visible telescopically as red flames arching up over the blackened solar disk. When they occur on the face of the sun, they can be observed (with proper filters) as dark filaments. During the Skylab missions, the astronauts were able to observe these eruptions whenever they occurred, using such instruments.

One prominence, measured during one of the missions, extended a half-million miles above the surface of the sun (Fig. 14.10).

Prominence matter generally appears first high above the solar surface and falls gracefully back following the magnetic lines of force (Fig. 14.11). The material originally comes from the sun but becomes visible only after it has cooled and condensed high above the surface. This usually happens above regions of sunspot activity.

When sunspots are at a minimum, other disturbances on the sun (flares, prominences, etc.) are also at a minimum. During such a period of the "quiet sun," its radiation is about as intense as would be expected from a body of the sun's temperature. This is true at all wavelengths of the electromagnetic spectrum (see the description of Planck's law on p. 136). Sun-

14.10 A huge solar prominence reaching 500,000 miles above the solar surface. This was photographed by the Skylab 3 astronauts in 1973. The size of the eruption can be visualized by comparing it with the small white dot which represents the size of the earth (National Aeronautics and Space Administration).

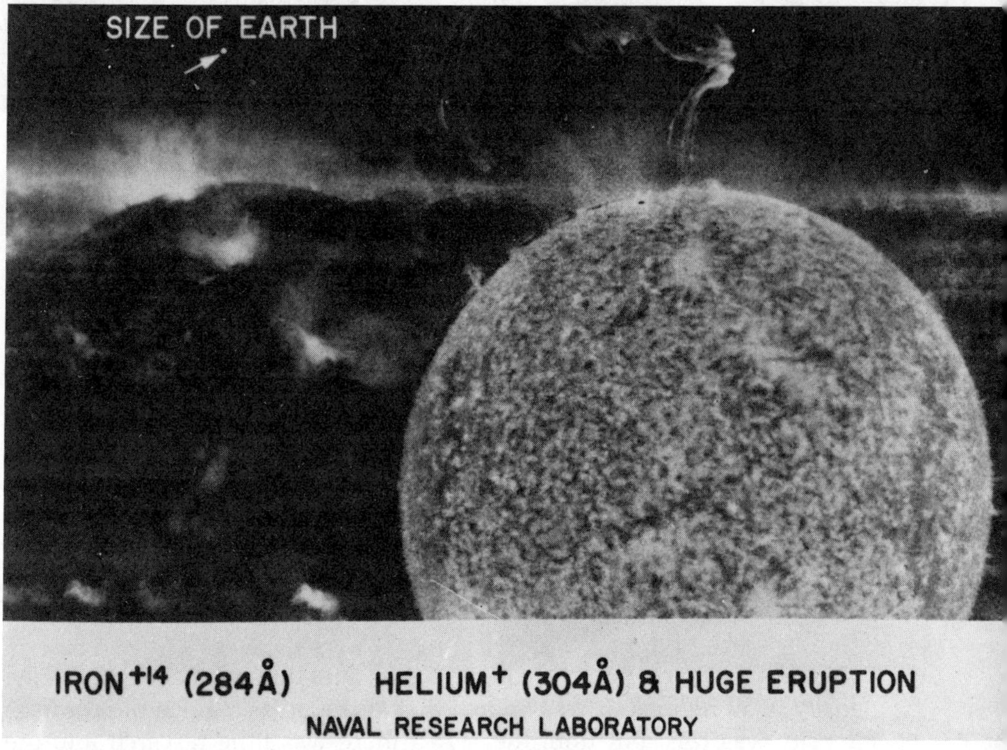

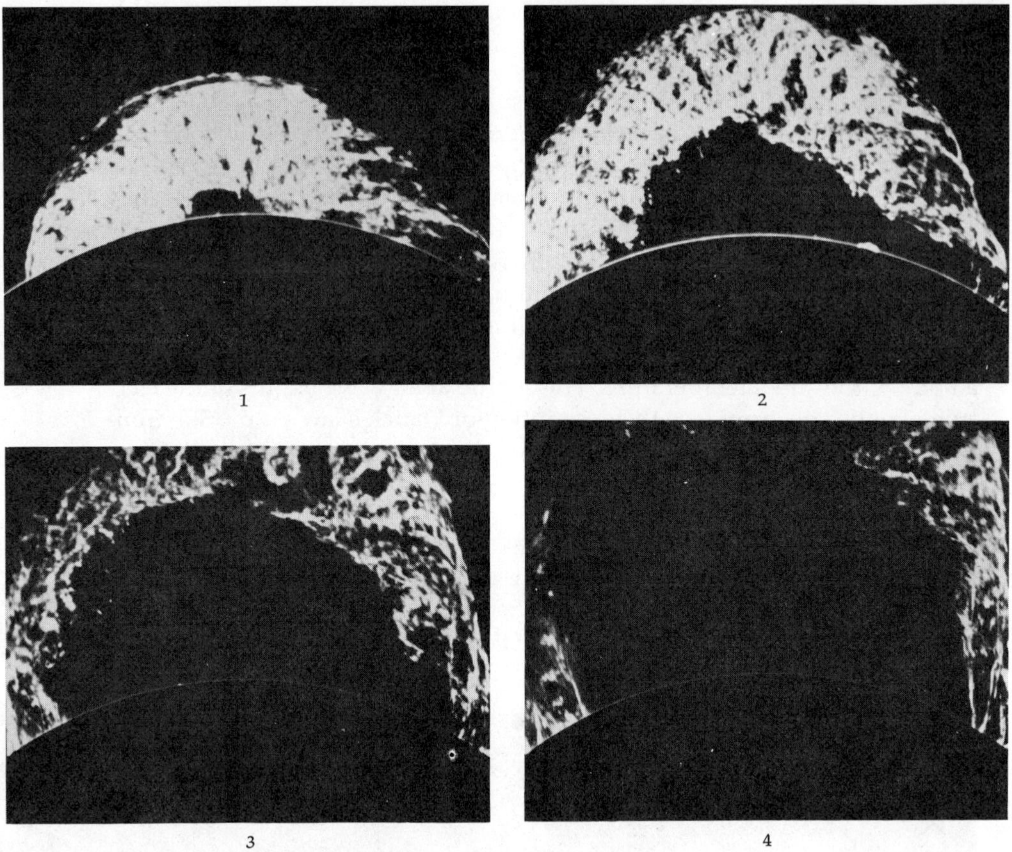

14.11 *A large solar prominence. This entire explosion took only one hour from beginning to end (photograph from the Yerkes Observatory).*

spots can therefore be used as an index of solar activity and disturbance. This is important because there is a definite correlation between the sunspot cycle and certain phenomena on earth. The sunspots are not the cause of these solar effects upon the earth, but rather are indications of the basic solar disturbances themselves.

Coronal Holes

Except at the solar poles, the corona was assumed to be of about the same density in all regions. It was known that at the polar regions the corona's intensity was less, but until the 1970's there was little research into the possibility of other *coronal holes.* By the time of the Skylab missions, their importance had been recognized and an intensive program for studying them was included.

The astronauts were able to detect (with special instruments) enormous gaps in this layer of the solar atmosphere. Not only were these gaps found to be regions of very low density compared to the rest of the corona (about one-tenth that of the normal corona), but they were also much cooler as well. It was also found that magnetic field lines in these holes did not arch back down to the solar surface again but instead extended straight out into space.

Then in 1973, astronomers found that high velocity streams of charged particles were emitted through these coronal holes. This accounted for the origin of previously unexplained streams which had been detected in the solar wind, moving at about twice the speed of the normal wind. Such velocities are possible because the magnetic fields in the coronal holes are relatively weak and matter is not impeded as it is spewed outward. When these high-speed winds reach the earth they create geomagnetic storms here, temporarily changing the strength and direction of the earth's normal magnetic field.

These holes, other than the polar ones, usually last about four to six months and are always close to major sunspot groups, so they too are related in some way to the amount of solar disturbance and activity. If the polar holes are included, these coronal holes may cover some 20% of the solar surface at one time.

Solar Pulsations

Besides all of these local disturbances on the solar surface, the sun's surface has also been found to pulsate in various ways, especially at sunspot maximum. One of these is an up-and-down vibration of about five minutes in duration. During this time, the region involved may move vertically about 1000 miles. Each portion of the surface pulsates on its own, but the entire surface is constantly in motion. It has been compared to a choppy sea.

There are also pulsating motions that involve larger areas of the sun and have longer periods and effect the entire solar globe. Similarly, the solar atmosphere also has been found to be oscillating on an enormous scale. The cause of all of these motions and their effect on the sun and ultimately on the earth is not known, but the study of them (much like earthquake research here on earth) may provide clues as to the sun's structure.

The Missing Neutrinos

One of the biggest mysteries about the sun might be called "the case of the missing neutrinos." When hydrogen is transformed into helium in the inner core of a main sequence star, besides the photons of energy that are thus created, a very special elementary particle is also released in this process. This is the *neutrino,* an uncharged subatomic particle which interacts very weakly with matter. That means it can pass almost completely unimpeded through the entire solar body. It has one chance in 10 billion of being absorbed even once during this journey. If it passes through the earth, it has one chance in 1000 billion of being absorbed.

Therefore, if some method of detecting these elusive particles could be devised, we could measure the rate at which the sun is generating energy in its interior long before that energy finally reaches the solar surface. Astronomers devised an experiment to attempt to detect them. In Lead, North Dakota, there is a deep mine in which astronomers installed a tank of 100,000 gallons of commercially available cleaning fluid. This installation is called the *Brookhaven Detector.* The chlorine in the cleaning fluid can capture neutrinos by reacting with them and producing a radioactive substance whose presence can be measured. According to calculations based on accepted models of the sun, the Brookhaven Detector was supposed to detect about one neutrino a day. Thus far, the rate has been considerably lower than this, much to everyone's surprise. The big question is, "Why?" The method of detection seems to be correct and should give reliable results. Is the sun's rate of creating energy slowing down? Or are the theories of nuclear transformation in error?

All these new findings raise important questions about the variations in solar radiation and their causes and effects on the earth. The sun has always been thought of as stable and dependable and now there are serious questions about this basic assumption. New and more exact measurements of the solar constant (see p. 135) are being made to see if any fluctuations can be detected. Only a 2% decrease in this all-important quantity of radiation would result in total glaciation of the earth. Thus far, however, no significant fluctuations have been found.

Discussion Questions

1. How does the fluctuation of solar activity affect the earth?

2. Describe the solar atmosphere. What characteristics does each of the layers have? Under what circumstances is each layer visible?

3. When viewed with high magnification from the earth, the surface of the sun appears something like oatmeal. Explain the cause of this appearance. What is it called?

4. Why do sunspots appear dark on the face of the sun? What causes them?

5. What is meant by the sunspot cycle? Why can it be considered either 11 or 22 years long? What does the "butterfly diagram" describe?

6. How did Galileo use the sunspots to demonstrate the sun's rotation?

15 *The Earth in Space*

One of the most philosophically important pictures that have been taken in the space age is that of the earth as it is seen from outer space (Fig. 15.1). How different man's concept of himself and his world would have been had such a picture been available even 500 years ago. We can now directly observe that the earth is round and that it rotates.

We know that the earth, the third planet from the sun, is actually one of the smaller members of the planetary system. If the sun were the size of a lemon, a small seed 40 feet away might represent the earth, and a pea 200 feet away would be the giant planet, Jupiter. The farthest planet, Pluto, would be ⅓ mile away, and the nearest star would be another lemon 2000 miles distant!

The earth is our home, and it is from this base that we view the rest of the universe. Because the motions of our planet cause constant daily changes in our viewing angle, we must first understand these motions so that we can allow for the apparent changes they produce.

On a large sheet of paper, draw a circle the size of a bowling ball to represent the earth. To scale, draw the following things on or around the circle:

Equatorial radius of the earth: 6378 kilometers.

Average height of Skylab orbit above earth's surface: 435 kilometers.

Deepest known part of ocean (Challenger Deep near Guam): 11 kilometers.

Height of highest peak of Rocky Mountains: 4.3 kilometers.

Height of Mount Everest: 8.8 kilometers.

By what percentage does the earth's surface differ from a perfectly smooth one? If a bowling ball had such imperfections, would they be easily visible?

15.1 *Half-earth as viewed from space by the Apollo 12 astronauts (NASA photograph).*

Rotation

In terms of the effect it produces (day and night), the most obvious motion of the earth is its rotation. A consequence (but not a proof) of this motion is the nightly swing of the constellations around the north pole of the sky (see Fig. 5.2).

It was not until the nineteenth century that simple, direct proofs that the earth turns on its axis were devised. The French physicist Jean Foucault constructed a clever device to demonstrate this motion (Fig. 15.2). By

15.2 *A Foucault pendulum as seen from above. Note that the outer circle measures degrees whereas the inner circle indicates hours of the day (courtesy Museum of Science and Industry, Chicago).*

means of a long wire, he attached a heavy lead ball to the high ceiling of the Pantheon in Paris in such a way that the ball was free to swing as a pendulum in all directions (this was done by means of a universal joint). Beneath the ball he placed a layer of sand so that as the ball moved back and forth, a point attached to its bottom marked a path in the sand. As the pendulum swung back and forth, the paths of successive oscillations did not coincide. The plane of oscillation appeared to be slowly shifting clockwise (Fig. 15.3). Since the only force acting upon the pendulum was

15.3 Schematic diagram of a Foucault pendulum. The pendulum is started at the north point (A) and swings toward the south. Because the earth is turning (rotating) under it as it swings, it ends its swing not at the south point but slightly to the west of it (B). It then swings back, not to A but to C, maintaining a very slight clockwise shift in its plane of oscillation. The apparent shift is caused by the earth's rotation. In this diagram the shift of the pendulum is highly exaggerated.

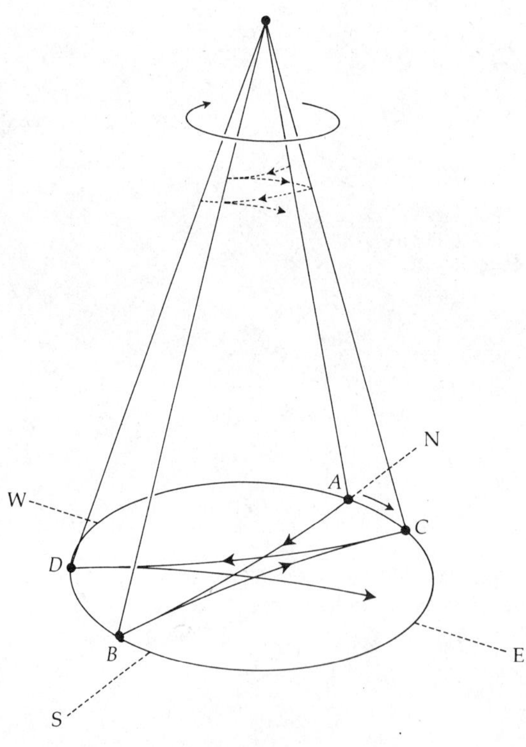

Many museums have Foucault pendulums. If you can visit one, find out how many degrees the pendulum shifts each hour. How does the museum keep the apparatus accurate, i.e., telling the correct time (usually hours are marked along the circumference of the circle in which the pendulum swings and, except for a pendulum situated at the earth's pole, it would not cover all 360° in 24 hours like a clock. Why is this so?

---

the downward pull of the earth's gravitation, there was no force to explain the pendulum's apparent shift. The earth's rotation beneath the freely swinging pendulum was the only other thing that could cause the pendulum to appear to change its direction. Foucault found the shift to be eleven degrees per hour in Paris. Later, it was discovered that similarly constructed pendulums in other places had different rates of shift. The farther away from the equator the pendulum was, the faster the shift. At the poles the shift amounted to fifteen degrees per hour (Fig. 15.4). These differences in the rate of shift of the plane of the pendulum's swing can be only explained by the rotation of the earth. No motion of the celestial sphere would have this effect.

Fill a large pie pan with sand. Place it on a turntable such as a Lazy Susan. While turning the table at a steady rate, draw a pencil through the sand in a straight line from center to rim. What is the shape of the line made by the pencil? Compare this with a cannon ball shot from north to south in the Northern Hemisphere. If you draw the pencil from the rim to the center of the pan while the Lazy Susan is turning, what is the resulting line in the sand?

Wind currents that flow across the land or sea act in a manner similar to a large Foucault pendulum. In the northern hemisphere, they are always deflected to the right in the same way that the bob of the pendulum appears to be deflected to the right, and for the same reason (Fig. 15.5). This can clearly be seen in many of the photographs of the earth taken from space.

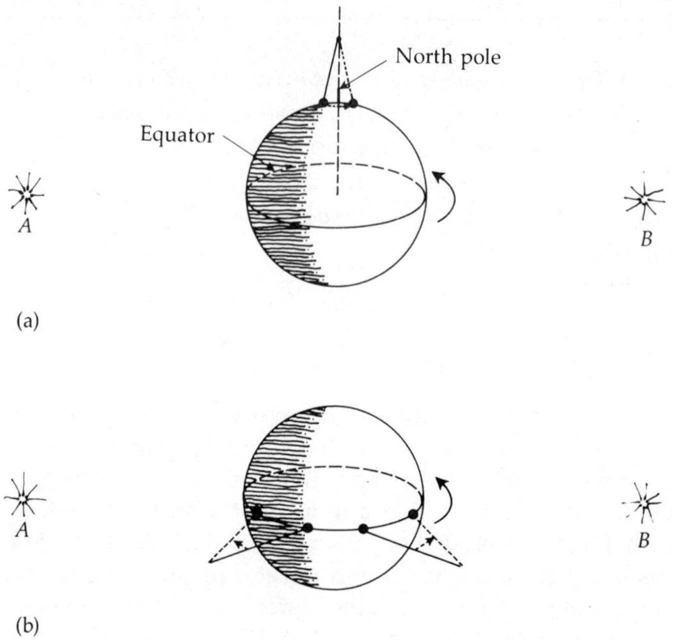

15.4 *The Foucault pendulum.*
(a) When the pendulum is placed at the north pole, the plane of the pendulum stays lined up with stars A-B *while the earth rotates beneath it, making the pendulum appear to be slowly shifting opposite to the earth's rotational direction, making a complete revolution in 24 hours (or 15° per hour).*
(b) When the pendulum is placed on the equator, the plane of the pendulum still stays lined up with stars A-B *as the pendulum is carried around the equator. The pendulum support is carried along with the motion of the earth but there is no observable shift in the swing of the pendulum itself. The pendulum's plane remains the same in relation to the stars and to the ground above which it swings.*

Timekeeping

The length of our ordinary day, of course, is based upon the rotational period of the earth. This is the time it takes for the earth to turn once on its axis relative to the sun, i.e., from one meridian crossing of the sun to the next. This is called the *solar day*.

Sunrise is the time when the sun's upper edge (not its center) first appears above the eastern horizon. *Sunset* is the time of the last appearance

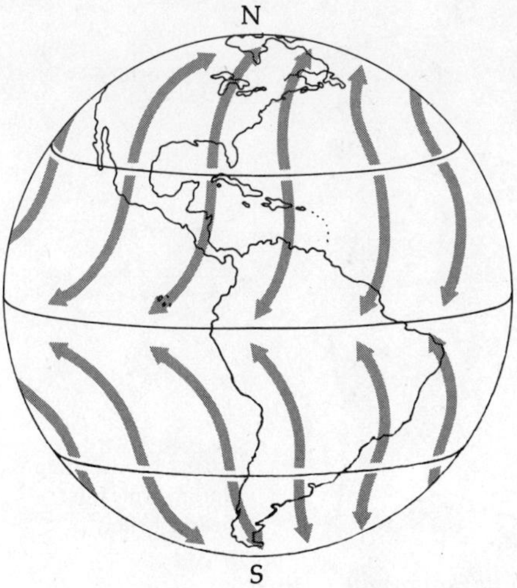

15.5 The rotation of the earth causes the flow of air masses to be deviated to the right in the northern hemisphere and to the left in the southern hemisphere.

of that upper edge on the western horizon. Since our atmosphere refracts or bends the sun's rays much like a lens, the sun at the horizon appears a little higher than it really is. We can actually see the entire disk of the sun when it is in fact still below the horizon; the refraction bends the sun's rays around the curvature of the earth.

When the sun is at the equinoxes, the length of the day and night are supposed to be equal—12 hours for each. This would indeed be the case were it not for the earth's atmosphere, and if we used the center of the sun to mark sunrise and sunset. However, because we use the sun's upper edge and because of refraction, there are differences in the length of day and night even when they should be equal.

Besides these variations in sunrise and sunset, the solar day itself is not uniform. As we found in Chapter 13, the earth's orbit is elliptical and the earth therefore does not move uniformly around the sun. In January, when it is about three million miles closer to the sun, it moves faster than it does six months later (Kepler's second law of planetary motion—see Fig. 15.6).

Also, the 23½° tilt of the earth's axis to its orbit makes any projection of the sun's apparent motion onto the celestial equator longer at the solstices than at the equinoxes (Fig. 15.7). Therefore, because of the uneven velocity and the 23½° tilt, the period of the solar day is irregular.

For timekeeping purposes, a *mean solar day* is used. This is based on an imaginary *mean sun* that uniformly moves approximately one degree per

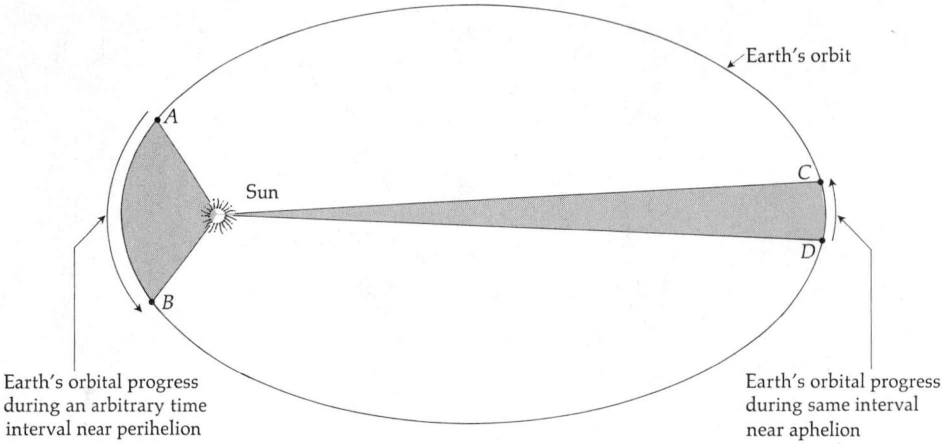

15.6 The sun does not appear to move smoothly (uniformly) through the sky because the earth moves in its elliptical orbit at a varying speed, conforming to Kepler's law of equal areas. The scale in this diagram is exaggerated.

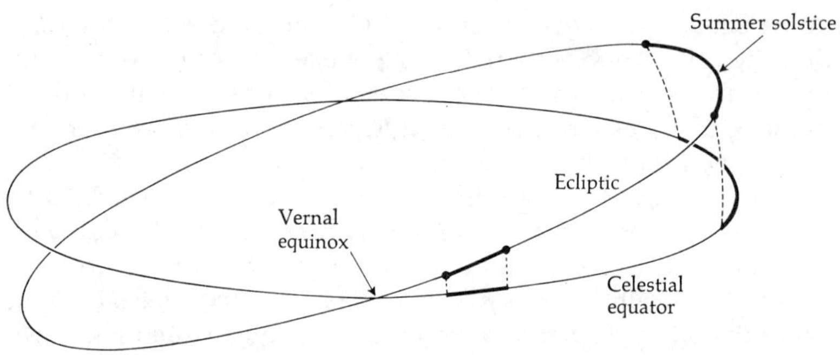

15.7 Any projection of the sun's apparent motion onto the celestial equator covers a larger angular measurement at or near the solstices than at or near the equinoxes. This is because of the 23 1/2° tilt of the earth's axis (which determines the plane of the celestial equator) to its orbit (which determines the plane of the ecliptic).

day toward the east along the celestial equator (not the ecliptic). The difference between the "apparent" and "mean" solar time is called the *Equation of Time.* This difference can be as much as almost 16½ minutes (in November) (see Fig. 15.8). When reading a sundial, the Equation of Time correction must be used to get the mean solar time. Sundial time is the local apparent solar time.

The time of the day at any instant is based upon the *hour angle* of the sun, that is, the angular distance it is from the meridian (measured in hours, rather than degrees). As we learned in Chapter 5, *antemeridian*

Time the meridian crossing of the sun each day. Use a compass to locate the exact north-south direction of your meridian. How much does the meridian crossing time vary from day to day? How does this relate to the Equation of Time?

15.8 *The Analemma, which is found on many world globes, indicates for any date the latitude at which the sun appears directly overhead at noon as well as the equation of time for that date.*

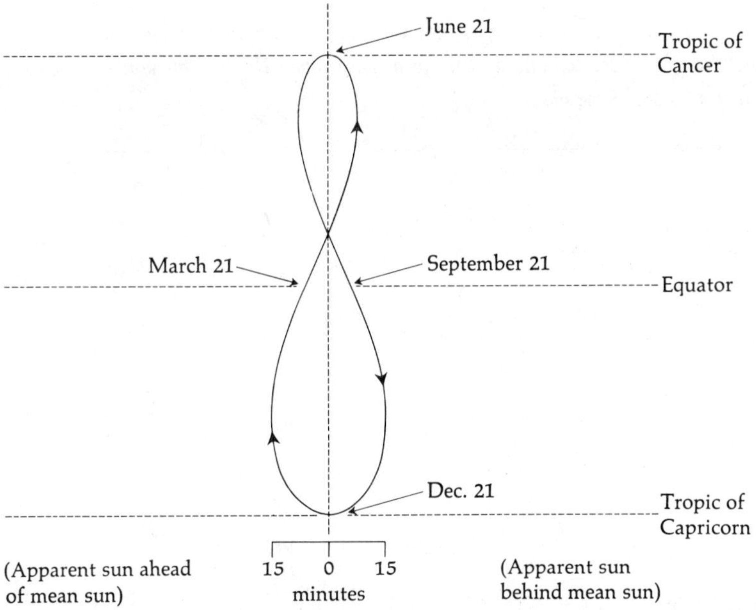

(a.m.) is the period before the sun reaches the meridian and *postmeridian* (p.m.) are the hours after this time. For convenience, we start our day at midnight so that solar time (if measured on a 24-hour basis, as is done in astronomy and in the Armed Forces) is actually the sun's hour angle plus 12 hours (Fig. 15.9).

Although mean solar time makes timekeeping uniform, it still varies continuously with longitude. The apparent sun will actually cross the meridian at Boston 12 minutes before it reaches the meridian at New York. The imaginary local mean sun will also cross the meridian 12 minutes earlier in Boston than New York. Therefore (again for convenience), the entire eastern section of the United States adjusts its clocks to show noon when the sun crosses the 75th meridian, near Philadelphia. This is called Eastern Standard Time. Similarly, cities like St. Louis, Chicago, and Kansas City have their noon when the sun crosses the 90th meridian, which is roughly in the center of the Central Standard Time zone. "Daylight Saving Time" is the standard time of the next time zone to the east. Thus Central Daylight Saving Time is nothing more than Eastern Standard Time.

The zero meridian at Greenwich, England is used not only for terrestrial orientation but also as a basic reference point in astronomy. The standard mean solar time of Greenwich is referred to as Universal Time (UT) and is used internationally when recording celestial events.

Measure the altitude of the sun at its meridian crossing at various times of the year. Record your observations and draw a diagram of them. Can you determine your latitude from these observations?

15.9 *Local apparent time is the hour angle of the sun past the meridian.*

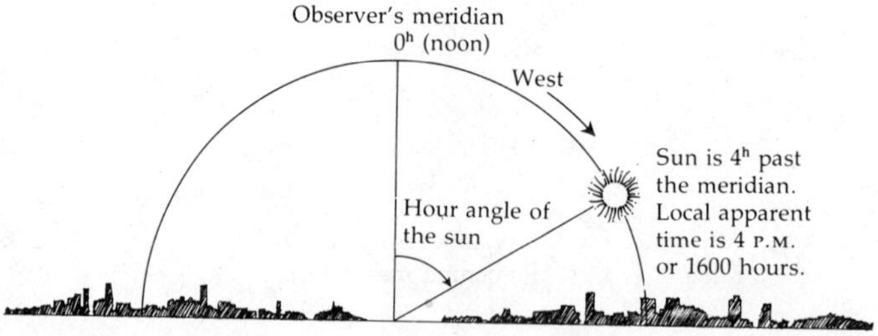

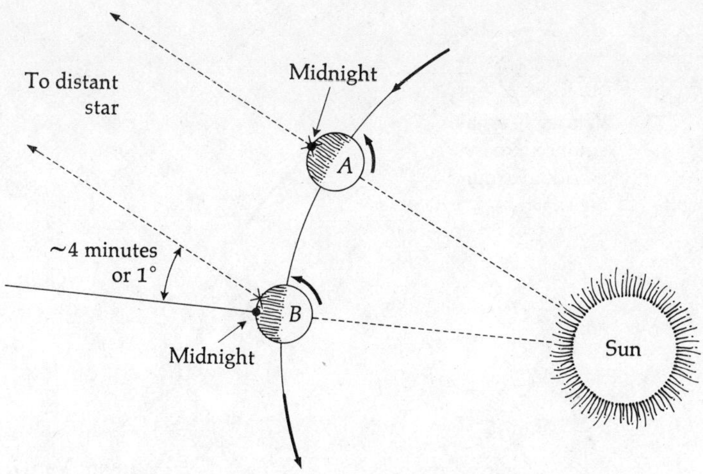

15.10 Sidereal and solar days. Point X *on the earth makes one complete revolution with respect to the stars (sidereal day) as the earth moves from* A *to* B. *It will take almost 4 minutes more (solar time) to make one complete revolution with respect to the sun (solar day).*

Sidereal Time

Because the earth is also revolving around the sun as it rotates on its axis, it makes one complete rotation relative to the stars in a slightly shorter time period than one solar day (Fig. 15.10). That is, a given star will cross our meridian about four solar minutes earlier each night.[1]

Time measured by the rotation of the earth relative to the stars is called *sidereal time*. The *sidereal day* is measured from one meridian crossing of the vernal equinox to the next. This can occur at any time of the day or night, depending upon the time of the year. In the fall, the vernal equinox crosses the meridian at night, while in the spring its crossing occurs during the day (Fig. 15.11). When the sun is near the autumnal equinox, the two time systems momentarily agree. A whole year must elapse before they will coincide once again, because of the four-minute difference in their daily time periods. In this interval, there will be one more sidereal day (366¼) than solar day (365¼).

---

[1] It was this shift in time each night that gave Karl Jansky the clue to the source of his mysterious radio hiss (see Chap. 12). Any object beyond our solar system would have this daily rate of shift in the time of meridian passage.

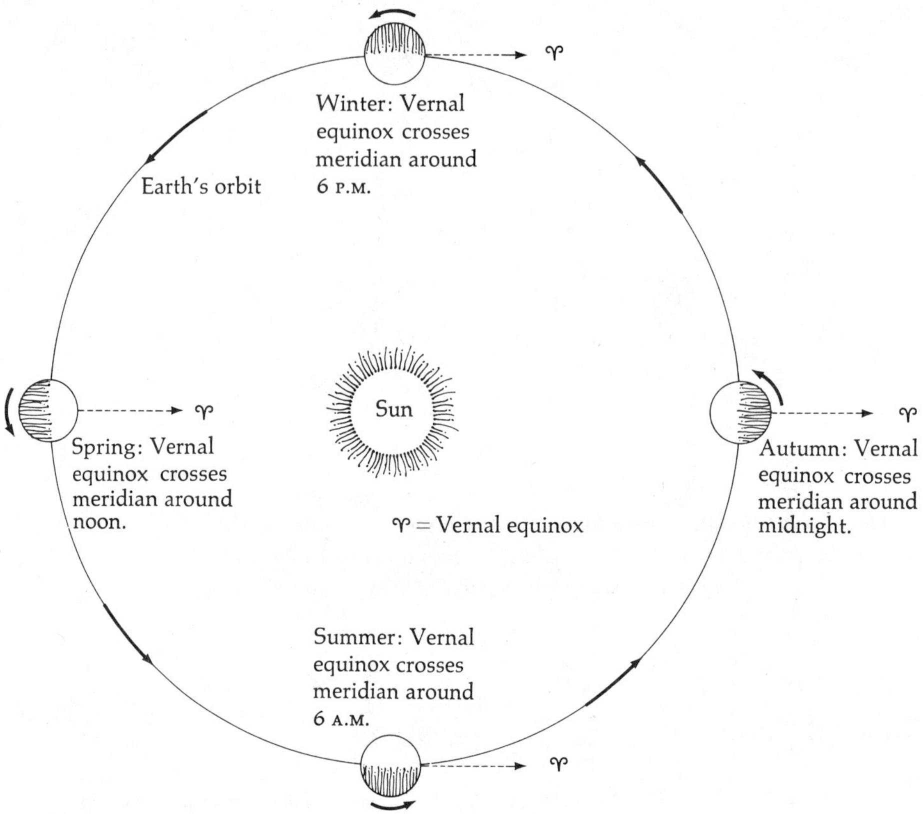

15.11 The "sidereal day" starts when the vernal equinox (♈) crosses the meridian. This can occur at any time of the day or night, depending upon the time of the year.

Make a simple sundial. Take a soda straw or a swizzle stick and stick it into soft earth in a sunny location so that it is inclined toward the north pole about 45° from the horizon. Each hour mark the ground where the shadow of your stick falls. On the next few sunny days, compare your markings with the hourly shadows. How much do the shadow locations vary over a period of time? When will your sundial again agree with your watch?*

*The inclination or number of degrees from the horizon should equal your latitude. For most of the United States, 45° is a close approximation to this amount.

The sidereal day is divided into 24 hours, each with sixty minutes, and each minute having sixty seconds, but each hour, minute, and second of this day is shorter than our ordinary time intervals. Sidereal time is used extensively in astronomy, and every major observatory has a sidereal clock which is divided into 24 hours instead of 12 (Fig. 15.12).

You are shipwrecked on an unknown island. With only a radio, a wristwatch, and a pencil, how can you locate your position in longitude and latitude? (Hint: What is the difference between your longitude and the longitude of a radio station whose broadcast you receive? If you set the watch by the radio signal, how can you compare it with your own local time?)

15.12 *Solar and sidereal clocks. Note that whereas the solar clock reads 13 hours 51 minutes 56 seconds ($13^h\ 51^m\ 56^s$) in Universal Time, the sidereal clock reads $15^h\ 6^m\ 50^s$ in Sidereal Time. On what date was this picture taken? Both clocks have hour, minute, and second hands as well as a "sweep second" hand which makes one revolution per second, enabling a reading in hundredths of a second (courtesy The Adler Planetarium, and Mr. Colling).*

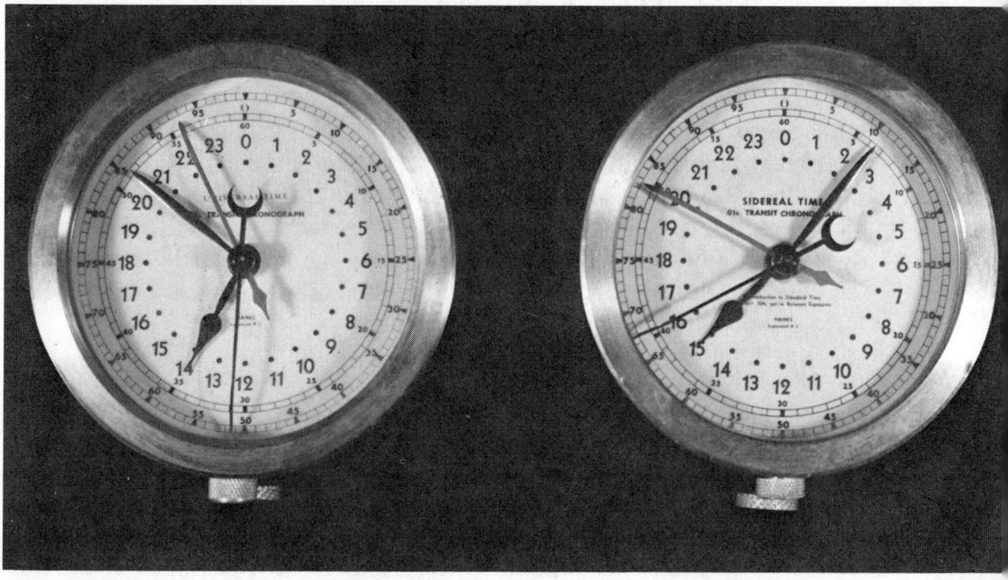

Revolution

The orbiting of the earth around the sun is undoubtedly its second most important motion in terms of the effects it produces on us. We have seen how difficult it has been throughout history for man to accept the concept of a moving earth. The proof of such motion was not attainable until very precise measuring instruments could be developed. Tycho Brahe rejected the theory of a revolving earth because he could not measure any stellar parallax (see Chap. 6).

Besides the use of stellar parallaxes, astronomers can also point to the changing radial velocities of the stars as the earth in its orbit moves toward or away from each one of them (Fig. 15.13). In fact, in radial velocity studies, this effect must always be corrected for.

As we discussed in Chapter 5, the revolution of the earth causes the different constellations to progressively come into view throughout the year. However, this is not proof of the earth's motion. Ancient civilizations were quite aware of the seasonal shift in constellations and devised their own earth-centered explanations.

Besides giving us our time period of one year, the earth's revolution, coupled with its 23½° tilt to its orbit, causes the seasonal changes we experience. The direction of the earth's axis in space stays essentially the same from one year to the next, changing only very minutely due to precession. Therefore, as the earth revolves about the sun, the northern and southern hemispheres are alternately presented more directly to the sun's rays (Fig. 15.14).

When the northern hemisphere is slanted toward the sun, the solar rays hit that half of the earth more nearly vertically, so that each unit area of terrestrial surface receives a maximum amount of radiation. Summertime temperatures result. At the same time of the year, the southern hemisphere is slanted away from the sun and receives much more oblique rays. The same amount of sunlight is spread over a much larger surface of the earth, affording minimal wintertime heating. Six months later the situation is reversed. Note that our seasons have nothing to do with the varying distance of the earth from the sun, since we are actually farther away from the sun during summertime in the northern hemisphere than we are during wintertime.

The Calendar

The yearly calendar can be thought of as a clock that uses large time intervals—days, weeks, and months. The year of the seasons is called the *tropical year*. This is the interval between successive passages of the sun across the vernal equinox. Spring begins in the northern hemisphere when

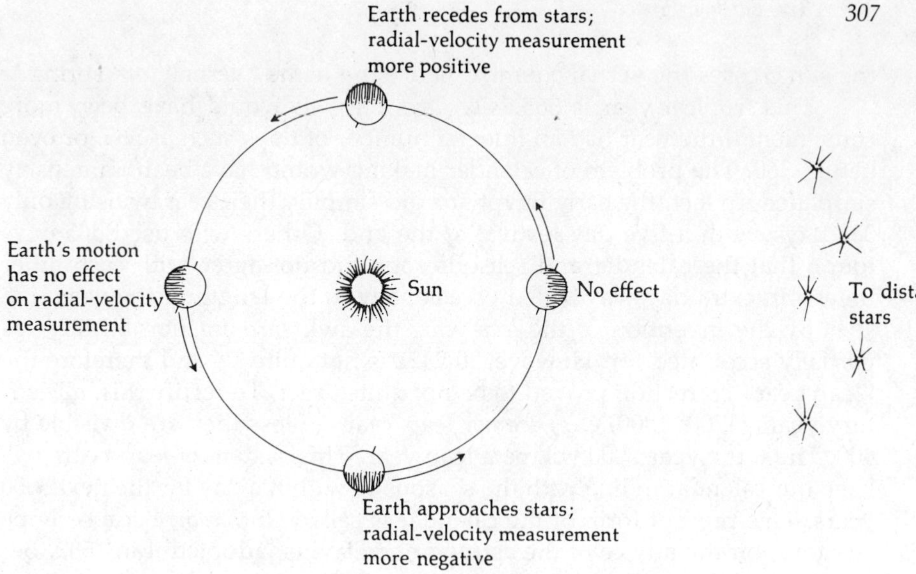

15.13 Radial-velocity measurements of stars fluctuate as the earth revolves about the sun.

15.14 Seasonal changes as the earth revolves around the sun, caused by the 23½° tilt of the earth's axis to its orbit. The same amount of radiation falls on a larger area in the northern hemisphere in winter than in summer and therefore does not provide as much heat per square centimeter. The seasons are reversed in the southern hemisphere.

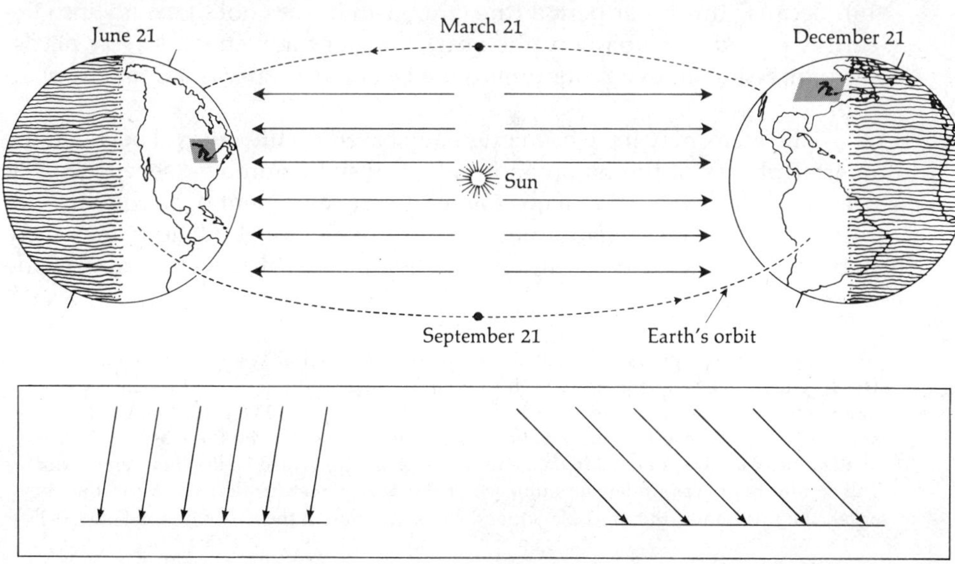

the sun crosses the vernal equinox, hence the name "vernal" or "spring".

This tropical year is 365.2422 days long. It would have been more convenient if the year had an integral number of days such as 365, or even better, 360. The problem of calendar making would have been immensely simplified. In fact, the early Egyptians did simplify their year by using only 360 days, with a five-day festival at the end. Others who used 365 days found that the calendar and seasons soon did not agree with each other unless an extra day was added occasionally to the length of the year.

By the invention of the *leap year*, the awkward fraction 0.2422 was partially accounted for. However, 0.2422 is not quite 1/4 and therefore the "leap year" correction proved to be not quite exact. To rectify this, all century years (1700, 1800, etc.) are *not* leap years unless they are divisible by 400. Thus, the year 2000 will be a leap year. This system of leap years will keep our calendar in line with the seasons to within a day for the next 3000 years. This present form of the calendar is called *Gregorian*, because Pope Gregory, on the advice of the astronomer Clavius, adopted it in 1582. Because it was Catholic in origin, however, England refused to adopt it for almost 200 years. They finally did so in 1752.

The history of the development of the calendar year and its various divisions finally resulting in our current calendar is a good example of man's constant attempt to organize his life with respect to nature. Now, by international agreement, we also have the "leap second." Because the earth's rotation is slowing about three milliseconds a day (see p. 316), about once a year a leap second must be added to keep time as measured by the earth's rotation in line with atomic time standards.

The calendar month was originally based upon the moon's period of revolution with respect to the sun, i.e., in terms of its phases (see Chap. 16). Because this lunar period (the "moon-th") does not fit evenly into the earth's period of revolution (the year), constant adjustments were necessary. Inserting an extra month into the calendar year was a frequent practice.

The week may have been originally based on the interval between the quarter phases of the moon and on the mystical number "seven." Seven celestial objects played a major role in the development of astronomy and astrology and in the naming of the days of the week.[2] The names were derived from the sun, moon, and the five planets that were visible to the

---

[2] The week very probably originated in the old astrological system of Planetary Hours. The seven celestial bodies in order of their assumed distance from the earth (which was then thought to be at the center of the universe) were: Saturn, Jupiter, Mars, Sun, Venus, Mercury, and the Moon. (Those that moved the fastest were believed to be the closest.) Each day, divided into 24 hours, derived its name from the planet that ruled its first hour after sunrise. The second hour was under the influence of the next planet (or other body) in the above order, the third hour, the third planet, and so forth. Thus, if the first day was Saturn's Day

naked eye. The order of the names was determined by a mixture of astronomy and astrology.

Although our calendar today has many disadvantages, especially in the business world, proposals for calendar reform have not gained much support. Figure 15.15 shows one such proposal, which would make the

15.15 The World Calendar. This example of an unchanging calendar not only makes the calendar the same each year, but also repeats itself every three months. Thus the months of January, April, July, and October are exactly the same. A "Year Day" occurs at the end of the year as the 365th day. Every four years another day is added just as we do during our Leap Year.

January

S	M	T	W	T	F	S
1	2	3	4	5	6	7
8	9	10	11	12	13	14
15	16	17	18	19	20	21
22	23	24	25	26	27	28
29	30	31				

April

S	M	T	W	T	F	S
1	2	3	4	5	6	7
8	9	10	11	12	13	14
15	16	17	18	19	20	21
22	23	24	25	26	27	28
29	30	31				

July

S	M	T	W	T	F	S
1	2	3	4	5	6	7
8	9	10	11	12	13	14
15	16	17	18	19	20	21
22	23	24	25	26	27	28
29	30	31				

October

S	M	T	W	T	F	S
1	2	3	4	5	6	7
8	9	10	11	12	13	14
15	16	17	18	19	20	21
22	23	24	25	26	27	28
29	30	31				

February

S	M	T	W	T	F	S
			1	2	3	4
5	6	7	8	9	10	11
12	13	14	15	16	17	18
19	20	21	22	23	24	25
26	27	28	29	30		

May

S	M	T	W	T	F	S
			1	2	3	4
5	6	7	8	9	10	11
12	13	14	15	16	17	18
19	20	21	22	23	24	25
26	27	28	29	30		

August

S	M	T	W	T	F	S
			1	2	3	4
5	6	7	8	9	10	11
12	13	14	15	16	17	18
19	20	21	22	23	24	25
26	27	28	29	30		

November

S	M	T	W	T	F	S
			1	2	3	4
5	6	7	8	9	10	11
12	13	14	15	16	17	18
19	20	21	22	23	24	25
26	27	28	29	30		

March

S	M	T	W	T	F	S
					1	2
3	4	5	6	7	8	9
10	11	12	13	14	15	16
17	18	19	20	21	22	23
24	25	26	27	28	29	30

June

S	M	T	W	T	F	S
					1	2
3	4	5	6	7	8	9
10	11	12	13	14	15	16
17	18	19	20	21	22	23
24	25	26	27	28	29	30

September

S	M	T	W	T	F	S
					1	2
3	4	5	6	7	8	9
10	11	12	13	14	15	16
17	18	19	20	21	22	23
24	25	26	27	28	29	30

December

S	M	T	W	T	F	S	
					1	2	
3	4	5	6	7	8	9	
10	11	12	13	14	15	16	
17	18	19	20	21	22	23	
24	25	26	27	28	29	30	Y

(Saturday), its first hour was astrologically dominated by Saturn. While Saturn "ruled" over all of Saturday, the second hour after sunrise was specifically ruled by the next planet in the above order, namely Jupiter. Counting down the list (repeating it over and over) to 24, we find that the twenty-fifth hour (which is the first hour of the next day) is the Sun's hour and therefore Sunday follows Saturday. Repeating this process each time, we get: Moon's Day (Monday), Mars' Day (Tuesday), Mercury's Day (Wednesday), Jupiter's Day (Thursday), and Venus' Day (Friday).

calendar year the same each year. However, it is hotly contested by some religious groups, because the "extra day" at the end of the year would destroy the seven-day sequence which they still consider sacred.

Orbital Changes

Besides providing us with convenient timekeeping periods, the various terrestrial motions are found to have profound effects upon our very existence. It has been discovered that very large changes in the climate of our planet (e.g., ice ages) may be due to relatively small variations in the earth's orbit.

There are believed to be three cycles which may drastically change the climate. These involve the time of the year that the earth is at its nearest and farthest point from the sun, the tilt of the earth's axis to its orbit, and the eccentricity of the earth's orbit itself (how much it differs from being completely circular). All these factors change over a period of many thousands of years, the three cycles overlapping each other and creating different effects on the earth.

By studying geological records dating back over 450,000 years, a correspondence between former ice ages and these changes has been found. A true ice age would take thousands of years to develop, but the early stages could effect us by, for example, shortening our growing season or causing unseasonably cold temperatures that would require tremendous usage of fuel. For these reasons, climatologists are studying the evidence very carefully in an attempt to foresee future weather conditions.

Discussion Questions

1. Which of the following time periods were probably based upon astronomical observations?

Day	Second	Year	Month	Decade
Minute	Week	Hour	Century	

2. Why must the earth's rotation be taken into account when firing long-range rockets?

3. The earth's orbit around the sun is an ellipse. The earth is farther away from the sun when it is summer than when it is winter in the northern hemisphere. Obviously, therefore, this distance does not determine seasonal change. Why then are there seasons on the earth? Can other planets have seasons?

4. In the Northern Hemisphere, the apparent deflection of the Foucault pendulum in its direction of swing is to the right. In which direction would it be in the southern hemisphere? Why?

5. Describe the various motions of the earth and the effects each has on our daily lives.

6. How can the Doppler effect be used to prove that the earth revolves around the sun?

7. What three changes in the earth's orbit are believed to be responsible for dramatic changes in the earth's climate?

8. What is the "equation of time"? Why is it necessary?

9. Define: sidereal time, apparent solar time, local mean solar time, standard time.

10. Each standard time zone covers a region of 15° of longitude. The Eastern Standard Time zone keeps the mean local time of the 75th meridian. When it is 6:00 UT, what time is it in New York City? At approximately what time of the day or night would the Rose Bowl football game be viewed in the Azores?

11. If a star is seen on the meridian at 11:00 p.m. tonight, when will it be on the meridian two months from now? (Hint: A given star crosses our meridian about four minutes earlier each night.)

12. Will the year 2100 A.D. be a leap year according to our present calendar system?

16 *The Double Planet*

Observed from space, the earth-moon system may well be described as a "double planet." Although the moon is not the largest satellite in the solar system, it is by far the largest in comparison to its primary planet, the earth. It has one-fourth the diameter of the earth. Because of this relatively large size, the moon's effect upon the earth is far greater than that of any other moon on its planet in the solar system.

The moon is usually described as orbiting the earth. Actually, however, they both revolve around their common center of mass which, because of the earth's much greater mass, is located about 1000 miles below the earth's surface on the side facing the moon. This is known as the *barycenter* (Fig. 16.1). As the earth rotates on its axis, the barycenter is constantly changing its position during each 24-hour period with respect to the continents on the earth.

Since it is the barycenter rather than the physical center of the earth that describes the ellipse around the sun, the center of the earth describes a slight wobble as it orbits the sun. For most purposes, it is not necessary to distinguish between the barycenter and the center of the earth. Only in precise calculations is it necessary to take the minute wobble of the earth in its orbit into consideration.

Tides

The swinging of the earth around the barycenter causes the waters on the side of the earth away from the moon to be "thrown outward", much as the water in a pail that you swing around your head is forced against the bottom of the pail (Fig. 16.2). Actually, the entire earth is affected equally by this swing effect (called *centrifugal force*), but because the earth is solid (like the pail), the main effect is seen in the changing water levels or tides.

At the same time the moon's gravitational effect on the earth is pulling in the opposite direction. Its main strength, of course, is on the side of the earth closest to it, and once again it is the effect on the waters of the

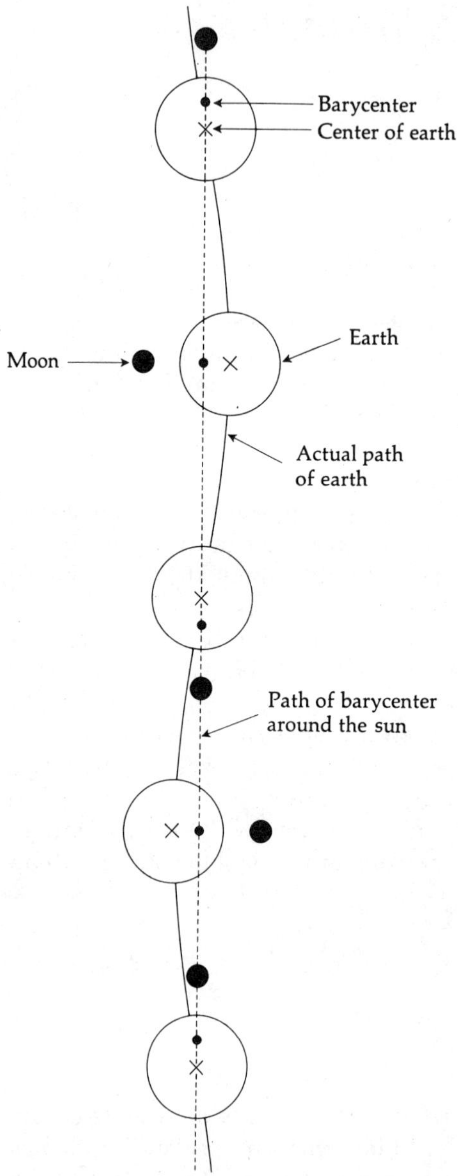

16.1 *The center of mass* (barycenter) *of the earth and moon revolves around the sun in an ellipse (although for simplicity it has been represented here as a straight line). One complete revolution around the barycenter takes one sidereal month. This causes the earth to appear to have a slight wobble in its orbit around the sun.*

16.2 The water in the pail is thrown in an outward direction as the pail is swung rapidly in a circle. Similarly, the swinging of the earth around the barycenter causes it to be "thrown outward". We are most aware of this effect in the constantly changing tides.

earth that is most noticeable.[1] These two forces, centrifugal and gravitational, are what cause similar tides on opposite sides of the earth.

Note in Figure 16.3 that at the center of the earth the two forces are balanced. If this were not so, the earth would not stay in its orbit around the barycenter. At the nearest point to the moon (*N*), the pull of the moon overbalances the effect of the earth's swinging about the barycenter, whereas at the farthest point from the moon (*F*), the latter effect is greater.

---

[1] The North American continent may rise as much as six inches when the moon is overhead. Aerial (atmospheric) tides can produce winds 1/20 mile per hour at the earth's surface.

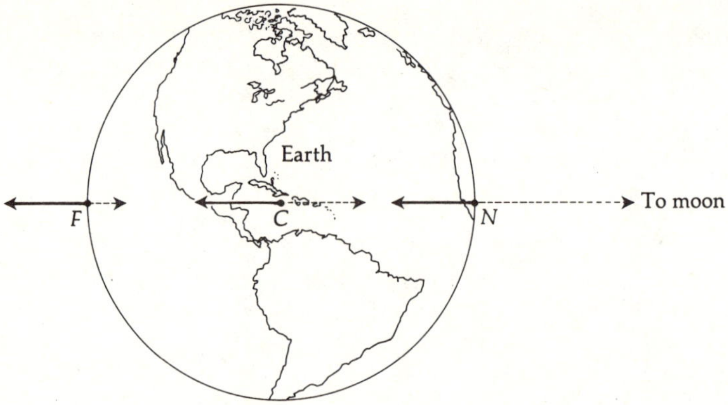

16.3 The dashed line represents the relative strength of the moon's gravitational pull on sections of the earth. The solid line represents the effect of the earth's swinging around the barycenter. The swing of the earth around the barycenter exactly balances the pull of the moon at the earth's center (C). *At the side nearest to the moon* (N), *the pull toward the moon overbalances the effect of the earth swinging around the barycenter, whereas on the other side of the earth* (F), *the latter effect is greater.*

Water, being fluid, is free to move and therefore flows out to produce bulges where the differential force is the greatest. These places experience high tide. Note that in Figure 16.3, all three solid arrows (representing the effect of the earth's swinging around the barycenter) are the same length.

The sun also produces tides, although it does not have so great an effect. However, when the sun and moon are in the same line (as at new or full moon), the tides are much greater than normal. They are then called *spring tides*. At full moon, although the sun and moon are on opposite sides of the earth, we still have extra high tides because both the sun and moon produce tides on both sides of the earth (Fig. 16.4). If the sun and moon are at right angles to each other (as seen from the earth), such as occurs at the first and last quarter of the moon, the tidal effects of the sun and moon tend to cancel each other out, and the tides are much smaller. These are called *neap tides* (see Fig. 16.5).

This constant flowing of tidal waters in and out over ocean floors creates a tidal friction which is very gradually slowing down the rotation of the earth. This effect was first brought to light when historical records of ancient eclipses were compared with modern computerized timings of these same events. It was found that the computer time and the actual solar time recorded 2000 years ago differed by about three hours.

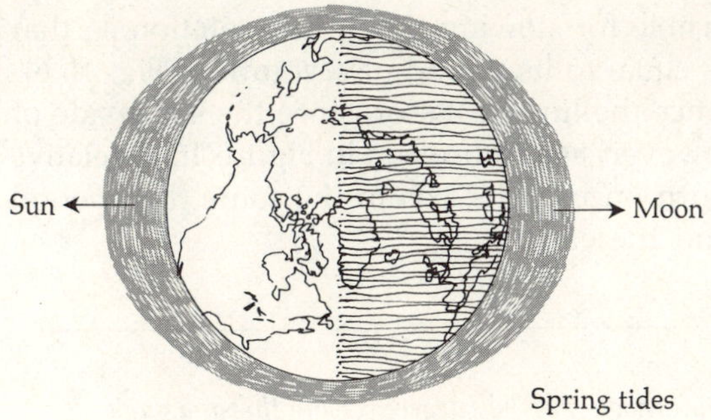

16.4 Tides are higher when the sun and moon are lined up with the earth. These are called spring tides.

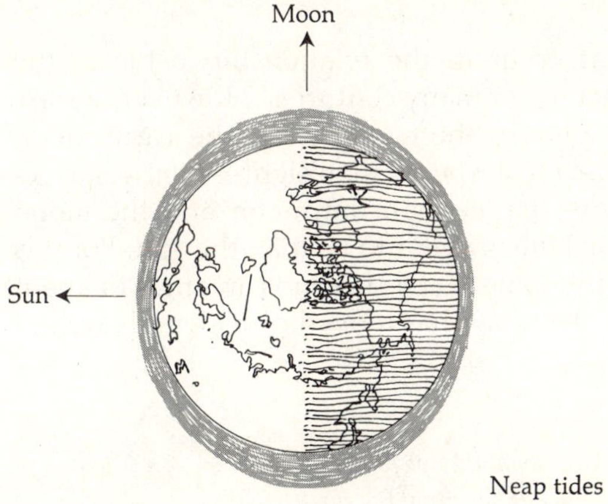

16.5 When the sun and moon are at right angles to each other (as seen from the earth), neap tides *occur. These are not as high as the spring tides.*

As the earth's rotational motion slows down, the moon's orbit increases and its orbital velocity decreases (Kepler's third law). This allows the angular momentum of the earth-moon system to be preserved. The month thus is becoming longer although at a slower rate than the day is lengthening (0.0016 seconds per century) and eventually both day and month will be 47 of our present days. When this happens the earth and moon will both keep the same sides continually toward each other, creating other effects here on earth.

Already the moon's rotation has been slowed sufficiently so that it always keeps the same side facing the earth. The earth's tidal forces are believed to have been responsible for slowing down this rotation so that the lunar period of rotation is equal to its period of revolution (Fig. 16.6). From the earth, we do not notice the lunar rotation since the same side of the moon always faces us. However, as can be seen in Figure 16.6, relative to the stars, the moon does turn once on its axis in the same time period that it takes to go once around the earth.

The full moon subtends an angle of about one-half degree. Using this as a guide, estimate the angular distance between two other celestial objects such as the pointer stars of the Big Dipper.

These minute changes that occur in the relationship between the earth and the moon will not effect us for many centuries. However, we are fortunate that at this particular time in the evolution of the earth-moon system we are able to observe a most spectacular sight—total solar eclipses. The sun is about 400 times larger than the moon and the moon happens to be at present about 400 times closer to us than the sun. For this reason, their apparent sizes are the same. Each subtends an angle of about ½° on the celestial sphere. Once the moon has moved considerably farther

16.6 *The moon rotates on its axis. If the moon did not rotate (a), we would see all sides of it. Since we see only one side from the earth, we know that the moon rotates as it revolves (b).*

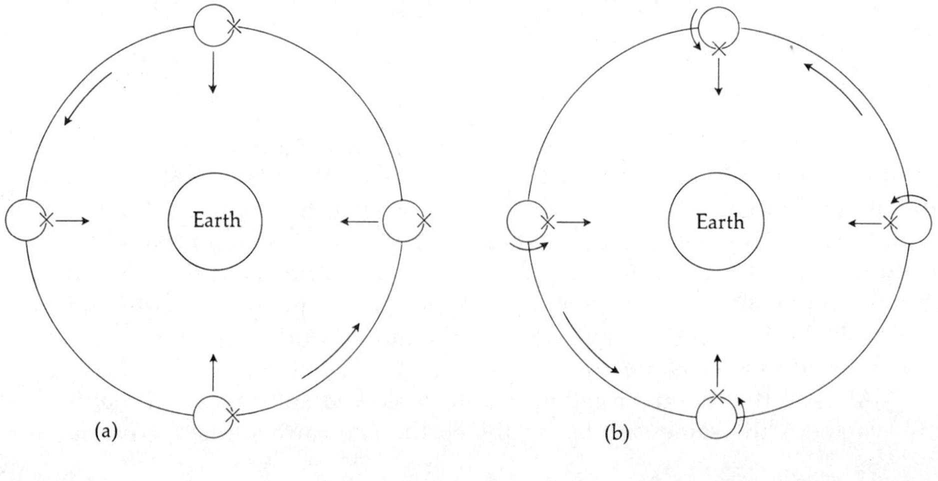

away from the earth, this relationship will no longer hold true. The moon will no longer be able to cover the entire disk of the sun during an eclipse and this spectacular sight will become a thing of the past.

Phases of the Moon

The side of the moon that we never see from the earth is called the *hidden* or *far side*. This must not be confused with the *dark side* of the moon, which is simply that hemisphere that is not illuminated at the moment by the sun. Only when we see the moon at its *full* phase is the far side the same as the dark side.

The full moon is seen when the sun and moon are at opposite sides of the earth. This means that a full moon can never be seen while the sun is high in the sky. In fact, a full moon rises at sunset (Fig. 16.7). Similarly, a crescent moon appears only when the moon is close to the sun on the celestial sphere (Fig. 16.8). Because the sun's glare is so strong, this crescent moon is visible only shortly after sunset or just before sunrise. The

---

Plot the path of the moon through the stars for one month. Indicate what phase it is at every few days. Indicate the time of each observation.

16.7 *The full moon rises at sunset. The moon is on almost the opposite side of the earth from the sun. As the earth rotates from west to east, the observer sees the sun set in the west as the almost full moon rises in the east. Dashed arrows indicate apparent motion of moon and sun due to earth's rotation.*

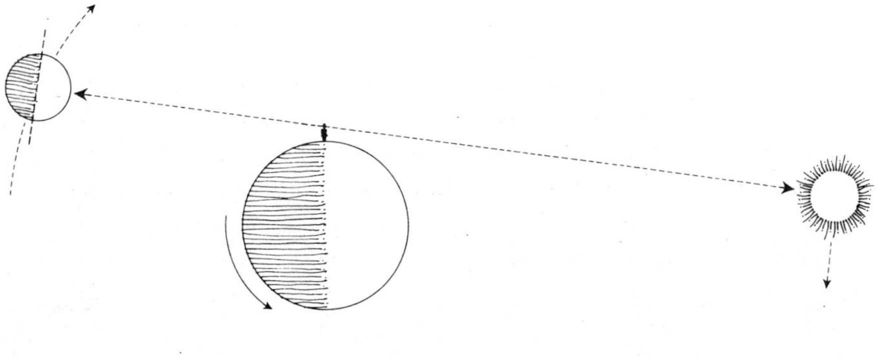

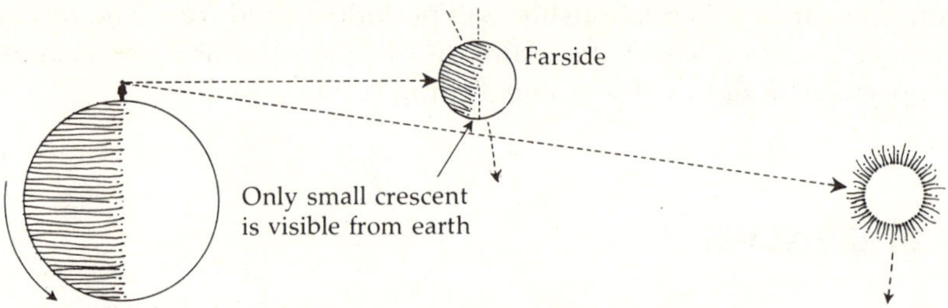

16.8 The "new" moon. When the sun and moon are on the same side of the earth, the angle between the observed position of the moon and the sun is relatively small. The observer sees the crescent moon only after sunset because the sun's glare obscures it during the daytime. Dashed arrows indicate apparent motion of moon and sun due to the earth's rotation.

greater the angular separation of the sun and moon on the celestial sphere, the more of the illuminated hemisphere we see. This is the cause of the familiar phases of the moon (Fig. 16.9).

Viewed from the moon, the earth also exhibits phases, but each phase is the exact opposite of the lunar phase as seen from the earth (Fig. 16.10). When the full moon is seen on earth, the dark night side of the earth faces the moon ("new earth").

The Motion of the Moon

Each night, with respect to the stars, the moon moves about 13° eastward on the celestial sphere. In the course of about 27⅓ days it makes an apparent path through all of the zodiacal constellations. This period of rev-

---

Note the position of the moon at about 9:00 in the evening. Where is it on the celestial sphere? What is its altitude? Its azimuth? Draw a sketch of the moon and the nearby stars. Be sure to label the north, east, south, and west positions. Two or three hours later, observe the moon again and draw a second sketch (or use the first one with a different colored pencil). How has the moon's position changed with respect to the stars? With respect to the horizon?

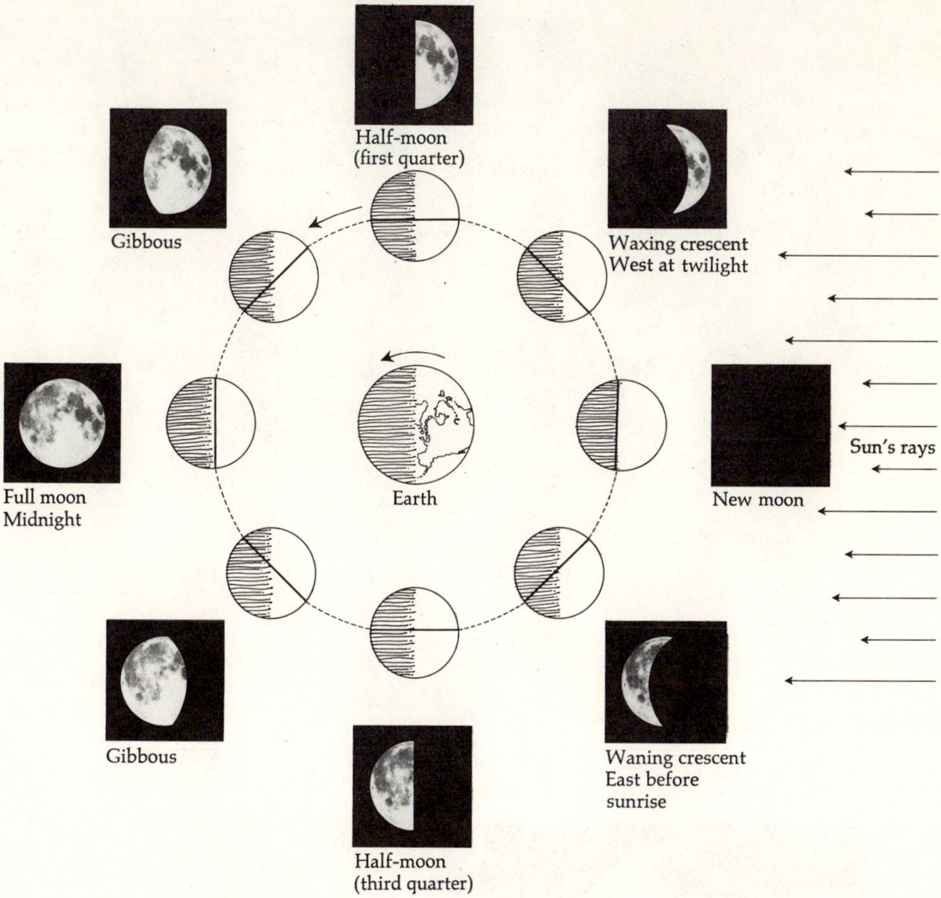

16.9 Phases of the moon. The solid lines divide the hemisphere of the moon facing the earth from the hidden or far side. Note the correspondence between the illuminated portion of the side facing the earth and the accompanying photograph of the moon's phases.

olution is called the moon's *sidereal period.* During this interval, however, the earth revolves about 1/13 (or 27°) of the way around the sun, making the sun appear to have moved eastward along the ecliptic by this amount. The moon must travel this extra distance in its orbit to come into line with the sun once again. Therefore, from full moon to full moon is 29½ days (Fig. 16.11). This is the *synodic period,* and is about two days longer than the sidereal period. We can compare these two periods with the concept of the sidereal and solar days (see Chap. 15). Our calendar month is based on the synodic rather than the sidereal period of the moon.

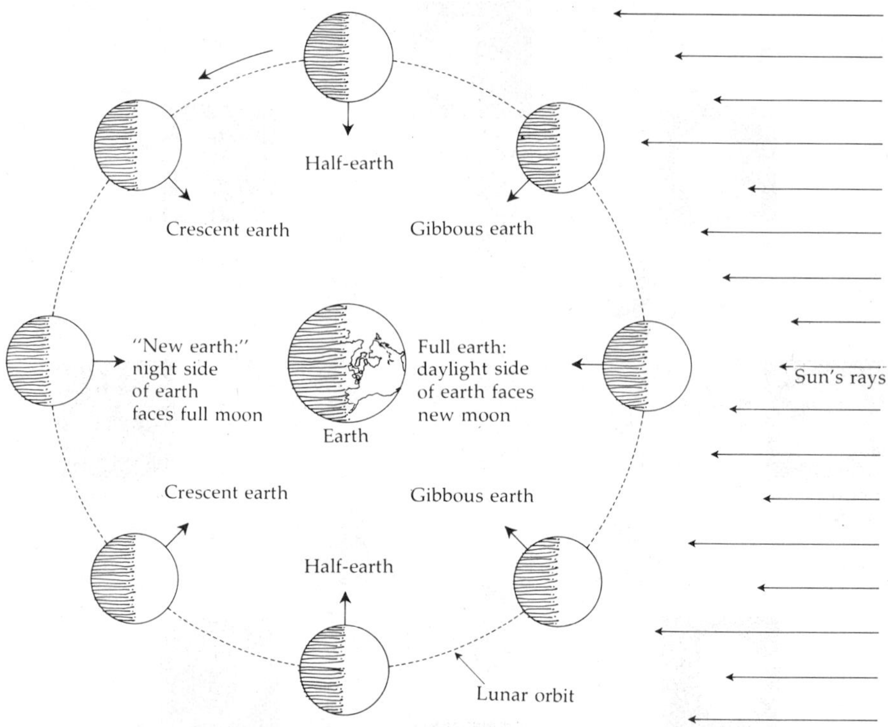

(a)

(b)

16.10 *(a) The phases of the earth as seen by an astronaut on the moon. Note that the phases of the earth are exactly opposite the phases of the moon. Compare with the lunar phases in Figure 16.9.*
(b) The crescent earth (photograph taken by Apollo 12 astronauts, NASA photograph).

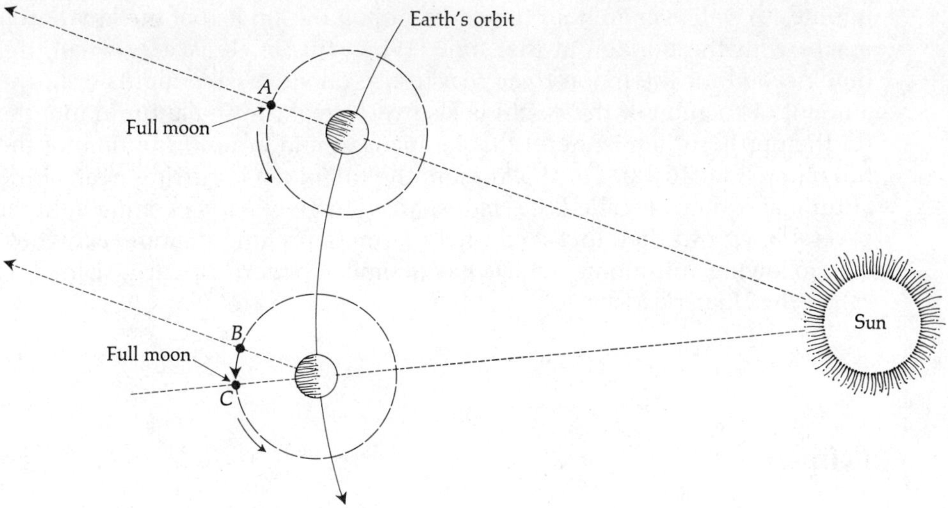

16.11 The full moon at point A *is seen on the celestial sphere against specific background stars. One sidereal month later (27 1/3 days), it is seen again against these same stars but is not yet at its full phase (point* B*). It must travel about two more days (synodic period = 29 1/2 days) to become full again (point* C*).*

The moon's 13° motion every 24 hours means that it moves about ½° per hour on the celestial sphere. Since the moon subtends ½° of arc on the celestial sphere, it moves eastward a distance equal to its own diameter every hour. The moon's diameter is about 2000 miles, which means that the moon is actually moving in its orbit about 2000 miles per hour. This *eastward* motion of the moon among the stars is easily observable by noting the position of the moon with respect to the stars early in the evening and then again several hours later. Of course, during that interval, the moon and stars will also appear to have all moved *westward* because of the earth's rotation. If the moon rises right next to a bright star, it will set somewhat after the star.

While the moon moves eastward 13° on the celestial sphere, the sun moves about one degree (a little less if it is to move 360° in 365¼ days). Therefore, with respect to the stars, the moon moves 12° eastward from the sun every 24 hours. This angular difference is the equivalent of about 50 minutes of time as reckoned by the sun—that is, the sun in its daily motion across the sky takes about 50 minutes to move 12° on the celestial sphere.

The moon, therefore, is seen to rise an average of 50 minutes later each evening. This time delay varies throughout the month from about 20 minutes to well over an hour depending upon the angle that the lunar orbit makes with the horizon at that time. When this angle is very small, the time needed for the moon to appear to rise on successive nights (i.e., the amount of rotation of the earth) is also very small. In the autumn months (in the northern hemisphere) this happens at and around the time of the full moon (Fig. 16.12). For this reason, the full moon occurring nearest the autumnal equinox is called the *Harvest Moon*. It gives early evening light for several days and therefore could help farmers prolong daytime activities. The following full moon, which has a similar pattern of early risings, is called the *Hunter's Moon*.

Eclipses

One of the most spectacular phenomena associated with the motions of the earth-moon system is an eclipse. These occur whenever the shadow of either the earth or moon falls upon the other. For this to happen, the earth, moon, and sun must all be nearly in line with each other, which is only possible when the moon is either at its new or full phase.

One might therefore wonder why there is not an eclipse at every new or full moon. The reason is that the monthly path of the moon through all the constellations of the Zodiac is not on exactly the same plane as the sun's apparent path (the ecliptic). The two paths are at a 5° angle to each other[2] (Fig. 16.13), and intersect at two points on the celestial sphere, called the *nodes* of the moon's orbit. Unless the moon and sun are at or close to one of the nodes at the time of a new or full moon, the moon will pass

---

Using a globe of the earth, a small ball, and a pocket flashlight, simulate the track of a total eclipse. This must be done in a darkened room. Show that a total eclipse always begins at sunrise somewhere on the earth and ends at sunset elsewhere on the earth.

---

[2] This closeness of the plane of the moon's orbit to that of the ecliptic is another unusual characteristic of the earth-moon system. Most of the other moons have orbits that lie close to the *equatorial* plane of their respective planet rather than close to the *orbital* plane like our moon. This factor and the relatively large diameter of the moon set the earth-moon system apart from the typical planet-satellite system.

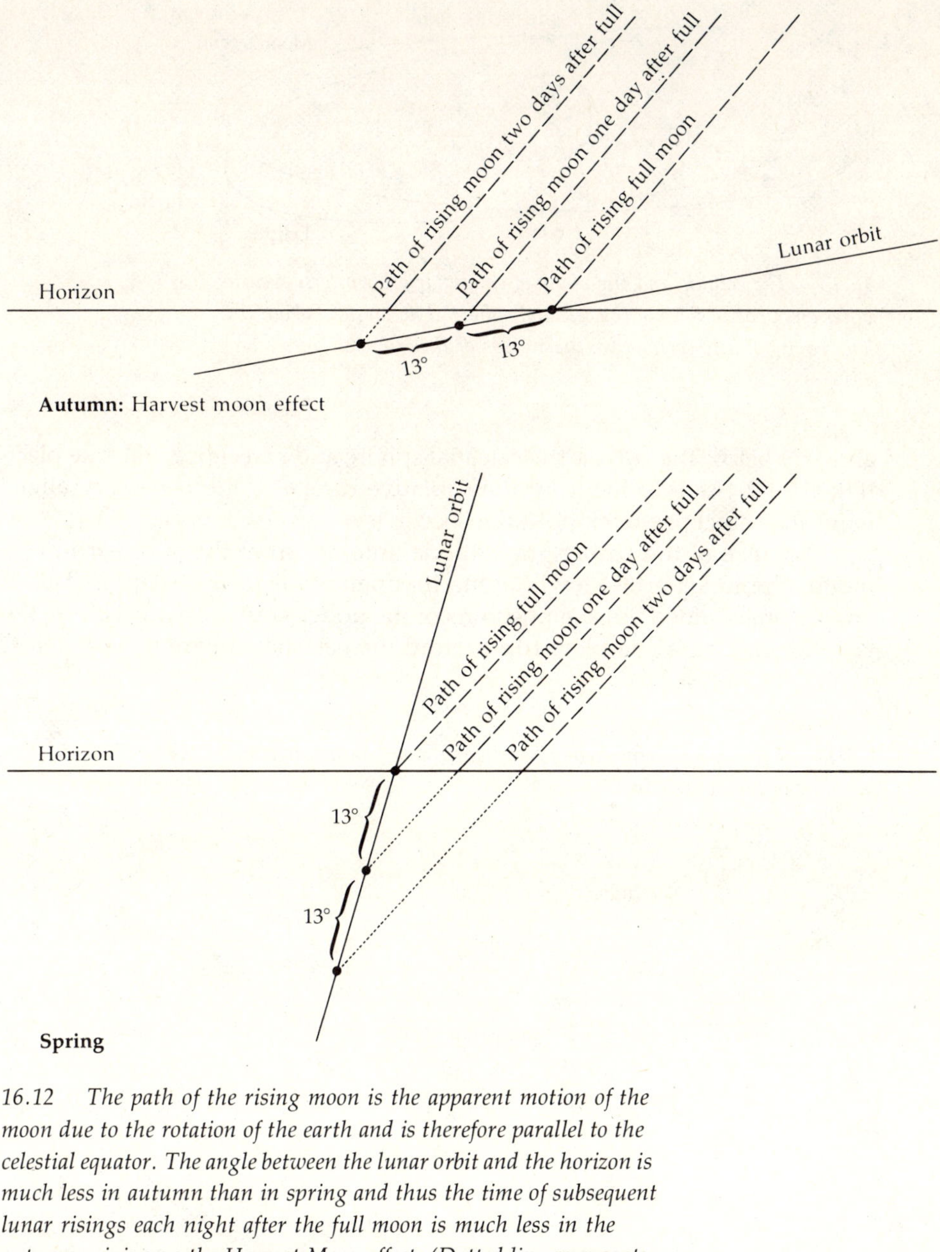

16.12 The path of the rising moon is the apparent motion of the moon due to the rotation of the earth and is therefore parallel to the celestial equator. The angle between the lunar orbit and the horizon is much less in autumn than in spring and thus the time of subsequent lunar risings each night after the full moon is much less in the autumn, giving us the Harvest Moon effect. (Dotted line represents time of delay of rising moon each night.)

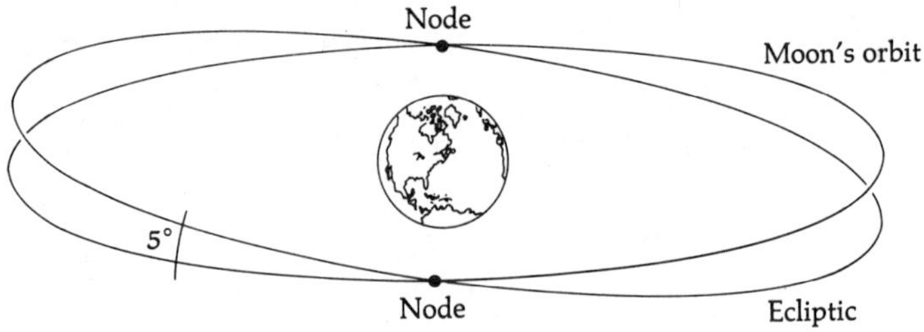

16.13 The ecliptic and the plane of the lunar orbit intersect each other on opposite sides of the celestial sphere at an angle of about 5°. The points of intersection are the nodes *of the moon's orbit.*

above or below the sun on the celestial sphere and no eclipse will take place (Fig. 16.14). As is evident by their relative rareness, the necessary alignment for an eclipse does not often occur.

Because of the strong gravitational interaction of the sun, earth, and moon, the lunar orbit slowly shifts its orientation to the ecliptic. The 5° angle is maintained but the positions of the nodes shift westward along the ecliptic, making a complete trip around the celestial sphere in 18.6 years.

16.14 Eclipses occur only when the moon is along, or nearly along, the line of the nodes (nn').

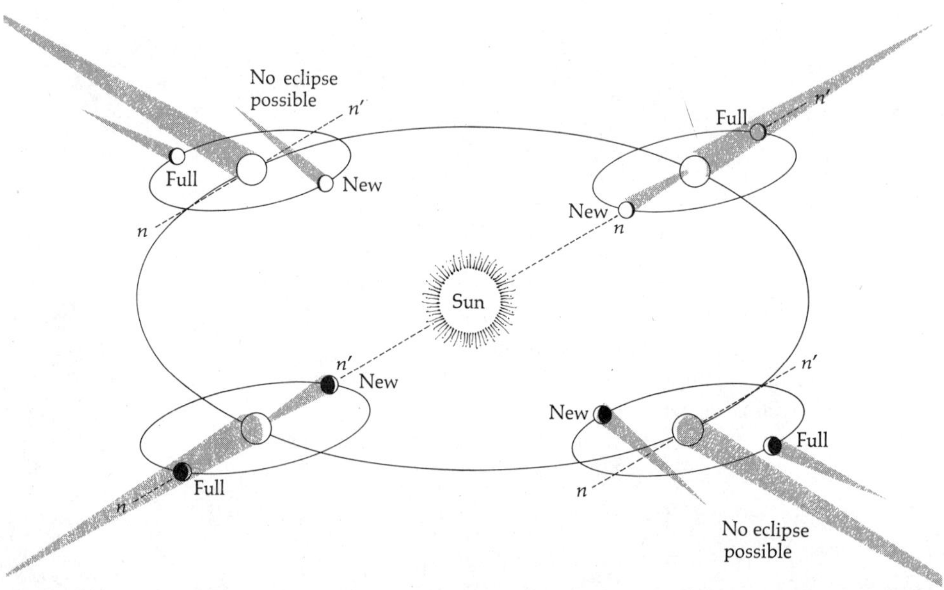

This constant changing of the position of the nodes is called the *regression of the nodes*. Because an eclipse can only occur when the moon and sun are close to a node, knowledge of the positions of the nodes at any given time is all-important in eclipse prediction.

Solar Eclipses

Of all eclipses, a total solar eclipse is by far the most spectacular. This occurs only in the very narrow band on the earth swept by the duncecap-shaped shadow of the moon (Fig. 16.15). For many hundreds of miles on either side of this totality, one can see partial phases of the eclipse, but if even a sliver of the sun's surface remains uncovered, the spectacle of a total eclipse cannot be observed. Although the period of complete totality lasts

16.15 *Total solar eclipse. (a) Only that area on earth covered by the darker shadow* (umbra) *experiences total eclipse. The lighter shadowed area* (penumbra) *has a partial eclipse. (b) For a total solar eclipse to occur, the moon's shadow and the earth must be within a certain distance of one of the lunar nodes.*

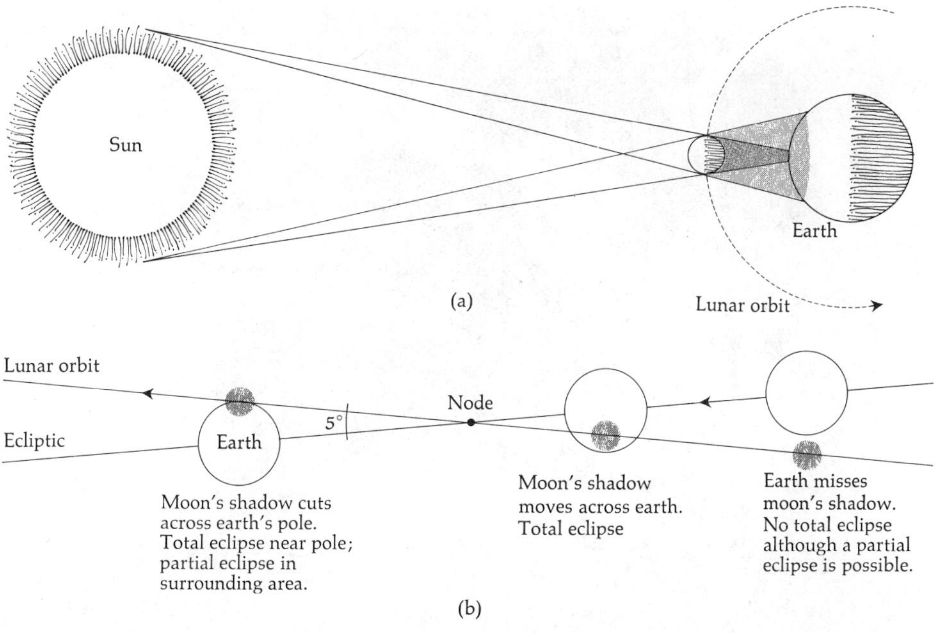

only a few minutes, much information about the sun has been obtained during these times (see Chap. 14).

Since only a very narrow path on the earth actually experiences the totality of any one solar eclipse, astronomers travel great distances to study it (Fig. 16.16). As totality approaches, the landscape takes on a strange appearance. The birds come home to roost and the cows start to return from pasture. The temperature drops and the wind picks up. Just before totality is reached, the last rays of the sun are seen shining between the mountain peaks on the rim of the moon, creating the *diamond ring effect* (see color plate J). Then the sky darkens and the solar corona flashes out (Fig. 16.17). Totality is reached and the brighter stars make their appearance. Totality lasts for several minutes, after which the first rays of the sun reappear and daylight returns.

Throughout the course of history, a solar eclipse has been an awesome spectacle, particularly in the days when its cause was not understood and it could not be predicted (Fig. 16.18). At such times, many thought that the end of the world was at hand. The early Chinese thought that a

16.16 *An astronomer prepares his instruments in order to observe the June, 1954 solar eclipse. To observe the total eclipse, this American expedition had to travel to Iran (photograph from Dearborn Observatory, Northwestern University).*

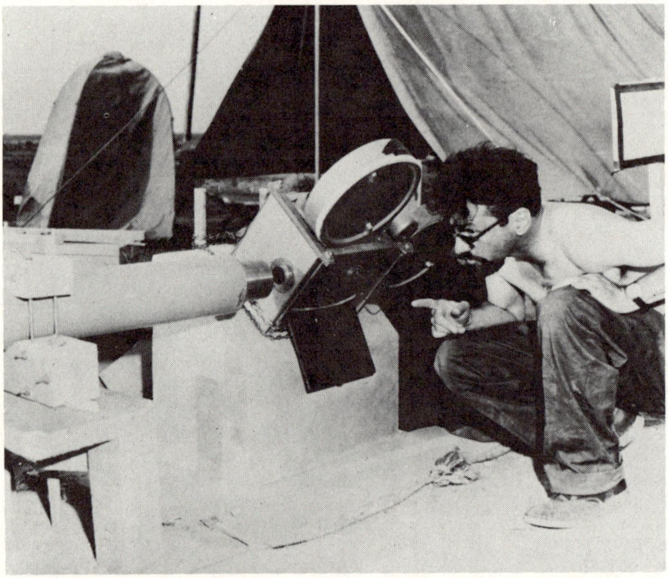

16.17 The totality *of a solar eclipse, during which only the sun's corona is visible (photograph from Dearborn Observatory, Northwestern University).*

dragon was swallowing the sun, and they made a great commotion in an effort to make the dragon disgorge the sun. Such noisemaking during eclipses was common to many early cultures and extends even to the present day among some tribes. Naturally, their drumbeating and cymbal clanging was always a complete success!

Viewed from any one spot along the narrow eclipse path, the total eclipse lasts but a few moments. However, after the solar eclipse has ended for one observer, it may just be starting for another, since the shadow sweeps eastward across the face of the earth (Fig. 16.19). Therefore, the entire total eclipse with respect to the earth may last many hours. Supersonic planes have been able to pace the sweep of the shadow and therefore extend the eclipse duration for a much longer time than that experienced by the ground observer.

Because the lunar orbit is an ellipse, rather than a circle, around the earth, there are times when the moon is too far from the earth to com-

16.18 In 585 B.C., at the height of a battle between the Medes and the Lydians, a solar eclipse occurred. This is said to have ended the fighting and brought about an armistice.

16.19 Nautical Almanac map of the total solar eclipse of March 7, 1970. This graphical representation of the area in which the eclipse was visible shows clearly the narrow path in which totality was experienced. Greenwich Mean Time is used. (From the Nautical Almanac Office, U.S. Naval Observatory.)

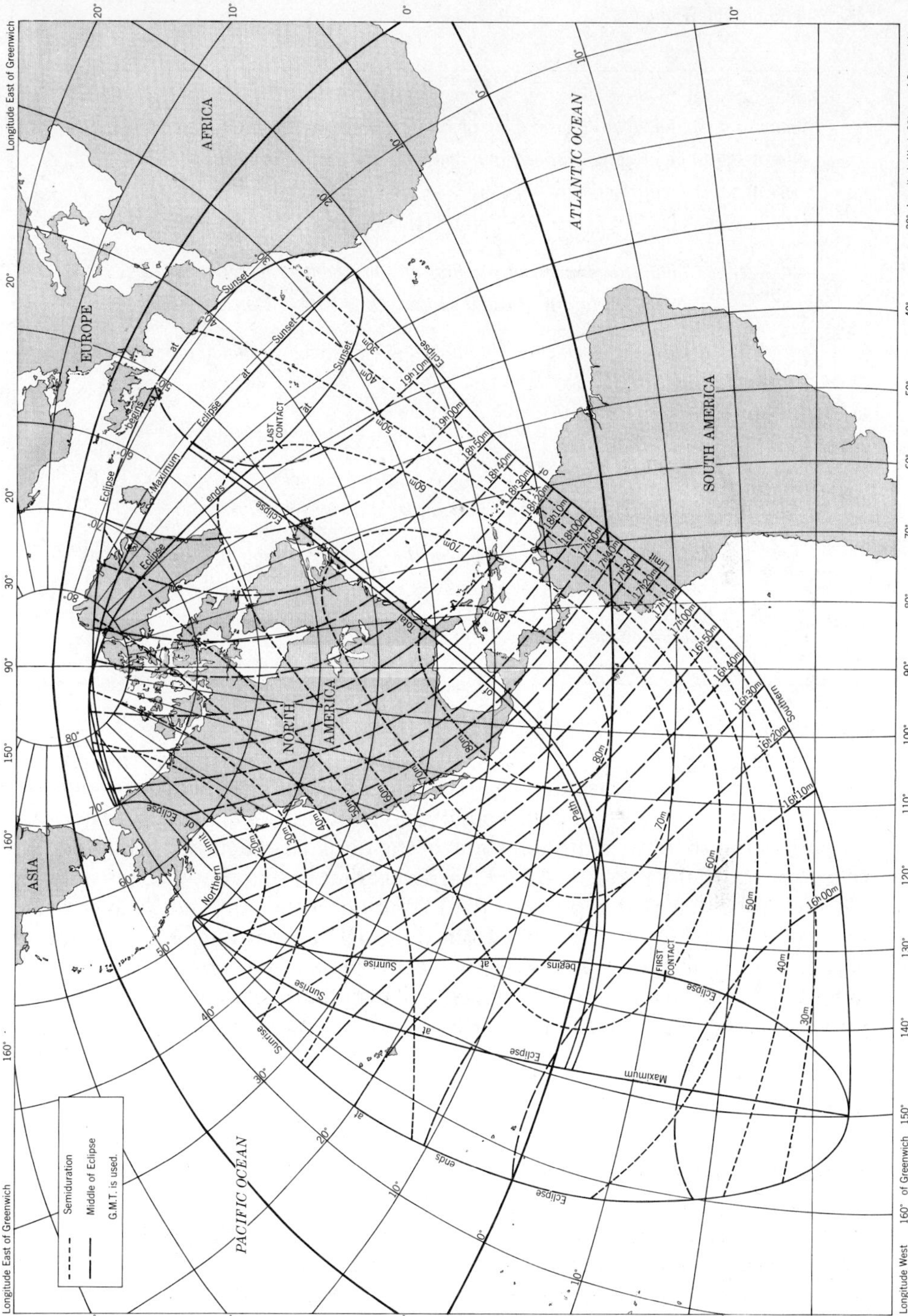
Longitude East of Greenwich
Longitude West 160° of Greenwich
Semiduration
Middle of Eclipse
G.M.T. is used.
PACIFIC OCEAN
ATLANTIC OCEAN
NORTH AMERICA
SOUTH AMERICA
EUROPE
AFRICA
ASIA
FIRST CONTACT
LAST CONTACT
Northern Limit of Eclipse
Southern Limit
Path of Total Eclipse
Eclipse begins at Sunrise
Eclipse ends at Sunrise
Maximum Eclipse at Sunrise
Eclipse begins at Sunset
Eclipse ends at Sunset
Maximum Eclipse at Sunset

Viewing a partial solar eclipse: Construct a viewing box through which the solar disk can be projected, or set up a telescope and focus the sunlight onto a piece of paper as in the Chapter 14 exercise. DO NOT LOOK AT THE SUN DIRECTLY AT ANY TIME DURING THE ECLIPSE!

1. *Time the moment of the first and last contact between the sun and moon on the celestial sphere. How long did the eclipse take?*
2. *Does the moon enter the sun's disk on the right-hand or left-hand side as we view it?*
3. *Estimate the maximum percentage of the solar surface that is covered.*
4. *Measure the drop in temperature during the eclipse.*
5. *Observe the shadows cast on the ground through the leaves of the trees during the eclipse; they will be crescent-shaped.*
6. *Find out what the path of totality was.*

---

pletely cover the sun's disk during an eclipse. We then experience an *annular eclipse* of the sun (Fig. 16.20). Totality never occurs, even along the center path of the shadow, although some of the unusual effects of a total solar eclipse are partially present. As we learned earlier in this chapter, it is only a coincidence that at this point in the evolution of the sun and its system the moon is at the right distance from the earth so that total eclipses can occur at all. The moon is slowly increasing its distance from the earth (about one centimeter per year), which will eventually place it too far away for its shadow to ever reach the earth.

Lunar Eclipses

A lunar eclipse occurs when the full moon enters the shadow of the earth (Fig. 16.21). For any one observer, this type of eclipse lasts longer in time than the solar eclipse and it can be seen by a larger portion of the earth. Since the shadow is on the moon, wherever the moon would normally be seen, the eclipse is visible. It is therefore visible simultaneously for all ob-

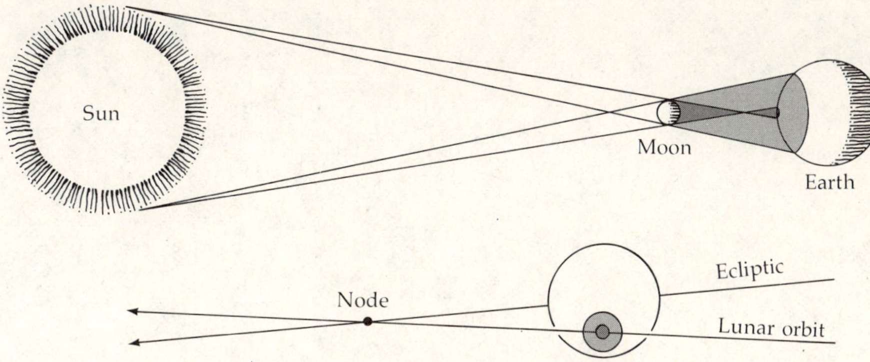

16.20 Schematic drawing of an annular eclipse. The moon is too far from the earth for the umbra part of its shadow to reach the earth. However, a small path on the earth's surface does experience an annular eclipse during which the moon is seen silhouetted against the sun with a ring of sunlight encircling the moon.

Viewing a lunar eclipse: Can be viewed directly, with binoculars, or a telescope.

It is difficult to detect the exact moment when the moon enters the penumbra, although eventually a darkening of the eastern limb will be seen. However, the entry into the umbra can be seen with the unaided eye and timed.

1. *How long is it from the moment of entry into the umbra until the moment of totality? From this measurement, how fast is the moon moving?*
2. *How long is totality?*
3. *How does the moon appear during totality?*
4. *Why does the moon not disappear completely when it is totally in the earth's shadow?*
5. *On what part of the earth is it possible to view the total eclipse?*
6. *Lunar eclipses were given as proof of the spherical shape of the earth. Why would only one such eclipse* not *be enough proof?*

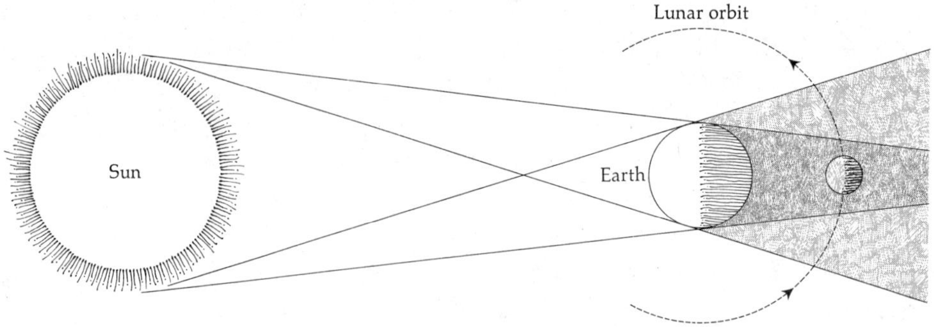

(a)

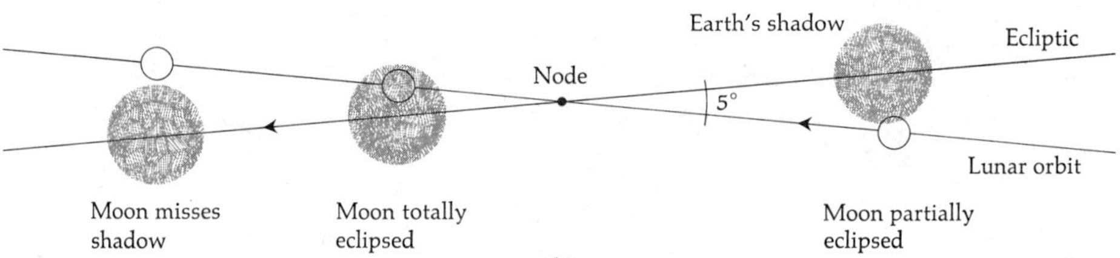

(b)

16.21 Total lunar eclipse. (a) Totality occurs when the moon is completely within the darkest part of the earth's shadow (the umbra). Before and after totality, the moon moves through the penumbra, causing the moon to be slightly darkened. (b) For a total lunar eclipse to occur, the moon and the earth's shadows must be within a certain distance of one of the lunar nodes.

servers on the hemisphere facing the moon. The observers can watch the progress of the moon as it passes through the earth's shadow, and because the shadow is larger in diameter than the moon itself, totality can last as long as one hour 40 minutes.

The phenomena of a lunar eclipse are very much less spectacular than solar eclipses and are of relatively little scientific value. Even when the moon is fully in the earth's shadow, the moon does not entirely disappear. Light from the sun passing through the earth's atmosphere is refracted and dimly illuminates the eclipsed moon (Fig. 16.22). It appears as a ruddy-colored ball, because the longer, red wavelengths of light pass through the earth's atmosphere more easily than the shorter, blue ones. (This is also why the rising or setting sun or moon often appears red near the horizon.) An observer on the moon would see the sun eclipsed by the earth while a total lunar eclipse was in progress for an observer here on earth.

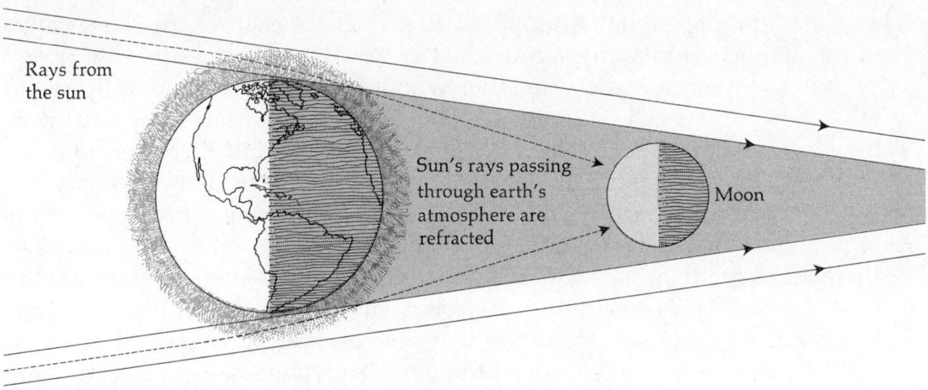

16.22 During a total lunar eclipse, the moon does not become completely dark. Some of the sun's rays are refracted as they pass through the earth's atmosphere, illuminating the moon.

The Saros Series

Over two thousand years ago, the ancient Babylonians noticed from long tabulations of eclipses that they occurred in a regular pattern. Their understanding of this pattern made it possible for them to predict future eclipses with reasonable accuracy. They found that any given eclipse recurs under nearly identical circumstances (duration, type, etc.) in a little more than 18 years. This was called the *Saros interval* and is related to the period of the moon's phases (synodic month), the period between successive returns of the moon to its nodes (nodical month), and the period between successive returns of the moon to the perigee (anomalistic month). It so happens that 223 synodic months, 242 nodical months, and 239 nomalistic months are all nearly 6585 days (Fig. 16.23). Consequently, 6585 days after a given eclipse, the sun, the earth, and the moon are almost exactly in the same geometrical relationship and an eclipse of the same character will occur.

Every eclipse, lunar or solar, belongs to a particular Saros series. Each series lasts many centuries and there are many Saros series current at any one time. The first solar eclipse of a new Saros series begins close to one of the terrestrial poles. About 18 years later, the second eclipse of that series occurs, a little farther from that pole, and is visible about ⅓ of the way

16.23 The Saros Cycle:

223 synodic months	= 6585.321 days
242 nodical months	= 6585.357 days
239 anomalistic months	= 6585.538 days

around the earth to the west. This is because the Saros cycle is really 6585 and *one-third* days long. After about 70 successive eclipses, the last eclipse of the series occurs many centuries later very close to the opposite pole.

In any one calendar year, there can be as many as seven eclipses (in which case there will be five solar and two lunar or four solar and three lunar). There must be at least two eclipses each year; if there are no more than two, they will both be solar. However, because the path of totality of a solar eclipse is so narrow, the chance of one occurring where you live is only about once in more than 300 years (Fig. 16.24). Therefore, although there are actually more solar than lunar eclipses, the lunar eclipses are the most frequently seen by most observers, since an eclipse of the moon can be seen from wherever the moon is visible on that night.

There is now evidence that the early Babylonians were not the only peoples who recognized the rhythm of the Saros series and were able to calculate future eclipses as well as other calendar events. Stonehenge, the ancient pillared temple in southern England, is now thought to have been an ancient observatory (Fig. 1.2). Its builders are unknown but they obviously had the astronomical sophistication to set up an elementary "computer" for keeping track of specific days of the year and for predicting eclipses.

The Egyptian pyramids and the Mayan temples are believed to be other examples of early astronomical "calendars". Recently scientists have found that the nomadic Plain Indians of North America also had a similar, if much more primitive structure—the Big Horn Medicine Wheel in northern Wyoming. Although not all of these "calendars" were used for eclipse prediction, they all were designed to indicate the summer solstice or the vernal equinox dates.

Date	*Where visible*
Feb. 26, 1979	Northwest U.S., Canada
Feb. 16, 1980	Central Africa, India
July 31, 1981	Siberia
June 11, 1983	Indonesia
Nov. 22, 1984	Indonesia, South America
March 29, 1987	Central Africa
March 18, 1988	Philippines, Indonesia
July 22, 1990	Finland, Arctic Regions
July 11, 1991	Hawaii, Central America, Brazil
June 30, 1992	South Atlantic
Nov. 3, 1994	South America
Oct. 24, 1995	South Asia
March 9, 1997	Siberia, Arctic
Feb. 26, 1998	Central America
Aug. 11, 1999	Central Europe, Central Asia

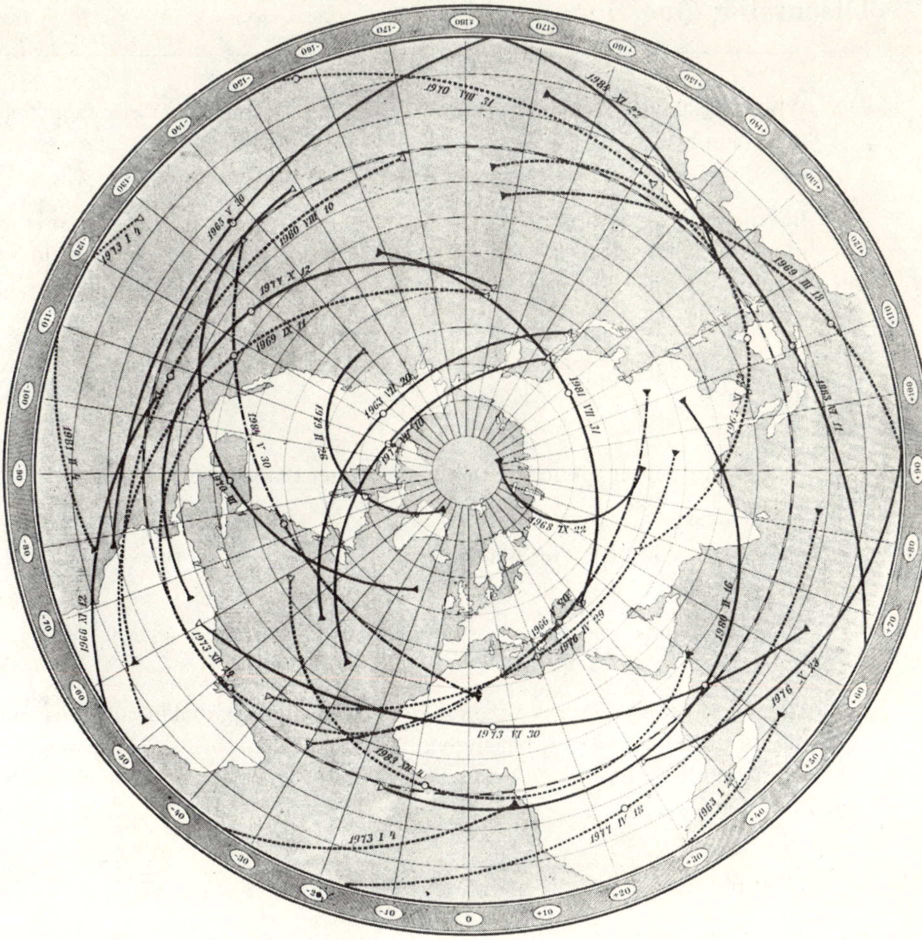

——— Total
.... Annular
._._._. Annular total

△ Begins
○ Noon
▲ Ends

16.24 *Diagram of the paths of all solar eclipses from 1963 to 1984 (photograph from Yerkes Observatory). See table on opposite page for some important future total solar eclipses.*

Discussion Questions

1. When the sun is rising and the moon is in its last (or third) quarter, where will the moon be seen in the sky?

2. In a diagram, show why neither the crescent nor full moon can be seen on the meridian at sunset. In what phase would the moon have to be in order to be seen on the meridian at sunset?

3. When a full moon is seen from the earth, what phase of the earth is seen from the moon?

4. Why is the synodic period of the moon longer than the sidereal period?

5. What causes tides? Under what conditions do the highest tides occur?

6. What conditions are necessary in order to have a lunar or solar eclipse?

7. There was a total solar eclipse on July 10, 1972, and a partial lunar eclipse on July 26, 1972. Make a diagram showing where the sun, earth, and moon were in relation to each other on these dates. What was the phase of the moon on each date?

8. What phase of the moon is seen when it is:

    a. Setting at 7:00 p.m. in the wintertime?
    b. On the meridian at midnight?
    c. Rising at midnight?

    Draw diagrams to illustrate these situations.

9. The moon's orbit and the ecliptic intersect at an angle of about 5°. What is the greatest declination that the moon can have?

17 *The Earth's Natural Satellite—The Moon*

The moon is unquestionably the most frequently viewed object in the sky. It is usually the first celestial object looked at through a newly-bought telescope and, unlike all but the very brightest stars, it is visible even to large metropolitan city dwellers (Fig. 17.1).

As is true with all celestial objects, for well over 99% of mankind's existence, the moon has been seen only with the naked eye. Yet because it is so conspicuous in the sky, this unaided view has had one of the greatest influences on philosophy and literature. References to the moon are found in poetry and prose, and it has variously been referred to as Cynthia, Selene, Diana, or Luna. The word "lunacy" comes from the supposed effects of the full moon on the behavior of man. The study of the moon's "geography" is called *selenography*. And what child has not seen the familiar "man in the moon" looking down upon him.[1]

Going to the moon has long been a dream of mankind, but throughout most history it was considered (at least scientifically) an impossible goal. In literature, there have been many imaginary journeys, among the most famous of which is Jules Verne's *A Journey to the Moon*. Johannes Kepler in his *Somium* also wrote of such a hypothetical journey. Because his story was based in part on the scientific principles of that day, it is probably one of the earliest attempts at science fiction. Certainly, man's first real step upon the surface of the moon was a major turning point in history and will always be acclaimed in the annals of scientific adventure.

The Distance to the Moon

Without a precise knowledge of the varying distance of the moon from the earth, pinpoint landings on the moon would not have been possible. The

[1] When the moon is seen through an astronomical telescope, the optics of the telescope cause the image to appear upside-down. The "man in the moon" is then not so easily recognized.

17.1 The full moon as seen through a telescope on the earth. Through an astronomical telescope, the moon (and everything else) appears inverted (i.e., "south" is at the top). However, by convention, the moon is usually shown as it appears to the naked eye, with the "north" at the top. In this photograph, this convention has been followed (photograph from the Hale Observatories.)

classical method for determining this distance is essentially the same technique as used by surveyors. An elementary form of this method was used as early as ancient Greek times (Fig. 17.2). The famous Greek astronomer Hipparchus calculated that the distance to the moon was 29½ times the earth's diameter. This result was remarkably close to the true value (about 30 times the diameter of the earth).

Although other more accurate measurements were made in the cen-

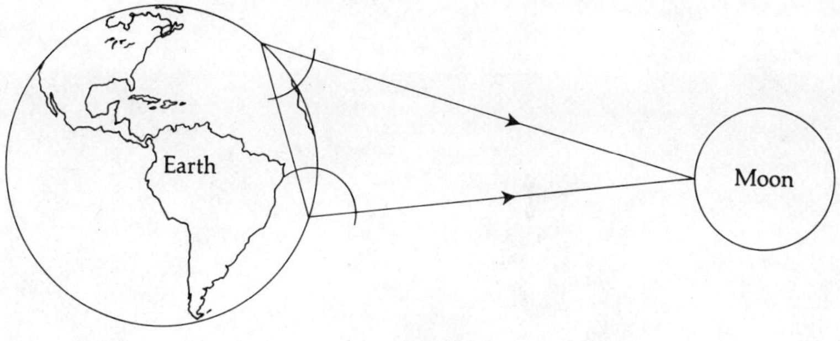

17.2 By sighting the moon from two different places on the earth, a triangle is formed. Using simple geometry, we can then calculate the distance to the moon.

turies that followed, it was not until the development of radar in the middle of the twentieth century that an entirely new and very precise method of measuring the moon's distance was provided. Project Diana in 1946 first used this new technique. Radar pulses, bounced off the moon, took roughly 2½ seconds to make the round trip. Since radar waves travel at the speed of light (186,300 miles per second), the distance to the moon could be immediately determined.

However, the most precise method for measuring the moon's distance was made possible by the Apollo astronauts. They placed special mirrors on the moon's surface so that laser beams from the earth could be reflected from them. Laser light stays in a very narrow pencil-like beam instead of spreading out like radar or ordinary light, and can therefore be pinpointed to specific locations on the moon or earth. Since the beam is highly concentrated, relatively little power is needed. So precise is this method that the distance between a point on the earth and a point on the moon can be determined to within inches.

From the center of the earth to the center of the moon, the average distance is 384,393 kilometers (238,857 miles). The elliptical orbit of the moon causes the distance to vary more than 6% either way. Because objects more distant will appear smaller, this change in the moon's distance affects its apparent size, as shown in Figure 17.3. On the right side is the moon when it is closest to the earth (perigee), while on the left side is the moon when it is farthest (apogee).

The Moon Illusion

The apparent difference in the size of the moon at perigee and apogee is not noticeable unless the two "half-moons" are compared as in Figure 17.3.

17.3 *Because the moon's distance from us varies, its apparent size also varies. Two photographs placed side by side show how great this effect is (courtesy of Willard A. MacCalla).*

Yet everyone is well aware of the apparently larger size of the full moon when it is rising. It seems considerably larger than when it is high in the sky, especially at the time of the *Harvest Moon*. The full and nearly full moon rises at about the same time on successive evenings and its unusually large size is often commented upon.

Until fairly recently, the explanation generally given for this apparition was that we were able to compare the moon to familiar objects when it was near the horizon, but not when it was overhead. Yet this same illusion is experienced on the open sea, where there are no such objects for comparison. Furthermore, this illusion applies to the constellations as well. The Big Dipper, when seen near the horizon, also appears noticeably larger. Yet if we take a series of photographs of the rising moon (Fig. 17.4) there is no evident change in size. The camera "sees" the moon as a constant size whether on the horizon or overhead.

There is much controversy and puzzlement over the cause of this optical illusion. One possibility is that because there is no depth perception

17.4 The moon rising, a series of exposures taken every 2½ minutes with a fixed camera (courtesy of Roger Hosfeld).

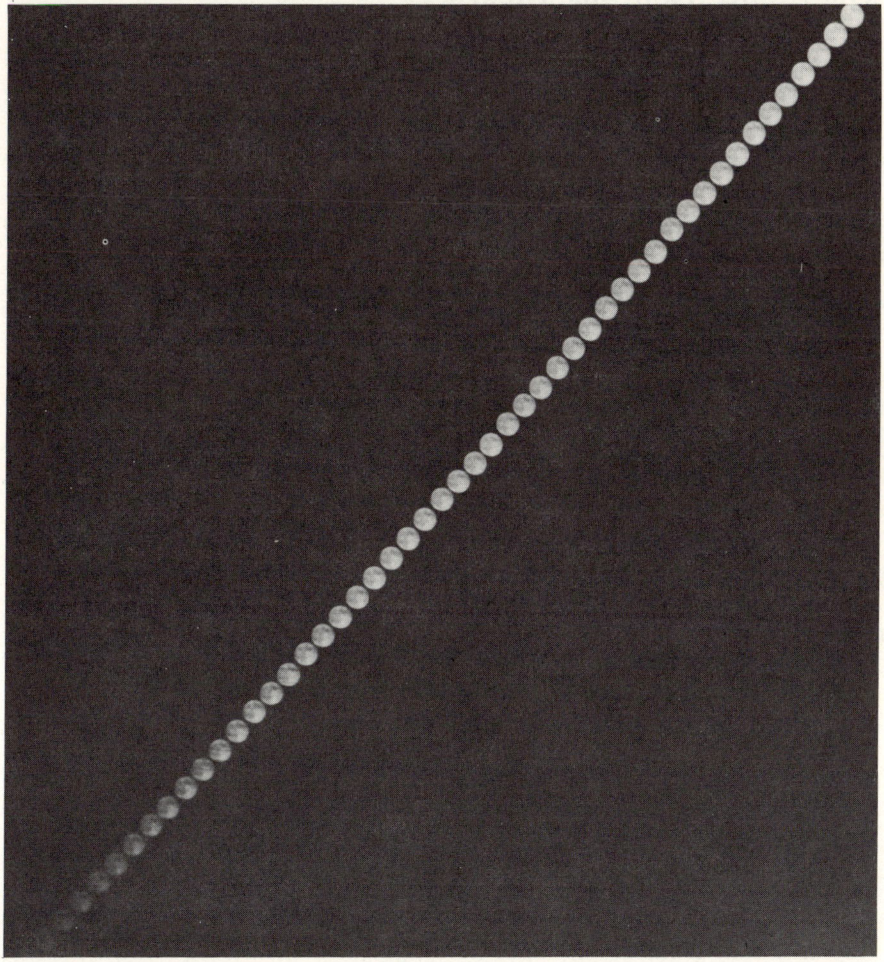

when we look at the sky, we have no concept of where the celestial objects really are in space. We tend to think of the celestial sphere as an inverted hemisphere whose rim describes our horizon. Actually, however, that rim is infinitely far away, a fact that our senses cannot perceive. Our perception of this part of the sky is distorted as is illustrated in Figure 17.5. Objects overhead, which are viewed directly and not on an angle, are not seen in this way. Note also that the individual stars are not enlarged at the horizon because, being so far away they appear as point sources. Any enlargement would be too minute to notice.

The Landscape of the Moon

When we look at the moon without a telescope, its light and dark features are easily seen. Through the telescope, however, these features become resolved into much greater detail. The large dark areas are the *maria*, so-called because they were originally thought to be seas.[2] Actually they are relatively smooth, low-lying flat regions with low reflectivity. The brighter areas are the highlands, the rough crater-covered mountains, and hills (Fig. 17.6).

Also seen clearly through a telescope are the *rilles*, which are canyon-like crevasses and valleys. These can be straight or meandering, the straight ones commonly over five kilometers wide. Hadley Rille, a good

17.5 The moon illusion. We think of the celestial sphere as an inverted hemisphere with its rim at our horizon on the earth. Actually, this "rim" is infinitely far away. We therefore perceive a distorted moon, a moon much larger at the horizon than overhead.

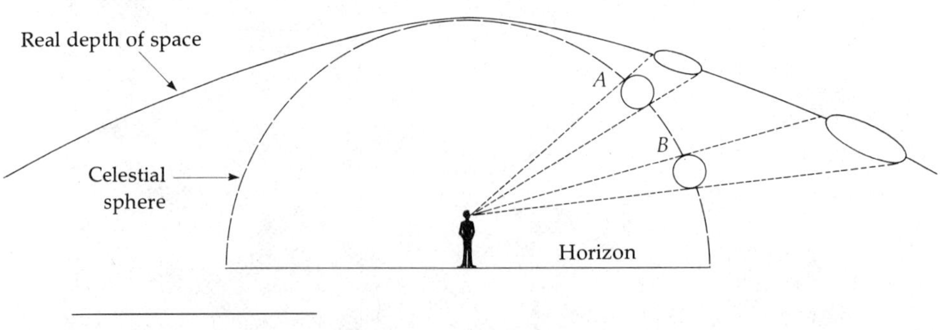

[2] *Maria* is the Latin word for "seas". The singular form is *mare*.

17.6 Major features of the moon (at full phase). The bright rays from the crater Tycho prevent further identification of features in that area when the moon is full. These rays are only surface deposits and cast no shadows. They are only observable at or near full moon. At other phases, dark shadows outline the numerous craters and other features.

example of the sinuous type (Fig. 17.7) averages only 1.2 kilometers in width.

When the full or nearly full moon is viewed through a telescope, bright rays emanating from several large craters become evident. These are streaks of rubble that have been ejected outward at the time the crater was formed by meteorite impact. They are so low in height that they do not cast shadows and therefore are not seen at other phases of the moon. In Figure 17.6, these rays are particularly obvious radiating out from the large crater Tycho. The Apollo astronauts confirmed that they are indeed long streaks of dust and rubble.

Observe the full moon both with the naked eye and with a telescope or pair of binoculars. Draw a map of its surface as seen under both conditions. Label the prominent features.

17.7 *A nearly vertical view of the Hadley-Apennine area, photographed from the Apollo 15 Command/Service Module while in lunar orbit, July 30, 1971. Hadley Rille meanders through the center of the photograph. The Apennine Mountains are on the right. The Apollo 15 LM landing site is on the east (right-hand) side of the rille's "chicken beak" in the center of the picture. The smooth, dark area on the left is the Marsh of Decay (NASA photograph).*

The very excellent photographs taken with the aid of telescopes here on earth have been superseded by the many pictures taken by the Ranger and Surveyor lunar orbiter spacecraft and during the Apollo moon flights (Fig. 17.8). First, the unmanned Ranger and Surveyor used television

17.8 Schroeter's Valley and the crater Aristarchus, which is the bright-appearing large crater on the left. The other large crater is Herodotus. Aristarchus is approximately 35 kilometers (about 21.75 miles) in diameter. Schroeter's Valley starts between these two craters and cuts a winding path through part of the Ocean of Storms. Photographed from the Apollo 15 CSM while in lunar orbit (NASA photograph).

scanning techniques to transmit pictures back to earth. Most of the first highly detailed mapping of the lunar surface was done in this way. Later, when manned craft orbited the moon, photographs of much higher quality than could be obtained by television transmission were brought back (Fig. 17.9).

Extremely detailed photographs of the vicinity of the landing sites were taken by the Apollo astronauts (Fig. 17.10). These photographs showed that the smaller features are similar to the larger ones. Tiny craters are found all over, even in the smooth-appearing maria. They have essentially the same form as their larger counterparts (Fig. 17.11). Miniature rilles are also found, and the same brown-gray barrenness pervades the close-up scene as it does the more distant view.

17.9 The large crater Kunowsky. Note the many craters on the floor of the larger crater. Photographed by the Apollo 12 crew (photograph from the National Space Science Data Center).

17.10 A small section of the lunar surface photographed by the Apollo 11 crew. Note the faint outlines of large craters in the dark maria as well as the many smaller craters in this area (photograph from the National Space Science Data Center).

Covering the entire surface of the moon is a fine, dustlike sand. It varies in depth from a fraction of an inch to several inches. Although it never was deep enough to be a hazard, the astronauts found that it clung to their space suits as they moved about on the lunar surface (Fig. 17.12).

With the use of a telescope or good pair of binoculars, identify and locate five major features of the lunar surface. What are the features that make up the familiar "man in the moon" face?

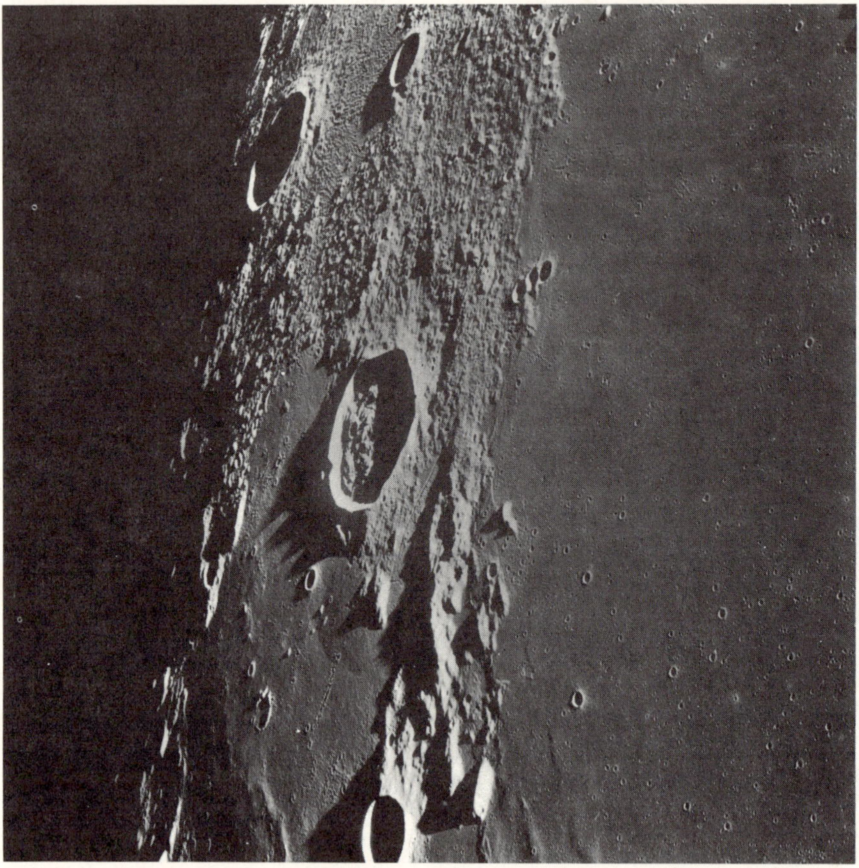

17.11 The lunar terrain photographed by Apollo 12 astronauts. Even the seemingly smooth maria are pockmarked with many tiny craters. The two large craters are Kepler and Encke (photograph from the National Science Data Center).

During the Apollo missions, the moon's far side was also extensively photographed and mapped from the orbiting capsules (Fig. 17.13). The pictures of the side of the moon which we can never see from the earth showed a multi-cratered terrain much like the near side. It is as barren and colorless. However, there are no large, smooth areas such as the maria of the near side. Nor are there as many large basins which are believed to have been caused by tremendous meteoritic impacts in the early formation of the moon. These asymetrical aspects of the lunar surface have yet to be fully explained.

17.12 An Apollo 12 astronaut examines the Surveyor III, an unmanned space craft which had been sent to the moon in 1967, nearly three years before. Note the lunar dust clinging to the white suit. The Apollo 12 LM is in the background (photograph from the National Space Science Data Center).

Surface Features

The maria, which are found only on the near side of the moon, are the largest single features of the lunar surface. Their contrasting darkness against the lighter surface features creates the "man in the moon" image. There are 14 maria or lunar "seas", the largest of which is *Mare Imbrium* (the Sea of Showers) which has a diameter of about 1100 kilometers (660

17.13 A small section of the far side of the moon as photographed by the Apollo 11 astronauts (photograph from the National Space Science Data Center).

miles). The other maria have equally romantic names (Tranquillity, Serenity, etc.), all of which are in Latin, which was the scientific language of the seventeenth and eighteenth centuries.

In the very early formation of the moon, huge meteorites are believed to have struck the recently-hardened crust, creating large basins in the surface. Most of these basins were created on the near side. Later, at a very slow rate, the basins filled with lava from the lunar interior. This filling process may have taken a billion more years. This lava flow created the maria on the near side. It is not known why there was no lava flow on the far side.

Both the near and the far side of the moon are well covered with craters of all sizes. Whereas the maria were all given romantic names, the craters were named after famous people, mainly scientists and philosophers (Tycho, Plato, Copernicus, Kepler). This custom is still observed today. In 1970, a special international conference was held to decide upon names for those craters not previously identified except by number. Many craters on the far side, however, had already been named by Russian scientists immediately after their lunar space craft had photographed this side in 1959. As a result of the Apollo missions, many of the prominent smaller features have also been given names.

Even the largest craters are considerably smaller than the maria. Clavius and Grimaldi, the two biggest craters, are both about 240 kilometers in diameter (150 miles) although most craters are much smaller. From the earth, about 30,000 craters can be detected, the smallest about one kilometer in size. Photographs taken on the moon showed craters much smaller than this, of course (Fig. 17.14).

Large craters are notably less numerous in the maria areas, although

17.14 The lunar terrain. Note the peaks in the center of the large crater in the upper left. Also note the hundreds of tiny craters that have pockmarked the grosser features. Photographed by the Apollo 11 crew (photograph from the National Space Science Data Center).

close examination of photographs reveals many *inundated craters*. This strongly suggests that these craters were at one time filled in by the same lava which produced the maria. There are other regions where the craters are so numerous that many overlap each other (Fig. 17.15). When one finds a perfectly formed crater superimposed on another, sometimes breaking that crater's walls as in Figure 17.16, it clearly represents more recent crater formation.

Crater walls sometimes reach 10,000 feet above the crater floor. They are much steeper on the inside than on the outside of the crater, where

17.15 The lunar terrain. Note the several places where the larger crater walls have been broken or appear to have been inundated. Photographed by the Apollo 12 crew (photograph from the National Space Science Data Center).

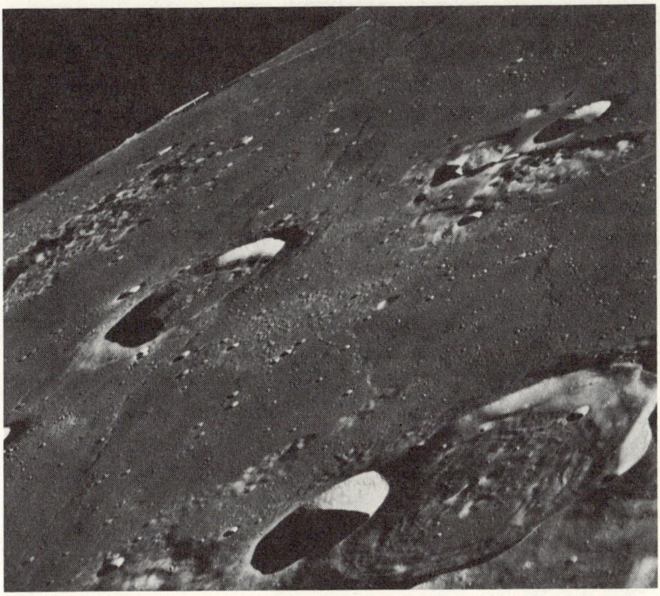

17.16 The lunar terrain photographed by the Apollo 12 crew from 60 miles above the surface. Note the crater wall broken by a more recently formed crater in the foreground (photograph from the National Space Science Data Center).

they usually slope more gently down to the surrounding land. Mountain peaks can be seen in the center of some of the larger craters (Fig. 17.14). These characteristics closely resemble the craters left by bombs and can be simulated by throwing stones at various angles and speeds into wet sand or mud. All of this strongly supports the theory that the majority of craters, particularly the larger craters, were formed by meteoritic impacts.

In addition to the craters and maria, the lunar surface also has a few mountain ranges. These are similar to those on earth, and some of them reach 25,000 feet or more, which is comparable to our highest mountains. In many cases, terrestrial names were given to these lunar ranges (Alps, Apennines, etc.) (Fig. 17.17).

With the aid of a telescope or binoculars, make a drawing of one section of the moon at different phases (or pick one crater or other feature to draw). Note especially the shadows.

17.17 A view of Hadley Delta mountain which rises about 4000 meters (about 13,124 feet) above the plain. On the horizon in the center is Silver Spur, named by the Apollo 15 crewmen who took this picture from their LM at the landing site. Note the many craters in the foreground and the layered appearance of Silver Spur (NASA photograph).

The Apollo Missions

The scientific aims of the Apollo missions were to obtain measurements and data that could not possibly be obtained from the earth. It was a most successful series of missions. The moon became a sort of scientific laboratory; many types of instruments were brought to the moon and some were left as semipermanent installations (Fig. 17.18). From the wealth of data collected, the many mysteries of our nearest neighbor are beginning to be solved. Photographs were taken not only of the moon but of the earth, sun, and stars as well. Seismometers were set up and left in place so that their signals could be recorded and analyzed back on earth. And about 840 pounds of moon rocks were brought back to the earth for study (see Fig. 17.19 and Plate E).

While they were on the moon, the astronauts amply demonstrated with their awkward kangaroo-like steps what had long been known about

17.18 The Apollo 11 Lunar Module (LM)—the first manned spacecraft to land on the moon (July, 1969) (photograph from the National Space Science Data Center).

the surface gravity of the moon: It is only 1/6 that of the earth. Objects fell much more slowly when they were dropped and traveled a much greater distance when they were thrown than they would have done here on earth. All communications between the two men while they were outside the Lunar Module had to be by radio intercoms, since without an atmosphere there can be no sound waves.

Although the Apollo astronauts kept earth time during their stay of

17.19 Moon rocks: a close-up view of one of the Apollo 12 lunar samples. This sample, a breccia, is highly compacted and is heavily coated with glass. Note the several pits in the surface. Numbers are centimeter measures. (NASA)

several days on the moon, the sun did not set. It moved only through a small arc in the sky. The earth remained virtually fixed high in the dark sky above them and shone with a brilliance almost 50 times as bright as the full moon on earth.

Astronauts described their journeys on the lunar surface in terms of north and south directions (Fig. 17.20). But what is north on the moon? A compass would be useless since the moon has no permanent, general magnetic field as does the earth. Still, north is quite a definite direction on the moon. Since the moon rotates on its axis, it has two poles. The one oriented most nearly like the north pole of the earth is, by convention, the north pole of the moon.

Composition of the Moon

The chemical composition of the moon is strikingly different from that of the earth. The analysis of the rocks brought back by the Apollo astronauts revealed that, for example, whereas the earth is about 36% iron, the moon is only about 8%. More than 50% of the lunar material is oxygen, most of which is chemically united with silicon, which is the second most abundant

17.20 Apollo 15 astronaut with the lunar rover at the edge of Hadley Rille (NASA photograph).

element on the moon. Some of the other lunar minerals are unknown on earth. Because of these differences, theories that the moon split off from the earth after it had been already formed must be discarded.

The determination of the ages of the various rock samples taken from the different landing sites gave many clues as to the events and processes at the time of their formation. When methods of isotope dating were used, some of the rock samples were found to be as old as 4½ billion years. Most, however, were less than 4 billion years, which would indicate the possibility of much lunar activity in the early formative days. Study of these rocks supports the theory that the moon and earth were probably formed at the same time, perhaps as a double planet.

The seismometers left on the moon have recorded thousands of moonquakes and meteoritic impacts. From these comes strong evidence of a differentiated, layered structure. The moon has a crust and a mantle much like the earth's. Beneath the mantle some investigators believe that there may well be a molten zone. Certainly the interior of the moon is not cold, but its actual temperature is still undetermined. To complicate the picture still further, the remnant of a magnetic field has been found frozen into the moon's outer shell. This small magnetism does not affect a compass but will require a much more complex theory of lunar evolution than has been proposed thus far.

When spacecraft first started circling the moon, astronomers noted that they were perturbed in their orbits whenever they passed over specific areas, most often the maria. These perturbations were found to be caused by concentrations of denser material beneath the surface. Because they are denser than the surrounding matter, these *mascons* (mass concentrations) exert a higher gravitational force and hence are able to affect the path of low-orbiting objects. The lava which rose from the lunar interior and filled in the holes or basins created by meteoritic impacts has a higher density than the original lunar crust found in the highlands. This is why the mascons are usually found in the maria regions. There are no mascons on the far side of the moon.

Origin of the Moon

These new data impose certain requirements, which any theory of the origin of the moon must meet. Several moon-history scenarios can be written. Oddly enough, a history involving a completely molten moon and a history of a moon formed by accretion of smaller masses are both compatible with the Apollo mission data. But the place of the moon's formation also remains unknown. Did the moon form entirely apart from the earth or in association with it? The radical differences in the chemical composition of the moon as compared to the earth imposes perhaps the greatest single obstacle to the theory that the earth and moon were formed in close association. It is difficult to see how this difference could have occurred if the earth and moon were formed originally as a double planet. Yet to hold that the moon had its origin elsewhere in space still leaves open the question of chemical differences and also the question of how the moon got where it is today.

The Apollo missions were only the very beginning of an era of discovery that may prove to be as important as the voyages of Columbus, and probably much more so. They were the pioneering steps that opened the moon as a future base and scientific laboratory (Fig. 17.21). Such a base offers immense opportunities for astronomical research. An observatory on the moon has many important advantages over telescopes in earth-orbiting satellites, although both make atmosphere-free observations possible. The moon is a stable platform and allows continuous observations for periods

17.21 Apollo 11 astronaut descends the ladder of the Lunar Module (photograph from the National Space Science Data Center).

equal to fourteen earth days. A lunar telescope, particularly if placed on the far side, would be free of all earth glare and other terrestrial interference. A radio telescope on the far side would be completely free of manmade electrical interference.

You are a member of a space crew scheduled to rendezvous with a mother ship on the lighted surface of the moon. Mechanical difficulties force your ship to land 200 miles from the rendezvous point. During the landing much of the equipment aboard is damaged, leaving only the following items in good condition. Which of these would you take for your 200-mile hike to meet the mother ship? Which items are useless and should not be taken? Why?

Box of matches
Food concentrate
50 feet of nylon rope
Parachute silk

Portable heating unit
Two .45 calibre pistols
One case dehydrated milk
Two 100 lb. tanks of oxygen
Stellar map (of moon's constellation)
Life raft
Magnetic compass
Five gallons of water
Signal flares
First-aid kit containing injection needles
Solar-powered FM receiver-transmitter

Not only astronomers, but physicists, chemists, biologists, and other scientists who greatly need a "vacuum" laboratory free from atmospheric and electronic interference would man this outpost. Certainly, a scientific colony on the moon available to all mankind would be an outstanding venture in international cooperation.

Discussion Questions

1. Discuss the problem of timekeeping on the moon. To what extent could a lunar colony keep earth time or follow an earth calendar? Would time zones have any meaning there? Devise a calendar for a moon colony.

2. When is the full moon near the vernal equinox?

3. If the Apollo 15 crew had stayed at Hadley Rille for several earth months, how often would they see the sun rise and set? The earth rise and set?

4. It has been said that the Apollo missions gave rise to many more questions about the moon than they answered. What were some of the unexplained findings?

5. If a civilization had arisen and remained on the hidden side of the moon, could they have discovered the existence of the earth? Discuss.

6. Do you think that a picture such as Figure 17.18 would have had the same effect on people who lived 150 years ago as the discovery of Jupiter's moons had in the early 1600's?

18 *The Terrestrial Planets*

Each of the nine worlds in our solar system is unique. Ocean-covered Earth, the ring system of Saturn, the moon-like terrain of Mercury, and giant belted Jupiter all illustrate the range of differences to be found. Yet with all their differences, the nine planets fall into two basic groups: the giant or major planets and the smaller "terrestrial" ones. Earth is one of the terrestrial group, along with Mercury, Venus, Mars, and Pluto.

The relative sizes of the sun, planets, and some of the larger moons are shown in Figure 18.1. With the exception of our moon, all the satellites (moons) are relatively tiny in comparison to their planets, yet four of these moons are probably larger than Mercury. We have already seen that a star can be the same size as the earth or smaller (see p. 168), and therefore must conclude that size alone does not determine these classifications.

In general, a star is thought of as a body that radiates its own energy. However, Jupiter radiates energy, and black holes (by their very nature) do not radiate anything. We usually think of a planet as a nonradiating body that revolves about a star, whereas a moon or natural satellite is one that revolves about a planet. Again we must be careful with this definition, since dust particles and comets which revolve around the sun are obviously not planets. Many astronomers consider Pluto an ancient "escaped" moon of Neptune.

Figure 18.2 summarizes the presently determined primary statistics of each planet. Note that the speed of the planets in their orbits decreases as the distance from the sun increases. This is in keeping with Kepler's third law of planetary motion: $p^2 = a^3$ (see p. 261). Since most of the planetary orbits are nearly circular, we can use the formula for the circumference of a circle ($2\pi r$) to determine the approximate distance it travels in one revolution. The r or radius of the circle is equal to the semimajor axis of the orbit (a) in Kepler's equation.

$$2\pi a \text{ (distance)} = V \text{ (velocity)} \times p \text{ (time)} \qquad \text{(Substitute } a \text{ for } r\text{)}$$

$$V = \frac{2\pi a}{p}$$

$$V^2 = \frac{(2\pi a)^2}{p^2} \qquad \text{(Square both sides of equation)}$$

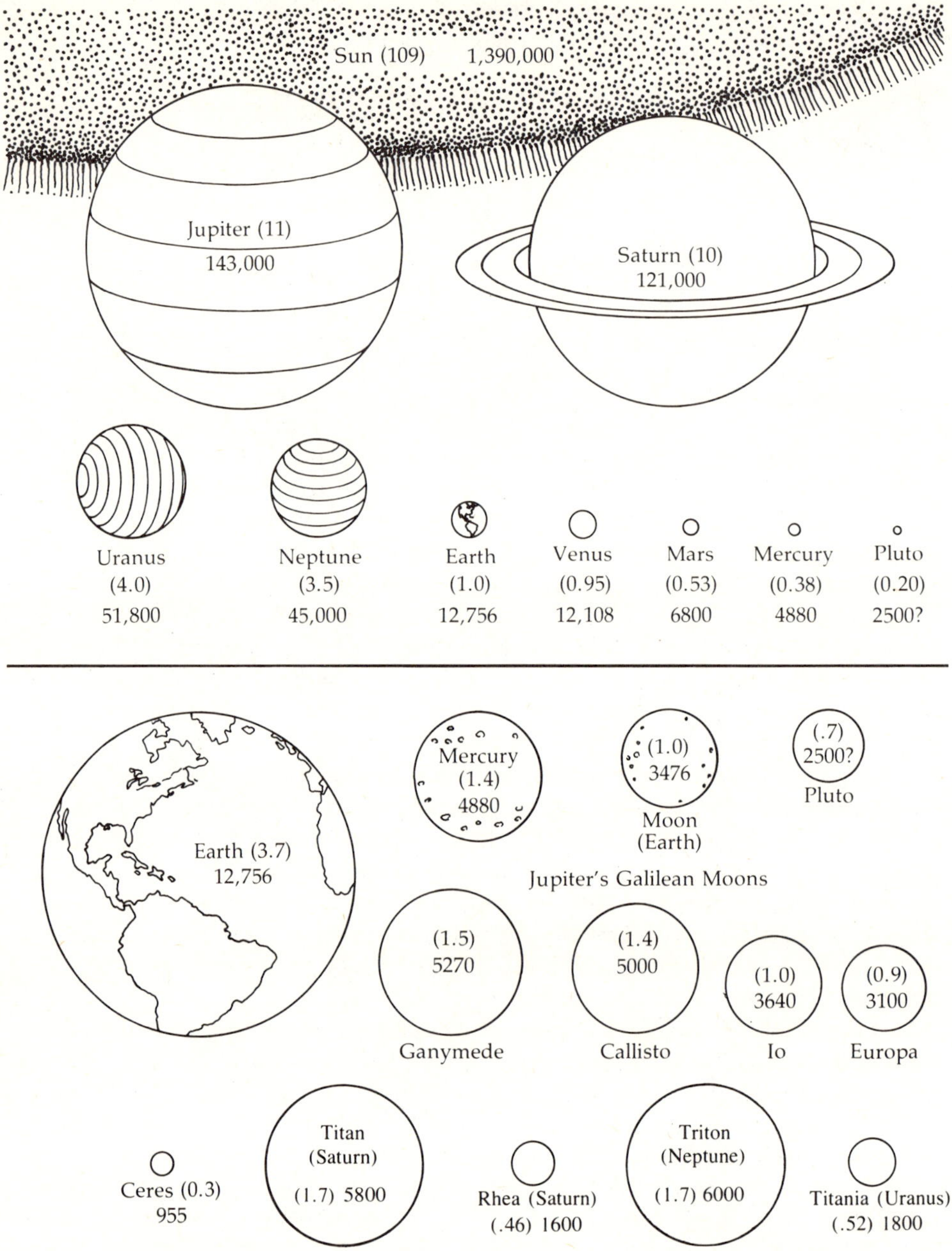

18.1 The relative sizes of some members of the solar system. The diameters are measured in kilometers. Numbers in parentheses compare the diameter of each object in the upper section with that of the earth. In the lower section, the comparison is with our moon. Ceres is the largest asteroid.

$$V^2 = \frac{4\pi^2 a^2}{a^3} \quad \text{(Substitute } a^3 \text{ for } p^2\text{: See Kepler's equation)}$$

$$V^2 = \frac{4\pi}{a}$$

$$V \sim \frac{1}{a}$$

Thus, the velocity of a planet in its orbit is inversely proportional to the square root of its distance from the sun.

As the planets move in their orbits, various terms are used to describe their positions relative to the sun and earth. Figures 18.3 and 18.4 illustrate the terms for planets closer and farther away from the sun respectively.

18.2 *The planets—their statistics.*

Name	*Distance from sun (AUs)*	*Size (Earth = 1.0)*	*Period of Revolution (days-yrs)*	*Orbital Speed (km/sec)*	*Period of Rotation (hrs-days)*	*Density (water = 1.0)*	*Surface Gravity* (Earth = 1)*	*Mean Temperature at visible surface: °C*
Mercury ☿	0.4	0.4	88 days	47.8	59 days	5.4	0.37	350 (day) −170 (night)
Venus ♀	0.7	1.0	225 days	35.0	243 days (retrograde)	5.2	0.88	− 33 (clouds) 480 (ground)
Earth ⊕	1.0	1.0	365 days	29.8	24 hours	5.5	1.0	22
Mars ♂	1.5	0.5	687 days	24.2	24.5 hours	3.9	0.38	− 23
Jupiter ♃	5.2	11	12 years	13.1	9.9 hours	1.3	2.64	−150 (clouds)
Saturn ♄	9.5	9.5	29.5 years	9.7	10.2 hours	0.7	1.15	−180 (clouds)
Uranus ⛢	19	4.0	84 years	6.8	23 hours ?	1.2	1.17	−210
Neptune ♆	30	3.5	165 years	5.4	22 hours ?	1.7	1.18	−220
Pluto ♇	39.5	0.2?	248 years	4.7	6.4 days	??	??	−230 ?

*The surface gravity of a planet, multiplied by 100, will give you the weight of a 100-pound astronaut on that planet.
The figures underneath the name of each planet are the internationally recognized symbols for each one.

Note that *east* and *west quadrature* are positions at right angles to the sun as seen from the earth. *Eastern* and *western elongation,* on the other hand, are positions on a line tangent with the earth and the other planet's orbit.

The giant and terrestrial groups of planets have other major contrasts besides their size. The smaller, terrestrial planets (with the exception of Pluto) were formed near the sun whereas the giants are much more remote. It has been suggested that the sizes of these giant planets are such because they had more "room to grow" far from the sun. Undoubtedly, the distance from the sun was a factor in determining the final composition of the planets. If the earth had been at the distance of Pluto, it would have had a much lower temperature. The velocity of the atoms comprising its original atmosphere would also have been lower, below the velocity of escape. If so, the original lighter gases such as hydrogen and helium would have been retained. As it is, the only helium that the earth has is locked as a gas in interior cavities. Once released, it soon escapes permanently from the atmosphere. Similarly, practically all the hydrogen on earth is locked into chemical compounds.

Because the giant planets were able to retain the original gases present at the time of their formation, they have an entirely different composition than the terrestrial ones. They are composed almost entirely of the

18.3 Positions of an inferior planet (i.e., one whose orbit is within that of the earth) in relation to the earth and sun.

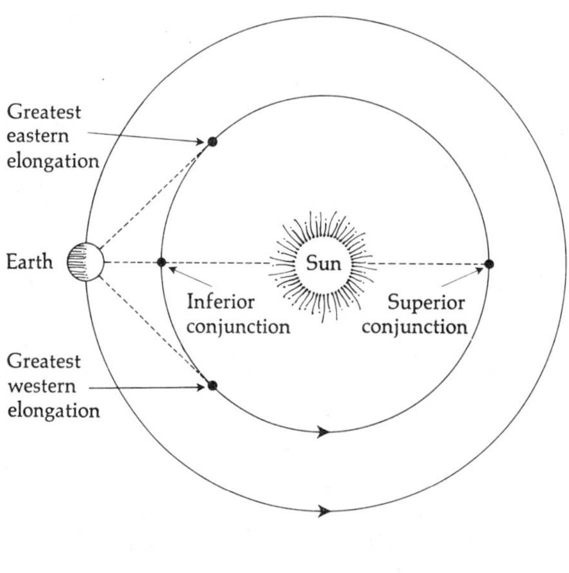

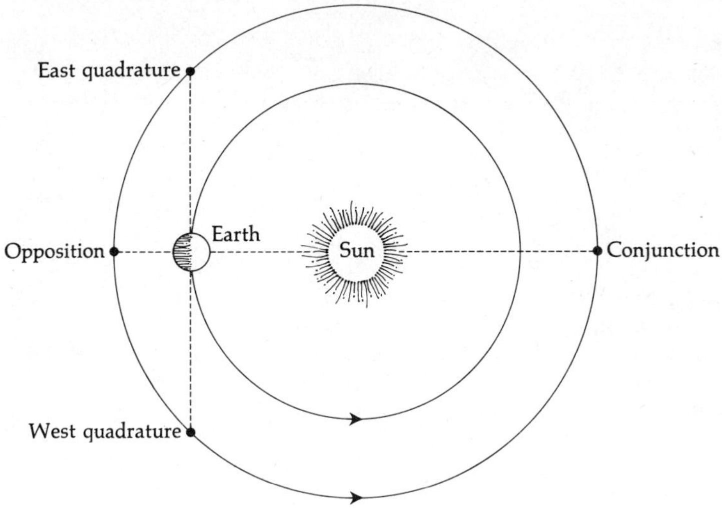

18.4 Positions of a superior planet (i.e., one whose orbit is outside that of the earth) in relation to the earth and sun.

lighter gases hydrogen and helium—the primary elements of the universe. Because of this they have relatively low densities. The terrestrial planets, on the other hand, are much denser and are composed mainly of the metals, having already lost most of the lighter elements. These smaller planets are thus quite different in composition not only from the rest of the solar system but also from the rest of the entire visible universe.

Besides retaining most of their original gases, the giant planets also managed to attract enough smaller globules around themselves to create what has been described as "miniature solar systems." Other than our moon, which does stand out as an anomaly in the solar system, there are only the two very tiny moons of Mars revolving around any of the terrestrial planets. On the other hand, the giant planets collectively have 30 of the 33 known moons, with Jupiter's 13 satellites topping the list. Also, four of the larger moons are actually bigger than Mercury. Figure 18.5 gives the statistical data for all these moons.

With the exception of the two smallest planets, Mercury and Pluto, all the rest have some form of atmosphere that can be detected from the earth. Three of the moons also have detectable atmospheres: Jupiter's moons Io and Ganymede, and Saturn's giant satellite, Titan. As we saw in our discussion on planetary formation (p. 250), a planet or a moon must have a sufficient surface gravity to hold an atmosphere. Both the mass and size of

	Name	Diameter (kms)	Mean Distance from planet (1000s kms)	Sidereal Period (days)	Apparent Magnitude
Earth	Moon	3476	384.4	27.3	−12.5
Mars	Phobos	25	9.4	0.3	12
	Deimos	13	23.5	1.3	13
Jupiter	Amalthea (V)	150	180.5	0.5	13
	Io (I)	3640	421.6	1.8	5
	Europa (II)	3100	670.8	3.5	6
	Ganymede (III)	5270	1070.0	7.2	5
	Callisto (IV)	5000	1882.0	16.7	6
	Himalia (VI)	120	11470.0	250.6	14
	Elara (VII)	40	11800.0	259.6	18
	Lysithea (X)	10	11850.0	263.5	19
	Leda (XIII)	8	12400.0	282.0	20
	Ananke (XII)*	10	21200.0	631.1	18
	Carme (XI)*	15	22600.0	692.5	19
	Pasiphae (VIII)*	25	23500.0	738.9	17
	Sinope (IX)*	15	23700.0	758.0	19
Saturn	Janus	350	157.5	0.7	14
	Mimas	400	185.4	0.9	12
	Enceladus	500	237.9	1.4	12
	Tethys	1000	294.5	1.9	11
	Dione	1000	377.2	2.7	11
	Rhea	1600	526.7	4.5	10
	Titan	5800	1221.0	15.9	8
	Hyperion	400	1479.3	22.3	14
	Iapetus	1200	3558.4	79.3	11
	Phoebe*	300	12945.5	550.4	14
Uranus	Miranda	550	123.0	1.4	17
	Ariel	1500	191.7	2.5	15
	Umbriel	1000	267.0	4.1	15
	Titania	1800	438.0	8.7	14
	Oberon	1600	586.0	13.5	14
Neptune	Triton*	6000	353.4	5.9	14
	Nereid	500	5560.0	359.9	19

*These satellites orbit their planets in a retrograde direction.

Note: Most of the sizes (diameters) of the moons are estimates, subject to change as new data becomes available.

18.5 Natural satellites of the planets.

a planet (or moon) determine its surface gravity.[1] This in turn determines the amount of gravitational pull on objects on its surface. This amount is commonly called "weight." Thus, a given mass will have different weights on different planets. Atoms of oxygen, for example, are held on the earth, but would be able to escape into space from the moon, which has only one-sixth the surface gravity of the earth.

A new era in planetary exploration opened on December 14, 1962, when for the first time an interplanetary spacecraft flew by and examined a planet—Venus. Since then, many space probes have provided us with a wealth of information about our solar system. Because so much of this data is still being analyzed, new findings are constantly being announced. Current scientific journals are the best source for these recent discoveries.

Mercury

Closest to the sun and the smallest of the planets, Mercury races around the sun in only 88 days. An astronaut on Mercury (if one should ever land there) would see the sun more than twice as large as it appears from the earth. It would be in the sky for 88 days and after that an equally long night would follow (Fig. 18.6). The astronaut would be bombarded with up to ten times as much radiation as he would receive while orbiting the earth, and meet with temperatures ranging from −300° F at night to up to +800° F during the long sunlit day. No other planet in the solar system has such a wide temperature range. This intense heat as well as Mercury's tiny mass explains why it has no atmosphere. It is composed mainly of rock and metals.

The astronaut on Mercury would weigh only about one-third as much as he does on the earth. This is twice as much as he would weigh on our moon and therefore his steps would not be as long or as kangaroo-like as they were on the moon. Like the Apollo astronauts, he would leave footprints in the fine dust that covers the surface and they would probably remain unchanged for millenia.

Although bright in the sky (when it is visible at all), Mercury is difficult to see from the earth because it appears so close to the sun. It is never more than 28° away from the sun on the celestial sphere. When observed through a telescope, it exhibits the same phases as the moon and

---

[1] Surface gravity or weight *(W)* is proportional to the mass *(m)* of the planet and inversely proportional to its radius *(r)* squared: $W \sim m/r^2$. If the earth were somehow squeezed into a volume of space with half its present radius, our weight would increase fourfold!

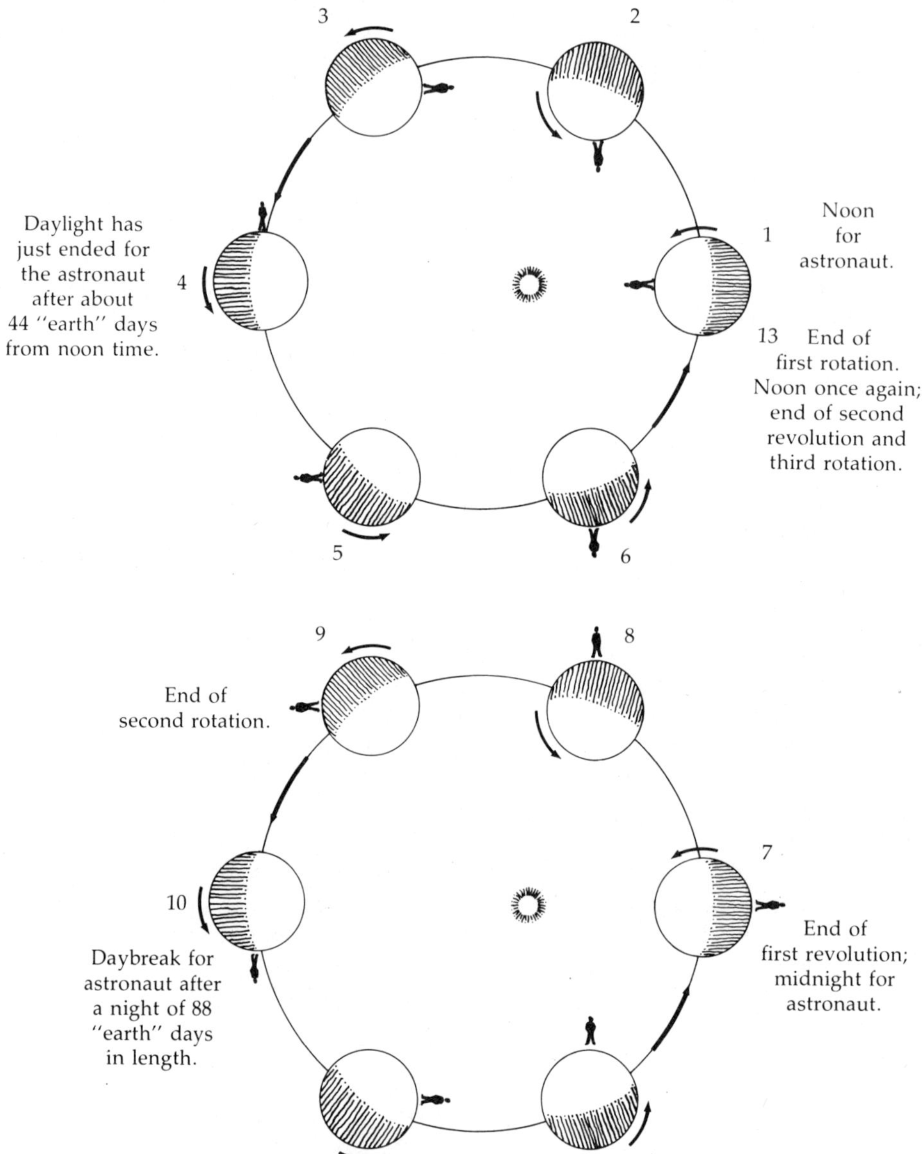

18.6 The rotation and revolution of Mercury. In two revolutions around the sun, Mercury rotates three times with respect to the stars, but to an astronaut on the surface those two revolutions or three rotations would constitute only one Mercurian day (from noon to noon). This "day" would equal 176 earth days.

Venus. However, like Venus, we can never see Mercury at exactly full phase.

Even when viewed with the most powerful telescopes on earth, only very faint markings are visible on its surface. On the basis of these it was once thought that Mercury's period of rotation was the same as its revolution—namely 88 days. This would have meant that it always kept the same side toward the sun. Mercury was therefore frequently pictured as being furnace-hot on one side and not far from absolute zero on the other.

It is said that Copernicus never saw the planet Mercury. Indeed, most people have never seen this planet. Yet for a period of about two weeks every four months Mercury can be seen shortly after sunset, shining more brightly than a first magnitude star. From an almanac, determine when Mercury will next be visible in the late evening twilight sky and locate it. If a telescope or binoculars are available, view it this way and draw a sketch of its appearance. Does it twinkle, and if so, why?

In 1965, through the use of the 1000-foot radio telescope at Arecibo, Puerto Rico (Fig. 12.16), it was discovered that the true period of rotation was only 58.6 days. This meant that Mercury spins on its axis three times in every two revolutions around the sun, thus showing nearly the same face with the same markings each time the planet is in the most favorable viewing position (Fig. 18.6). This undoubtedly was the reason for the earlier conclusion that Mercury kept the same face toward the sun.

Most of the information we have about Mercury comes from the Mariner 10 spacecraft, which was programmed to intercept and fly by Mercury three times in 1974–75. It passed within a few hundred kilometers of the planet at that time. The photographs it took and sent back to the earth showed that the surface of Mercury is greatly like that of the moon (compare Fig. 18.7 with Fig. 17.1). Numerous craters pockmark the landscape, many overlapping each other. There are also some smoother areas which resemble the lunar maria, although their origin may not be the same as the lunar feature. One unique feature of Mercury's surface is the *scarps*, or cliffs, some running for over 300 miles, cutting through large craters. These scarps were probably a thrusting up of the Mercurian crust as a consequence of the shrinking planetary core.

Besides sending back close-up pictures of the surface of Mercury, the other instruments on Mariner 10 revealed many unsuspected features. For

18.7 The surface of Mercury as photographed by Mariner 10 in 1974 from 200,000 kilometers (124,000 miles) away. The spacecraft's TV cameras took 18 pictures which were later computer enhanced and made into this photomosaic. The largest of the craters are about 200 kilometers in diameter (JPL-NASA).

18.8 A large circular basin on Mercury measuring 350 kilometers (220 miles). It appears to have been flooded with lava flows and then subsequently cratered. Compare this scene with those of the lunar surface in Chapter 17 (JPL-NASA).

example, a very weak magnetic field was detected (only about 1% that of the earth's) but strong enough to deflect the bombarding solar wind away from the planet. With Mercury's slow rotation, a magnetic field was never thought possible, since it was believed that faster rotation was necessary to produce such a field.

The existence of a magnetic field gave further evidence that although Mercury resembles the moon on its surface, it is not truly moonlike. Its density is found to be closer to that of the earth's and it is believed to have a large iron core, perhaps three-quarters of its diameter. Such a large core would be about the size of our moon. Above the iron core is a thin silicate mantle, very moonlike in appearance. Thus, with an interior that is very earth-like and an exterior that is moon-like, Mercury raises many questions about the evolution of the inner solar system.

The major axis of Mercury's orbit is known to slowly change its orientation relative to the stars so that the *perihelion* (point nearest to the sun) and *aphelion* (point farthest from the sun) are also constantly changing. Most of this rotation of the major axis can be accounted for by the gravitational effects of other planets, but a small portion has been found to be a relativistic effect, largely due to Mercury's very rapid orbital velocity. Thus, Mercury's orbital changes have been used as one of the classical proofs of the theory of general relativity.

Venus

Venus, because of its close similarity to the earth in mass and size, has sometimes been called our "twin sister" planet. In other respects, however, (surface temperature, composition of atmosphere, etc.) it is very dissimilar.

Venus is the object most frequently referred to as the "evening star" or "morning star"—especially at those times when it is visible as a brilliant object shortly after sunset or just before sunrise (Fig. 18.9). Many ancient peoples were not aware that this brightest of evening or morning "stars" was one and the same object. The Greeks, for example, designated it "Hesperus" in the evening and "Phosphorus" in the morning.

At its brightest, with a maximum apparent magnitude of −4.4, Venus is more than 100 times brighter than the star Altair in the constellation Aquila. Many people are sure it is a new and mysterious object in the sky at these times, sometimes even reporting it as a UFO (see Chap. 21). During both World War I and II many rounds of ammunition were fired at Venus, each side thinking it was a spy device of the enemy. When one knows exactly where to look for it, Venus can even be seen with the naked eye in the daytime when it is at or near its maximum brightness. It can easily be

---

Observe Venus in a small telescope. Sketch what you have seen. Repeat the observations a few weeks later and make a new sketch. Make diagrams showing the sun, planet, and earth at the times of your observations. Confirm Figure 18.9.

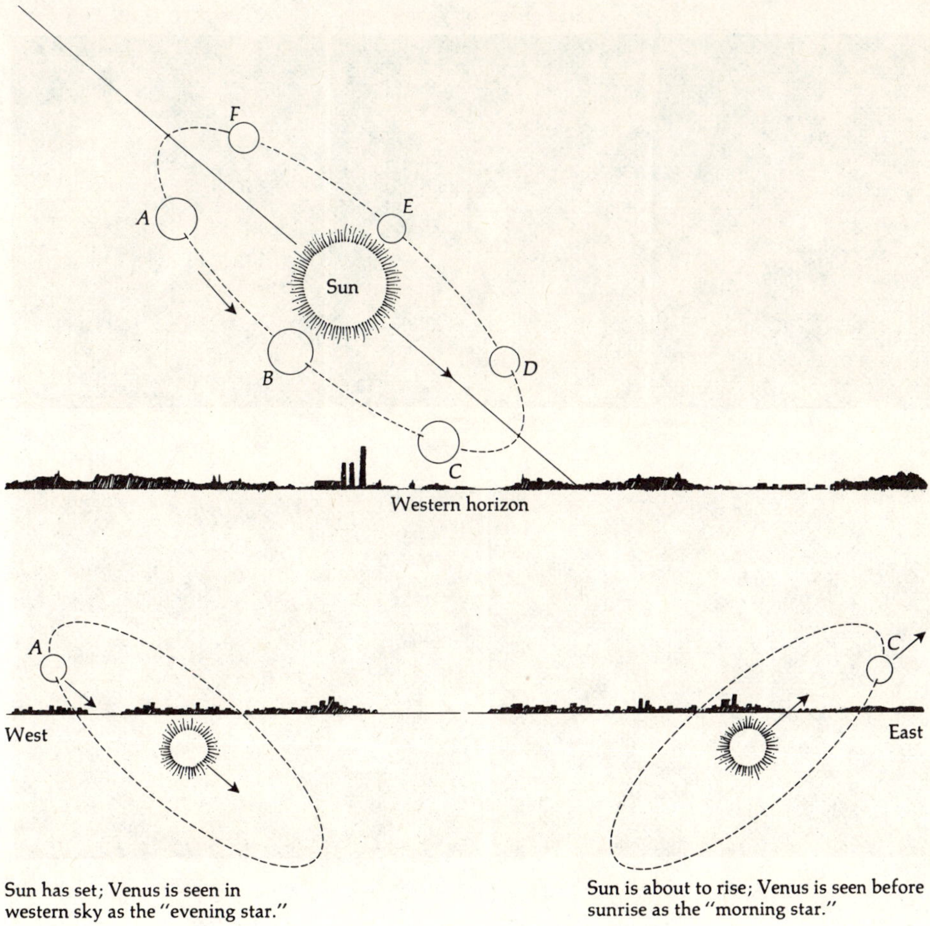

18.9 Venus takes 225 days to revolve around the sun. It is seen in different positions relative to the sun during this period. At both A *and* C *it appears at its greatest brilliance. At* A, *it will set after the sun and will appear as a crescent through a telescope. At* C, *it will rise before the sun and will again be a crescent when viewed through a telescope.*

seen through a telescope during the day except when it is very close to the sun.

Through a telescope, Venus, like Mercury, is observed to exhibit all the phases of the moon (Fig. 18.10). If the ancient Greeks had been able to observe these two inner planets with telescopes, it is unlikely that the

Five phases of Venus

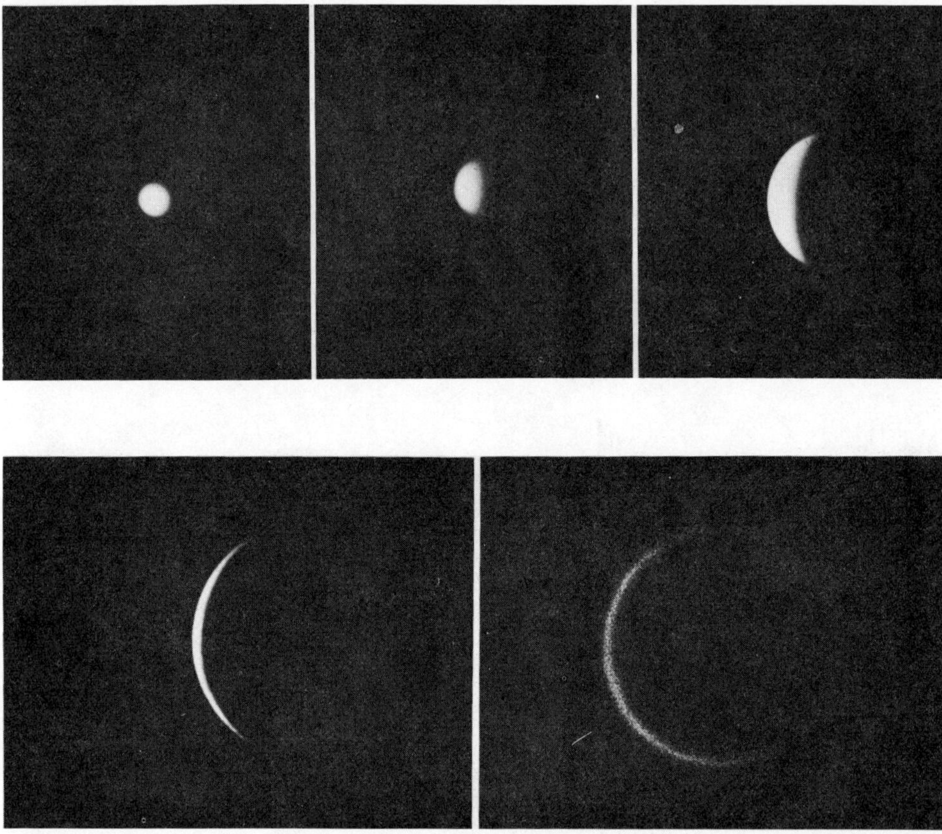

(a)

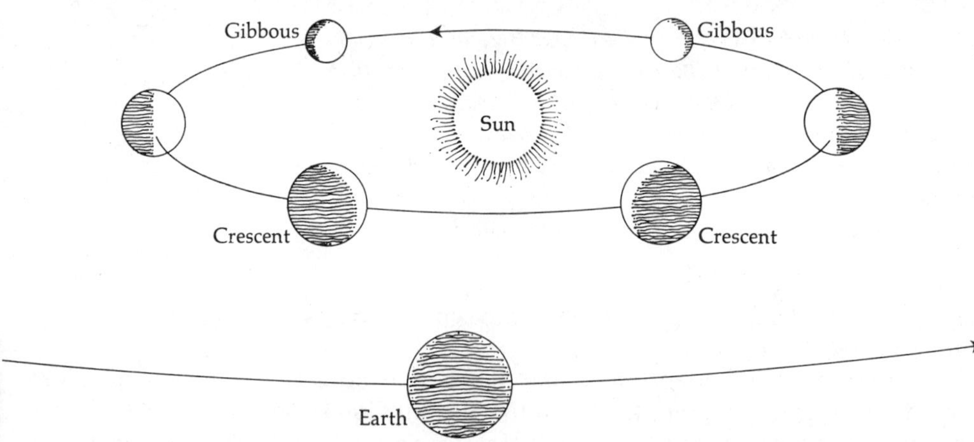

(b)

◄

18.10 The planet Venus exhibits the same phases as the moon. Because it is farther from the earth when it is at the full phase, its diameter appears smaller than when it is at the crescent phase. The planet Mercury goes through the same phases (photograph from the Lowell Observatory).

earth-centered Ptolemaic universe would have been accepted. Galileo was the first to see the great significance of the full set of phases of Venus. He recognized immediately that this would be impossible under the Ptolemaic system (Fig. 18.11) and announced his discovery cryptically: "The Mother of Love imitates the forms of Cynthia."

18.11 When Galileo viewed Venus through his telescope, he knew that the Ptolemaic system was incorrect.

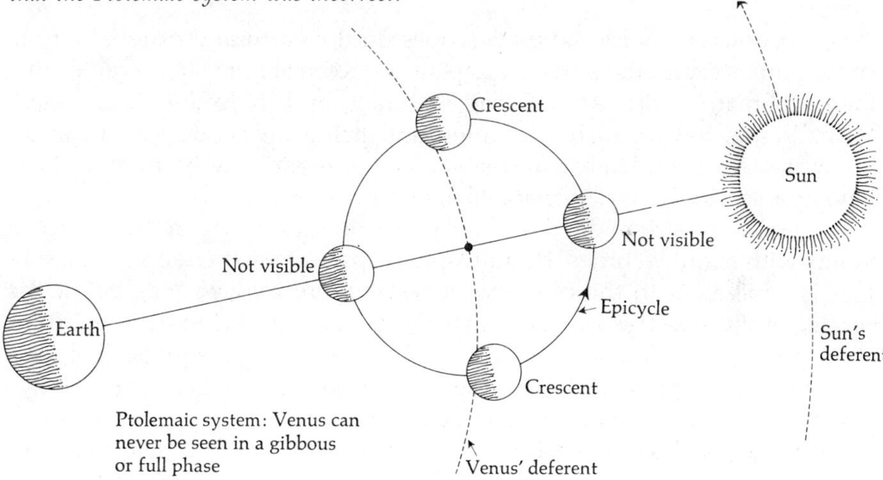

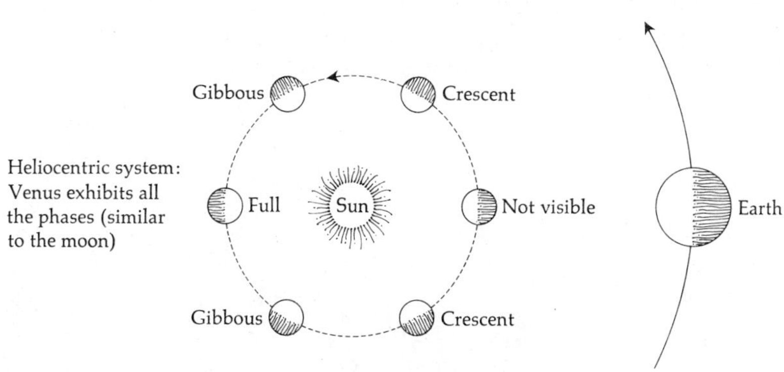

Construct a diagram similar to Figure 18.6 for the planet Venus.

The brilliance of the planet is due to the sunlight reflecting off the dense cloud cover which completely obscures its surface. Although Venus receives twice as much radiation from the sun as the earth, it absorbs only half, reflecting the rest back into space. Of the sunlight that does penetrate this thick atmosphere, only one percent reaches the surface, creating a lighting effect much like the earth's on a very dark, rainy day. And if entering Venus' atmosphere is difficult for solar radiation, leaving it is even worse. The solar radiation heats the lower regions of the atmosphere but is then trapped and unable to escape into space again. Carbon dioxide, which is the major component of its atmosphere, prevents longer-wavelength radiation (infrared heat waves) from leaving. This causes the *greenhouse effect* which appears to be the cause of the high surface temperature of Venus.

Observations with radio telescopes on the earth and especially from space probes found the surface temperature to be about 900° F even during the Cytherean[2] night. At 900° lead will melt, and therefore the old ideas about Venus having lush vegetation and being an abode for life as we know it seem completely impossible. Venus must now be thought of as having a searingly dry, furnace-like surface.

The heavy carbon dioxide atmosphere presses down on the surface of Venus with about 90 times the atmospheric pressure experienced on earth. This is equivalent to the pressure of water more than ½ mile below the surface of the sea. It is the radically different chemical composition of this atmosphere from that of the earth which accounts for the great difference in atmospheric pressure, despite the fact that the surface gravity of the two planets is nearly the same. This thick atmosphere is found to be 97% carbon dioxide and less than 0.2% water. No oxygen has been detected. Corrosive compounds such as sulfuric acid are also present. At one level in the atmosphere it is believed there may actually be sulfuric acid "rain." If so, it is probably the most corrosive naturally occurring fluid in the solar system.

Although Venus approaches more closely to the earth than any other planet (coming within 25 million miles at times), the optical telescope can

---

[2] Cytherea is another name for Venus.

probe no deeper than the outer cloud layers. They reveal nothing but a uniform, featureless, light yellow cloud mass. But radio, radar, and ultraviolet studies from the earth, as well as from space probes by both the United States and the Soviet Union, have provided a far better picture of the planet. On December 15, 1970, the Soviet spacecraft Venera 7 reporedly landed on the surface and actually transmitted data for 23 minutes. This was the first landing of a spacecraft on another planet. Since then other spacecraft have either passed close by or have entered the atmosphere and landed on the surface, sending back more data, including the first photographs of the surface.

The obscuring cloud layer which we see from earth does not extend all the way down but appears to lie in a band about 30 to 37 miles above the surface. For this reason, photographs taken on the surface of Venus by the Russian spacecrafts Venera 9 and 10 were able to reveal a barren terrain, smooth and sandy in some places and steep and rocky in others. Radar images show other features such as a vast trough-like depression and a huge dead volcano. Earlier radar images indicated that the general landscape was relatively flat, although shallow craters and other rough areas were detected (Fig. 18.12).

Radar studies also established the fact that Venus slowly rotates on its axis in a period of 243 days, turning in the opposite direction of the other planets (retrograde). Because of the combination of this slow backward rotation and the 225-day period of revolution, an astronaut on Venus (which is probably just as unlikely as the one on Mercury) would see the sun come up in the *west* every 117 earth days. However, because of the thick clouds overhead, it is doubtful if he would be able to see the sun at all. A year on Venus would last two Venus days! It is believed that Venus' apparent lack of any magnetic field is due to this slow rate of rotation. That makes the origin of Mercury's magnetic field all the more mysterious.

Ultraviolet studies from Mariner 10 (passing by on the way to Mercury) revealed a detailed cloud structure in the upper atmosphere of Venus (Fig. 18.13). Bands of clouds spiraling poleward move around the planet at speeds of 100 meters per second (over 200 miles per hour). Although they move in the same direction as the planet is rotating, they whip around Venus in four days, 100 times faster than the planet's rotational speed. (On earth, our jet streams have been clocked at only some 20 to 30 percent faster than the earth's spin.) The cause of these enormous winds in the upper atmosphere of Venus is not known. At the surface the breeze is so mild it would hardly even be able to raise any dust.

The possibility of life existing on Venus seems very unlikely because of its surface conditions and the composition of its atmosphere. Certainly the existence of a greater amount of water vapor would be a basic requirement of any life (as we know it) to exist.

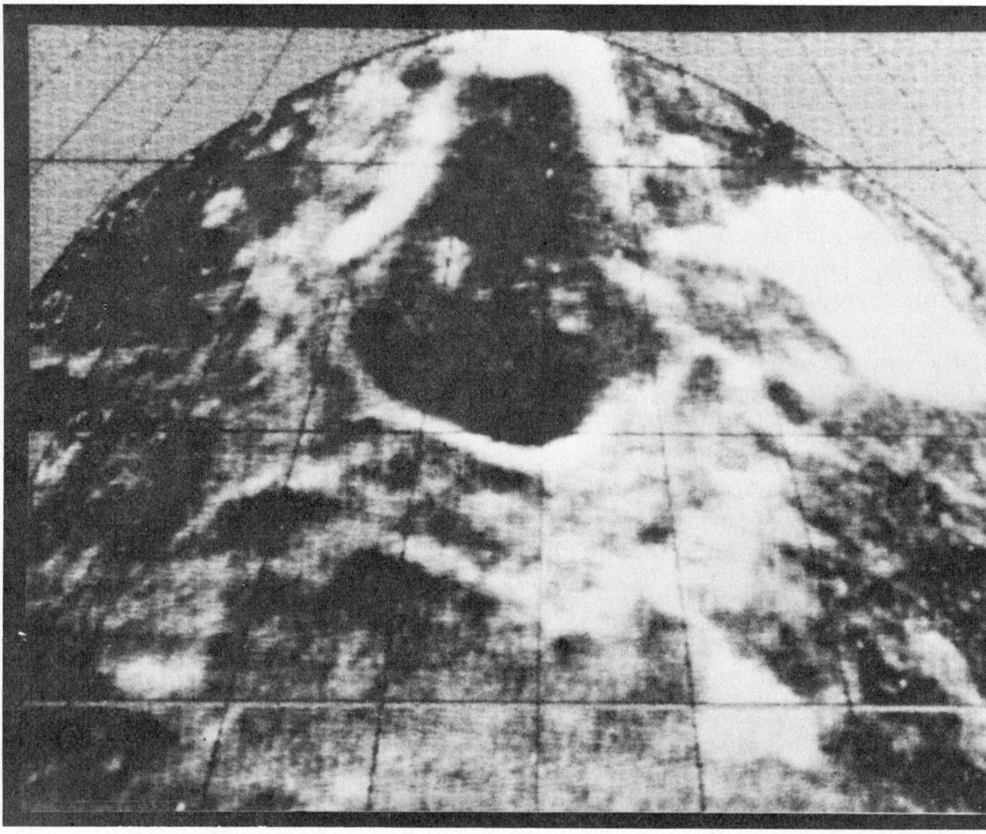

18.12 Radar map of Venus: a portion of the Venusian surface from which radar reflections are obtained. Features as small as 12 miles in size can be distinguished. The black area in the upper center is a large basin about the size of Hudson's Bay on earth. The oval shape of the radar picture is caused by the manner of processing the data (courtesy National Astronomy and Ionosphere Center).

Describe a day and night on Mercury and Venus. How long would the sun be in the sky? What kind of calendar could be devised for each planet?

18.13 Venus from 720,000 kilometers (450,000 miles) away as recorded by Mariner 10's TV cameras in 1974. Pictures were taken in ultraviolet (invisible) light and then computer enhanced and mosaicked to produce this composite. Both dark and light areas are thick clouds covering the planet (JPL-NASA).

Earth

For the first time in history, man has been able to see his planet from space. One cannot help but wonder how philosophical and theological ideas would have differed if such a view had been possible at an earlier date. Certainly, seeing our spaceship Earth from afar has already changed our way of looking at ourselves. It has especially emphasized our complete dependence on the condition of this "spaceship" and on the necessity of maintaining its life-support systems.

An observer from space would characterize the earth as a blue and white planet with a large pale yellow companion as its moon. If careful measurements were taken, our observer would find that the earth was not spherical but somewhat flattened at the poles and bulging at the equator.

While no changes would be observed on the moon, our observer would see large changes in the cloud formations on the earth. The swirling patterns of the clouds would show the effect of the earth's rotation (see page 294: Coriolis effect). Rate of rotation, size, mass, and surface temperatures could all be readily determined. It is highly unlikely, however, that our observer would be able to detect any visual signs of life, even though he could analyze the composition of the atmosphere and determine that it could support life (as we know it). The observer would discover that the earth is the only planet in the solar system to have an oxygen-nitrogen atmosphere. But the first clue that intelligent life existed on earth would come from detecting radio and television broadcasts.

Seasonal changes in the two hemispheres would be easily detectable, as would the land and ocean areas. It would be easily observed that the earth's surface is largely covered by water. It is of philosophical interest to realize how relatively little observers could learn of our planet if they were positioned at the distances of Mars or Venus. Obviously, as inhabitants of the planet Earth, we are not so limited.

The earth is believed to have been formed at the same time as the rest of the solar sytem. Matter from the primordial rotating cloud of dust and gas condensed into particulate material and gradually a solid planet accreted as it swept up smaller particles in its path around the newly forming sun. As it grew, it heated up, not enough for nuclear reactions to take place, but enough for its interior to melt. This allowed for *differentiation* into a core, mantle, and crust. The heavier elements such as iron and nickel sank to the center and formed the core. The lighter elements floated to the top, becoming the mantle and crust. Gaseous material also rose to the surface and all the gaseous elements except free hydrogen and helium were held there by the earth's gravity, eventually becoming the oceans and atmosphere. This differentiation of the elements has been noted in all the terrestrial planets (except Pluto) and in the moon.

It is thought that if the temperature on the earth had been higher, the chemical composition of the atmosphere would have been different and indeed we might have had a greenhouse effect as is found on Venus. Instead, we have a planet with a suitable atmosphere and temperature for the teeming life that exists on it. This life, which has been present for at least three-fourths of the earth's history, is protected not only by the earth's atmosphere but also by its magnetic field.

In the eleventh century A.D. the Chinese were probably the first to use a magnetic needle to show direction, although the ancient Greeks knew about magnetic materials. However, it was not until the seventeenth century that the existence of the earth's magnetic field was demonstrated. Today we know the strength and shape of the field but are not completely certain how it is formed. It is believed to be due to the still molten outer part of the earth's iron-nickel core. As the earth rotates, this fluid also moves, creating electrical currents much in the manner of a dynamo.

The magnetic field acts as a buffer to the solar wind and to cosmic rays (see p. 194). Near the earth, the field resembles the pattern taken by iron filings around a bar magnet (Fig. 18.14). But at higher altitudes it is distorted by the solar wind much like the water flowing around a speeding powerboat (Fig. 18.15). A bow wave forms on the sunward side while a long tail is found on the opposite side, extending to the moon and beyond. Trapped particles (mainly from the solar wind), enter this region, which is called the *magnetosphere,* and spiral around the magnetic field, following its lines of force. They enter the earth's atmosphere only above the earth's magnetic poles where, the field curves down to the surface. The interaction of these rapidly moving high-energy particles with the earth's atmosphere causes the magnificent auroras, *borealis* in the north and *australis* in the south.

The much weaker magnetic field of Mercury produces a similar magnetosphere but on a much smaller scale. In Mercury's case, the planet itself occupies most of the magnetosphere. However, a thin bow wave in front and a very long tail of charged particles behind have been detected.

18.14 *The magnetic field of the earth as compared with that of a bar magnet. Note that the geographic and magnetic axes do not coincide. The geographic axis points to the north and south celestial poles; the needle on a magnet points to the north magnetic pole.*

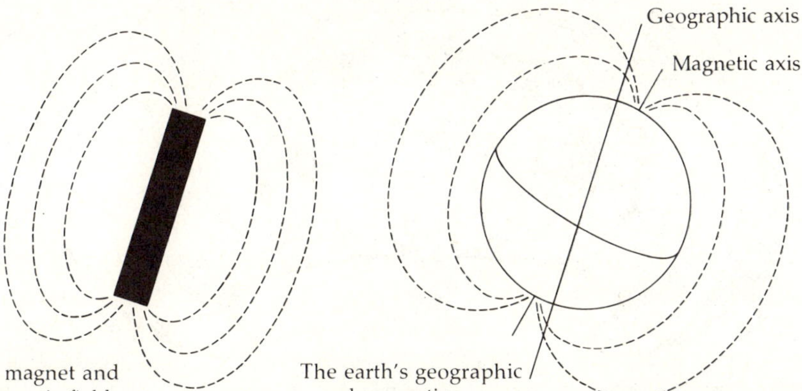

A bar magnet and its magnetic field.

The earth's geographic and magnetic axes.

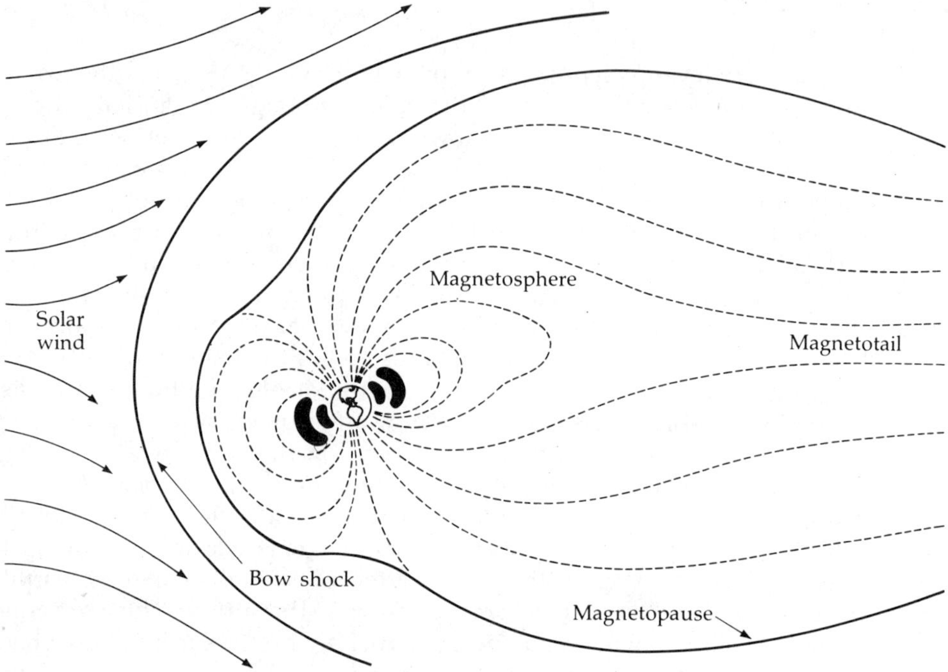

18.15 The earth's magnetosphere. Dotted lines represent the magnetic field lines which are distorted by the solar wind. Few of the solar wind particles cross the outer boundary of the magnetosphere, which is called the magnetopause. *The dark areas around the earth are the Van Allen Radiation Belts.*

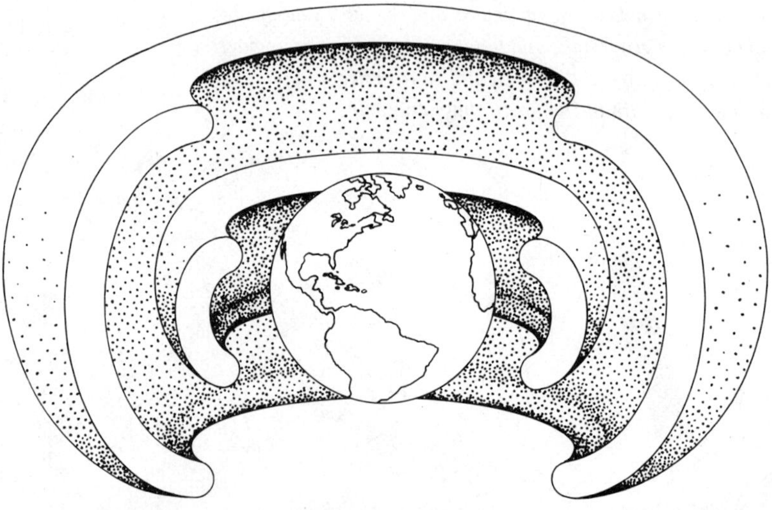

18.16 The doughnut-shaped Van Allen Radiation Belts surrounding the earth.

Within the earth's magnetosphere are two doughnut-shaped regions or belts in which the density of these charged particles is the greatest (Fig. 18.16). These are the *Van Allen Belts,* named after the American physicist James Van Allen, who discovered them in 1961 in one of the first experiments with artificial satellites. The inner belt lies about one earth radius from the surface while the outer one is about 3½ earth radii away. By comparison, the magnetic field on the sunward side extends some ten earth radii away.

The radiation within these belts is especially high and during periods of high solar activity such a large solar flare, the added influx of charged particles makes the level of radiation a great hazard to astronauts.

Mars

The "red planet" Mars, named after the god of war, has excited more interest, both factual and fanciful, than any other member of the solar system. At one time it was seriously considered to be an abode for intelligent life despite its greater distance from the sun. So certain were the believers of this theory that at one time an award was offered for proof of life on any planet *other* than Mars (and, of course, the earth.)

The Martian surface is readily seen through a telescope. Large dark and light areas can be distinguished and seen to change. These variations were at first believed to be seasonal growth and decay of vegetation as first one hemisphere and then the other received more direct rays from the sun. It is now known that the changes, although seasonal in some cases, are caused by the shifting, windblown dust and rubble on the barren surface.

If you could telescopically observe the earth from Mars, make a diagram of how the earth would appear (phases) at different places in its orbit. How would the moon appear? Put the moon in the diagrams to scale.

In 1887, when the Italian astronomer Giovanni Schiaparelli saw what appeared to be very thin lines on the Martian surface, he called them "canali," or channels. Unfortunately, the word he used was translated into "canals," which created the impression that intelligent life existed on Mars. Other astronomers, notably the American, Percival Lowell, spent long hours charting these illusory lines and expanded Schiaparelli's canali into a full-blown fantasy (Fig. 18.17).

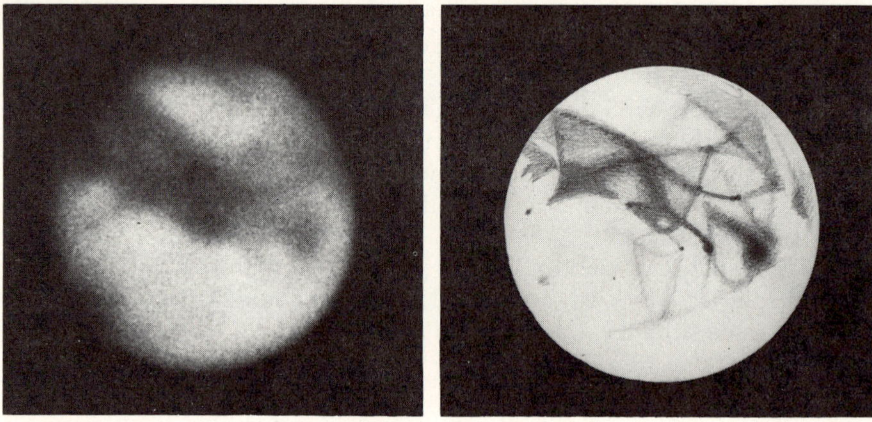

18.17 The planet Mars; a photograph taken from the earth and a 1926 drawing (photograph from the Lick Observatory).

The concept of the existence of Martians has remained much alive in the everyday imagination. Certainly the panic created in 1939 by Orson Welles' dramatic radio presentation of H. G. Wells' *War of the Worlds* indicated how pervasive this concept was in the thinking of many people. It is still very popular today, as witness the frequent linking of reported sightings of Unidentified Flying Objects (UFO's) with "Martians."

Perhaps, therefore, it was a great shock to many when the first Mariner spacecraft sent back pictures showing the first close-ups of the surface (Fig. 18.18). Not only were there no obvious signs of life, but there was nothing that even resembled "canals" or "lines." The most obvious features of these photographs were moonlike craters, crumbled ridges, and smooth desertlike terrain. The pictures showed a Mars that was quite different from the earth and indeed from anything that had been generally imagined about Mars.

Observe Mars through a small telescope. Make a drawing and describe your observation.

In November, 1971, the spacecraft Mariner 9 was put into orbit around Mars and became the first artificial satellite of another planet. It

18.18 The Martian surface, photographed by Mariner 6 in 1969. The large crater at the south edge of the photograph is about 15 miles across. The highly irregular terrain to the northwest of this crater is debris constituting the rim of a much larger crater over 150 miles in diameter lying to the southwest (courtesy Jet Propulsion Laboratory).

mapped almost the entire surface, revealing a great variety of land formations. Most provocative were landforms closely resembling terrestrial dry river beds, complete with deltas and tributary patterns and teardrop formations around higher terrain (Fig. 18.19). Most scientists agree that this is strong evidence that at one time Mars had flowing water. Only a flowing liquid could cause such land features and water is the most likely candidate for that liquid.

However, liquid water is conspicuously absent on Mars. Until 1976, when the Viking spacecraft reached Mars and landed on its surface, the possibility of finding any water seemed quite remote. The Martian atmo-

18.19 A Martian landscape: this photomosaic was made from 15 photographs taken by Viking 1 in 1976 from a range of 1680 kilometers (1040 miles). Evidence of ancient lava flows and water channels can be seen. It is also possible that the water may have "ponded" behind the ridges before cutting through to form "water gaps" (JPL-NASA).

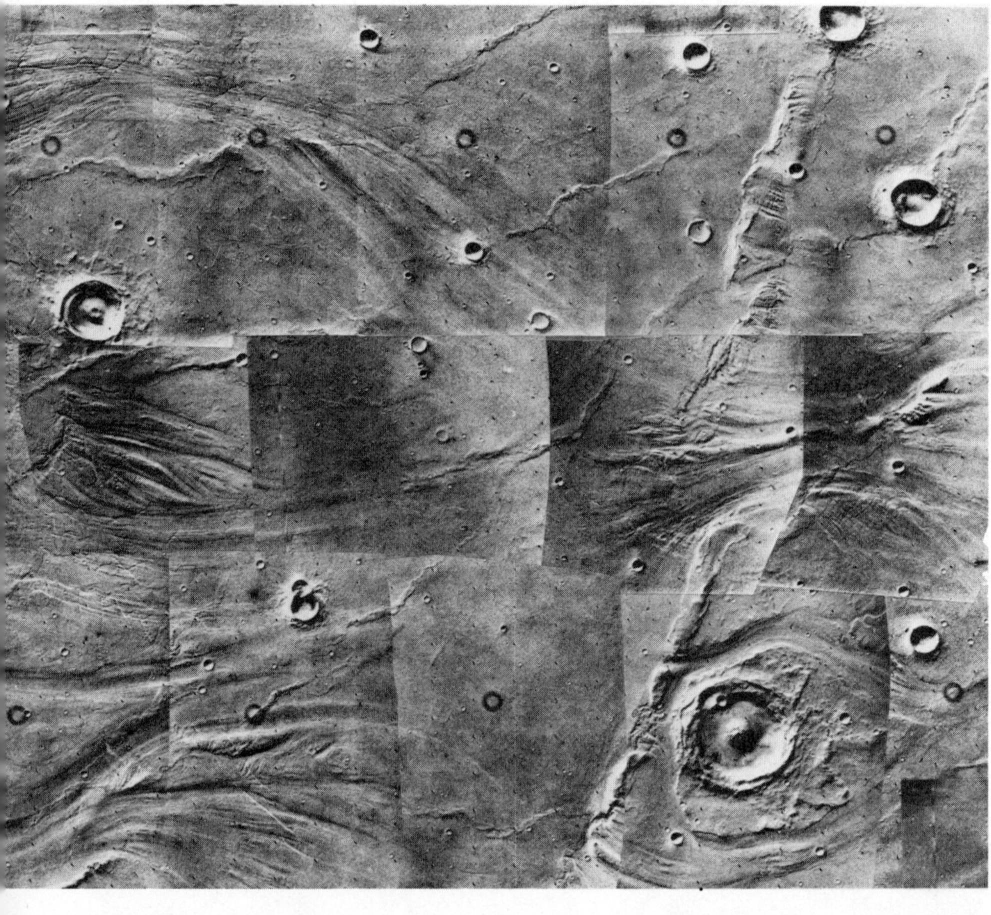

sphere, which is only about 1% the density and pressure of earth's (at sea level) has at most 0.1% water vapor. It is 95% carbon dioxide.

The Martian polar caps that wax and wane with the seasonal changes grow large in the Martian winter from the condensation of the carbon dioxide into a thin frozen layer that extends down to the middle latitudes. The opposite polar cap grows smaller at the same time as its carbon dioxide melts and evaporates into the atmosphere. But the Viking instruments found that the residual cap which never melted was ordinary ice. The depth of this "trapped" water is uncertain but has been described as "abundant." Such a reservoir exists at either pole.

The Viking lander also discovered that about 1% of the Martian soil was water. It was released when the Viking instruments heated the soil several hundred degrees. The soil is normally very dry, since the water in it is chemically bonded to other elements. It is possible that still more water exists below the surface in a permafrost condition.

Thus, water does exist on Mars but is either frozen or trapped in the soil. It is therefore quite possible, if not certain, that water did flow at one time on the Martian surface. Mars' orbit, like that of the earth's, goes through very long-period changes which could have tremendous influence on the planet's climate. If Mars' temperature was once warmer, the water would have been liquid and flowing. However, at that time the atmospheric pressure must have been much greater; otherwise water would have very quickly evaporated in such very thin air.

The photographs from the Viking landers revealed rocky terrain, sand dunes, and low hills in the distance. Of course, it must be remembered that the landing sites were chosen carefully for their relative flatness and smoothness to assure a safe landing. Far more spectacular were the features shown in the photographs of the Viking orbiters as well as in earlier orbiting spacecraft (Fig. 18.20).

There are a dozen very large volcanoes on Mars, as well as many smaller ones. All are believed to be dead. The largest, Olympus Mons (Fig. 18.21), is 26 kilometers high and 500 to 600 kilometers in diameter. This diameter is greater than the distance between Chicago and St. Louis and the height is three times higher than the highest mountain on earth! Another large volcano, Alba Patera, although only six kilometers high, is 1600 kilometers wide. Considering that Mars is about one-half the size of the earth, these are indeed enormous features.

The Martian landscape also has vast rifts or canyons, some of which have been compared to the continental rifts found deep beneath our oceans here on earth. The Valles Marineris, for example, is as long as the United States is wide. (Fig. 18.22). It stretches almost one-third of the way around the planet. In places it is 110 kilometers wide and several kilometers deep. Other rifts or canyons, although not as big as this one, are still huge by earth standards.

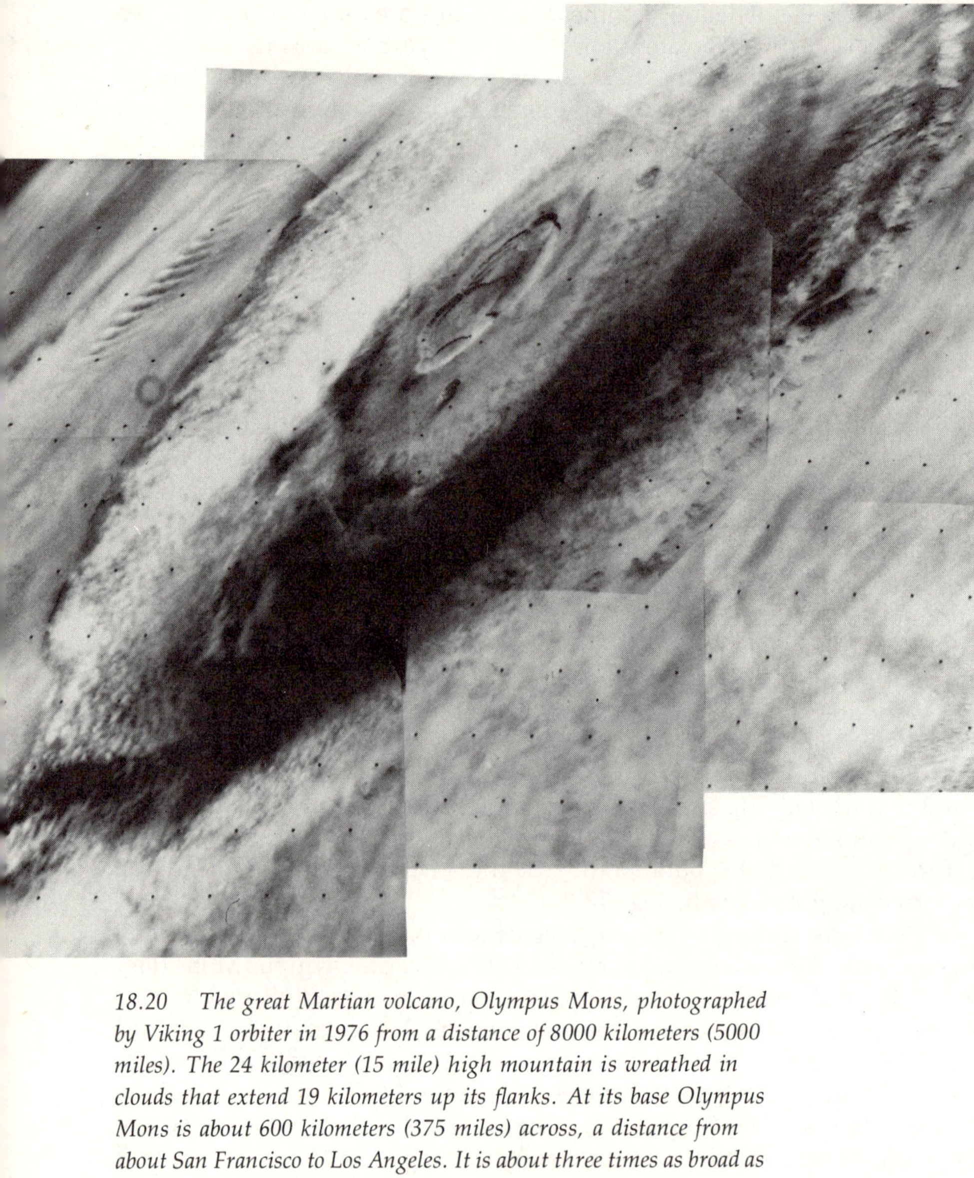

18.20 The great Martian volcano, Olympus Mons, photographed by Viking 1 orbiter in 1976 from a distance of 8000 kilometers (5000 miles). The 24 kilometer (15 mile) high mountain is wreathed in clouds that extend 19 kilometers up its flanks. At its base Olympus Mons is about 600 kilometers (375 miles) across, a distance from about San Francisco to Los Angeles. It is about three times as broad as Mauna Loa on Hawaii which is the largest volcanic pile on earth (JPL-NASA).

The surface of Mars is covered with large amounts of dust which completely obliterate the surface features during dust storms. These storms, with winds up to 100 mph, may develop very quickly and last

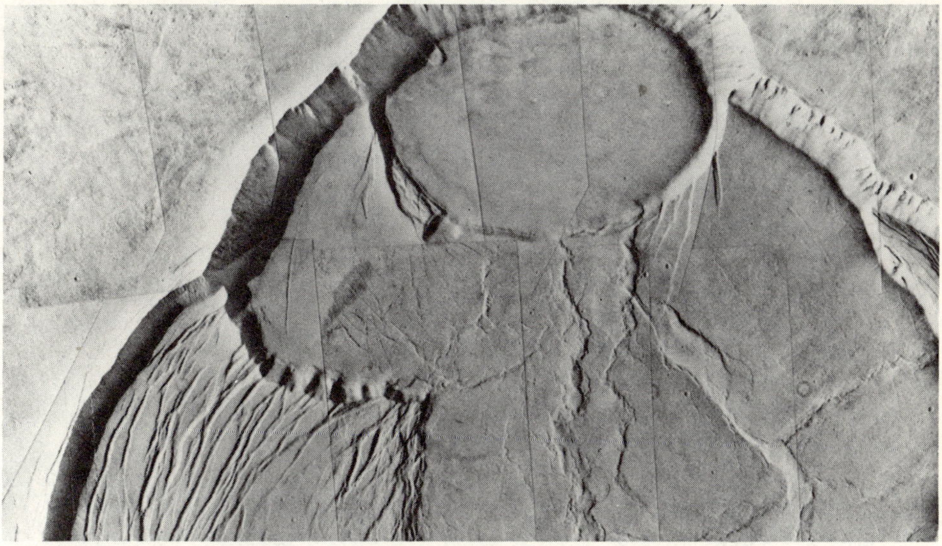

18.21 The caldera or summit crater of Olympus Mons. This mosaic of 13 high resolution photographs taken by Viking 1 shows features only 18 meters (60 feet) across. Such details will allow scientists to plot the millions of years of history of this tallest known mountain in the solar system. Easily distinguishable are the varied levels of frozen lava lakes from a series of ancient eruptions. The crater walls are about 2½ kilometers (1½ miles) deep (JPL-NASA).

Draw a map of the United States. Superimpose on it (to scale) an outline of the Martial volcano Olympus Mons and the Martian canyon Valles Marineris. Also to scale, superimpose a circle that shows the size of our moon.

---

several weeks. In the process, changes are made in some of the smaller surface features and in the surface coloration. When no storm is in progress, surface winds are very gentle, gusting to perhaps 19 mph. They do not move the dust very much at all. However, much of the dust stays suspended in the thin atmosphere and its reddish color causes the sky to appear pink as the sunlight is reflected off the airborne particles.

The redness of the dust and other surface rocks and soil is believed to be due to iron oxide (rust). How this was caused is not certain but the

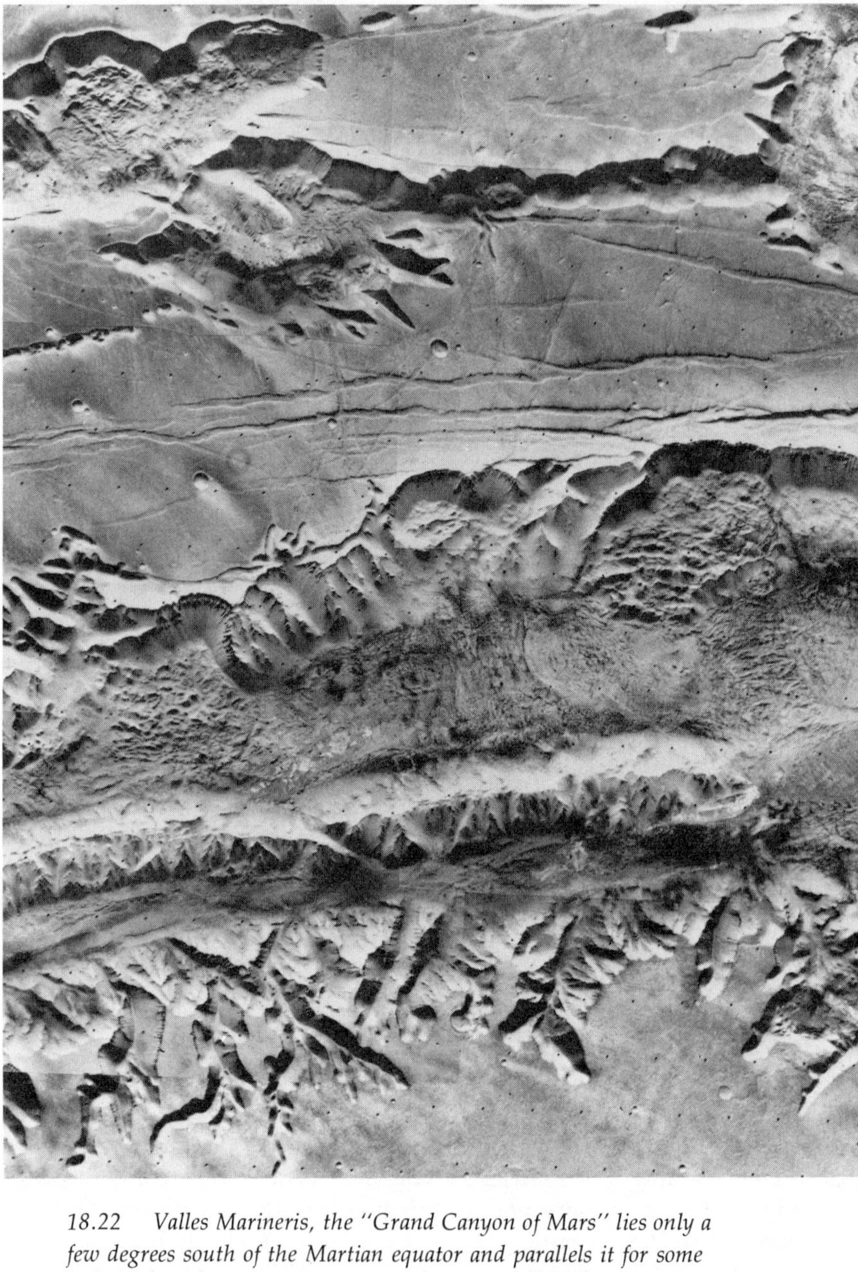

18.22 Valles Marineris, the "Grand Canyon of Mars" lies only a few degrees south of the Martian equator and parallels it for some 5000 kilometers (3100 miles). This photomosaic was made from pictures taken by Viking Orbiter 1 in 1976 from an average range of 4200 kilometers (JPL-NASA).

redness of the rocks and sky was clearly seen in the Viking photographs (see Plate C).

Mars resembles the earth only in the length of its day and in the inclination of its axis, which, as on the earth, produces the seasons. Its mass, however, is only about one-ninth that of the earth's, which means that its surface gravity is only about four-tenths that of the earth's. This would account for the thin Martian atmosphere, the lighter gases having escaped into space after they were released from the still-forming protoplanet mass.

Such a thin atmosphere gives little protection against the sun's radiation, even though this is only one-half the amount received by the earth. The ultraviolet solar radiation reaching the Martian surface is sufficiently intense to constitute a fatal dose by terrestrial standards.

The Viking I lander which landed on Mars July 20, 1976 recorded subzero temperatures both day and night, although other previous measurements have indicated that, at the equator at least, the temperature can approach that of a pleasant spring day here on earth. However, nighttime temperatures plummet, even at the equator, to −120 F and − 150 F.

The picture we get of Mars, then, is a cold, dry, barren land with an extremely tenuous atmosphere composed primarily of carbon dioxide. One of the major purposes of the Viking probes was to determine if life existed on this planet. Although the results of the various experiments proved controversial, thus far no definite signs of life have been detected, either at present or in fossil form. The photographs certainly show no such signs. The most complicated molecules found were carbon dioxide and water. No complex organic molecules were detected, although there is the possibility that the instruments used were not appropriate for such detection.

Scoops of Martian soil were subjected to heat, nutrients, water, and other chemicals in the many experiments aboard the landers. Some showed results of activity which resembled biological reactions, but not in the amount or duration expected from organisms on earth. Scientists do not know why such discrepancies occurred but they are not ready to completely rule out the possibility of life. Even if there is no life now, it is possible that very primitive life did exist at one time. Water may not have been present on the planet in a liquid form long enough for more advanced life to evolve.

In the seventeenth century, Johannes Kepler, in his belief in an orderly universe, declared that Mars had to have two moons, although no one had yet observed any. Reasoning that if the earth had one and Jupiter had four (only four were visible to the early telescopes), then by mathematical progression (1, 2, 4, . . .), Mars had to have two. Not until 1877 were its two tiny moons discovered and he was proven correct, although for very unscientific reasons.

The two moons were appropriately named Phobos (fear) and Deimos (panic), the two sons of Ares, the Greek god of war ("Mars" in Roman

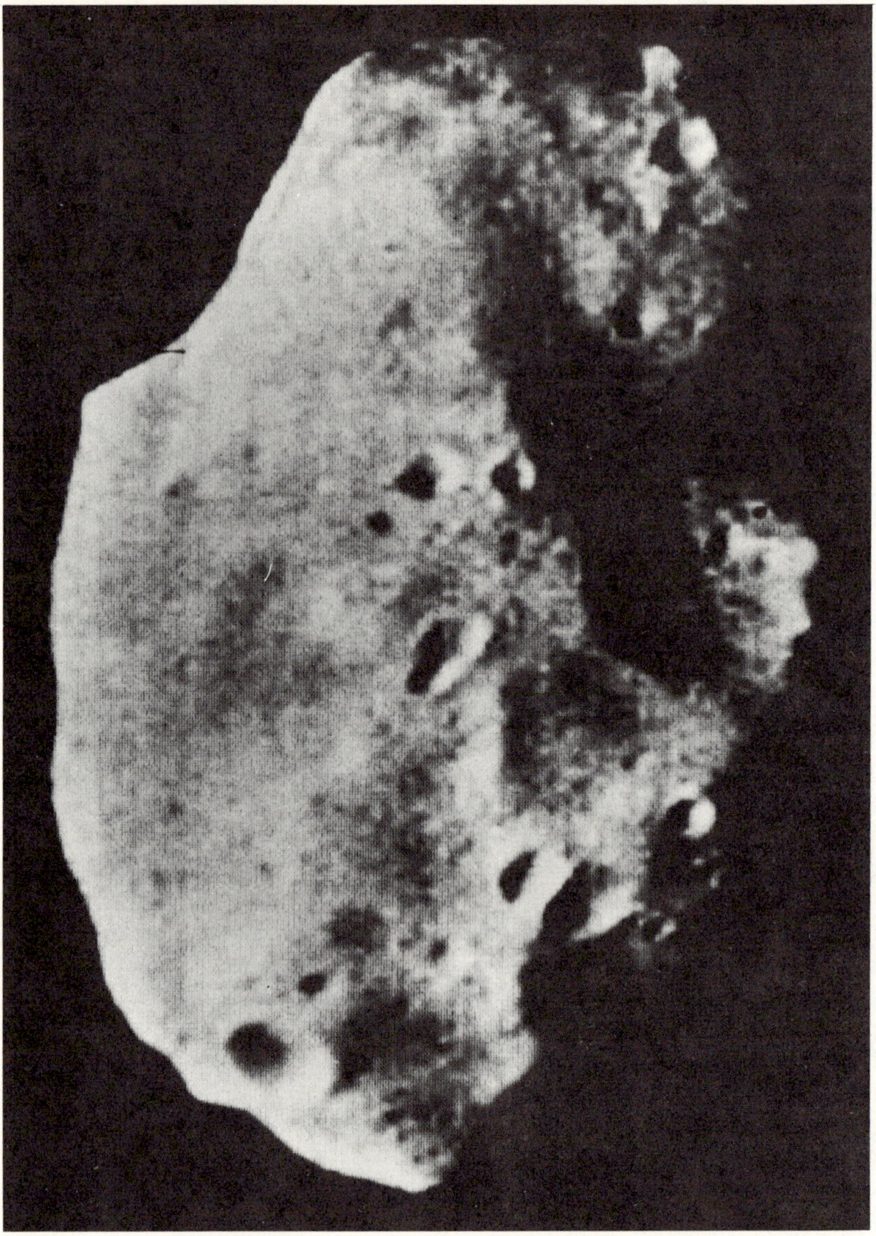

18.23 Phobos, one of the two tiny moons of Mars, photographed by Mariner 9 in 1971. The many craters suggests that Phobos is very old and has considerable structural strength (courtesy Jet Propulsion Laboratory).

mythology). Both orbit Mars in almost perfectly circular orbits very close to the Martian celestial equator. Neither is ever visible from the Martian poles. An observer on Mars would see Deimos, which is the farthest of the two, move slowly across the sky from east to west in about 60 hours. It would be as bright as Venus is in terrestrial skies. Phobos, on the other hand, would appear to rise in the west and move very quickly to the east, although it is actually orbiting in the same direction as Deimos (i.e., the counterclockwise direction of most revolution and rotation in the solar system). It revolves about Mars so much faster than Mars rotates on its axis (about eight hours against a bit over 24 hours) that it appears to move in a retrograde direction. It is brighter than Deimos in the sky but still would appear less bright than our crescent moon.

Photographs and other data received from the Mariner and Viking spacecraft revealed that both moons are cratered, nonspherical bodies resembling asteroids in many ways (Fig. 18.23) (see Chapter 20). Phobos also shows parallel grooves on one side, posing more questions as to the origin of these two relatively tiny objects. They are both almost identical in shape, with Phobos measuring 27 kilometers in its longest dimension, about the size of the asteroid Eros. Deimos' longest dimension is only 15 kilometers.

Discussion Questions

1. Suppose you were a future astronaut whose mission it was to land on Venus. How would your task and equipment be different from that of the astronauts who landed on the moon? How would a mission to Mars be similar or dissimilar to the Apollo missions?

2. What is the synodic period of a planet? How does this differ from the sidereal period? Draw a diagram illustrating the difference.

3. The closer a planet is to the sun, the faster is its orbital velocity. Why? Give some other astronomical examples of this principle.

4. What factors have determined the present composition of the planets? What factors determine whether a planet will have an atmosphere?

5. Which planets can never be seen near the moon when it is at full phase? Why?

6. How did Galileo use the phases of Venus to prove that the Ptolemaic system was incorrect?

7. The Moon, Venus, and Mars are all seen in the western sky one evening. Draw a diagram of the orbits of these bodies and indicate where each would have to lie in its orbit in order to be seen at this time.

8. Can Venus ever be seen at midnight from the earth?

9. Which of the planets can never appear at opposition?

19 The Giants of the Solar System

Revolving far beyond the orbits of the terrestrial planets, the giant planets Jupiter, Saturn, Uranus, and Neptune, are a marked contrast in the solar family. Each has its distinctive characteristics but they all possess a similar chemical composition: they are all almost entirely composed of hydrogen and helium.

Jupiter

When the Greeks gave the name of their mightiest god to the giant planet Jupiter, it could hardly have been more appropriate. With more than 300 times the mass of the earth, Jupiter is the largest, most massive of all the planets. It is almost 2½ times the mass of all the other planets combined. It is the center of a system of thirteen known moons, which is often referred to as a "miniature solar system."

Through a small telescope, one sees colorful atmospheric bands, prominent markings such as the Great Red Spot, and the four major moons. Galileo first saw these features with his elementary telescope in the early seventeeth century and ever since they have been standard astronomical showpieces for viewers with small telescopes and even with good, firmly held binoculars. In honor of their discoverer, the four major moons are often referred to as the *Galilean satellites* (Fig. 19.1).

The discovery of these four moons was the first proof that other planets could have satellites. It also helped to prove that the Copernican theory was right. Through the telescope, the four giant satellites move in an everchanging pattern. Because they all revolve nearly in one plane—the Jovian equatorial plane—they are visible as little white dots stretched out in a row on either side of the planet. Because of their relative nearness to Jupiter, their orbital periods are correspondingly short ranging from about 1½ days to 16½ days (Kepler's law). They are frequently eclipsed by Jupiter and just as frequently seen to cross *(transit)* Jupiter's disk, casting shadows on the clouds, and to an observer on Jupiter, producing solar eclipses at these

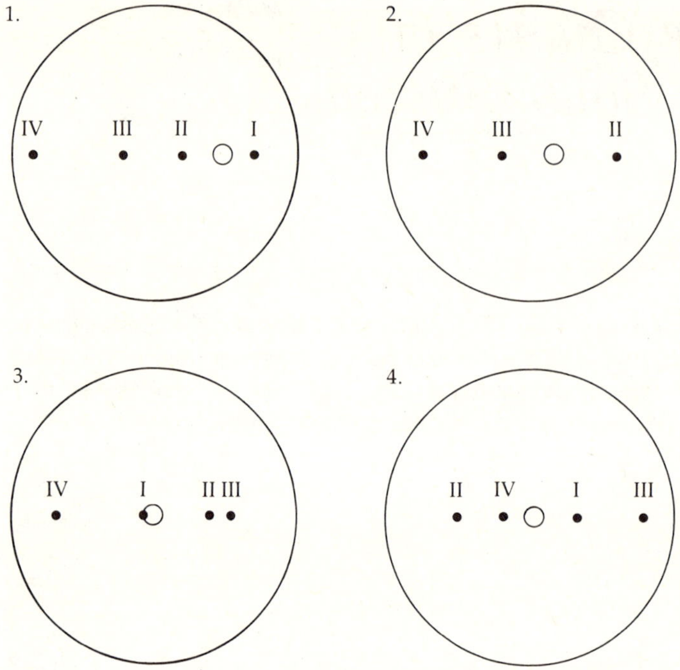

19.1 The changing configurations of Jupiter's four largest moons as seen from the earth on successive evenings. Note that on the second day, Io (I) is eclipsed by Jupiter, whereas on the following night it appears in front of the planet. All four moons pass in front of or behind the planet at one time or another, as seen from the earth.

points (Fig. 19.2). Named after Jupiter's lovers in Greek and Roman mythology, they are, in order from the sun, Io, Europa, Ganymede, and Callisto. Ganymede is larger than the planet Mercury; Callisto and Io are larger than our moon (see Fig. 18.1). All three are more massive than our moon.

When Pioneer 10 flew by Jupiter in 1973, it revealed that Io and Ganymede have highly rarified atmospheres, with perhaps a billionth the pressure of the earth's at sea level. Io's atmosphere was discovered to have

---

With a small telescope, record the positions of Jupiter's four largest moons over a period of a month or more. If enough observations have been made, try to calculate the period of each moon. Make diagrams or drawings of each observation.

19.2 The giant planet Jupiter, showing the Great Red Spot. The smaller darker spot near the top of the planet is the shadow of the moon Ganymede, which can also be seen in this photograph taken with the 200-inch telescope (photograph from the Hale Observatories).

a yellowish glow, arising from sodium vapor present in it. This contrasts with the satellite's reddish brown surface of salt and sulfur and still redder polar caps. The other major Jovian moons are believed to have a gravel-like soil and frozen surfaces.

From Io, Jupiter would loom very large in the sky, covering 18° on the celestial sphere—about the same angular size as the Big Dipper appears in our sky! Much attention has been paid to Io because it appears to trigger immense bursts of radio energy from Jupiter. These short but intense emissions represent the second most powerful radio source in the solar system, exceeded only by the sun. The emissions are probably closely connected with Jupiter's magnetic field, but how Io's presence sets off the bursts is not well understood.

The four Gallilean satellites are enormous compared with the other nine known moons of Jupiter. All but one of these relatively tiny satellites have orbits outside the orbit of Callisto. In 1975, the International Astronomical Union officially gave names to these small bodies, again using Greek mythology. Previously they had been known only by a Roman numeral in order of their discovery (see Fig. 18.5).

The outer eight of the tiny Jovian moons form two groups of four moons each with respect to the inclination and the direction of their orbits. The outer four moons have retrograde motion and all eight have highly eccentric, inclined orbits. This is in contrast to the Galilean satellites and to Almathea, all of which revolve close to the Jovian equatorial plane. Current theories hold that either each of the two outer groups was captured at the same time by the planet's gravitation or that some larger moon was fractured very early in its history, creating smaller fragmentary satellites.

There is some evidence of a 14th Jovian moon, but it has not yet been confirmed. Many planetologists believe that Jupiter has many more undiscovered moons, possibly too small to be detected from earth with current equipment. Future space probes may well reveal some of them.

Jupiter itself is eleven times the diameter of the earth, yet it has an equatorial rotational period of a little less than 10 hours. This rapid rotation combined with the giant size of the planet results in a speed at the equator of 35,500 kilometers (22,000 miles) per hour, contrasting with the earth's equatorial speed of slightly more than 1000 miles per hour. Such rapid spinning on its axis causes a very obvious polar flattening and an equatorial bulge. There is almost a 9000 kilometer difference between Jupiter's equatorial and polar diameters (the earth's equatorial diameter is a little less than 12,600 kilometers).

The colorful atmospheric bands, easily visible through a small telescope, are also caused by Jupiter's rapid rotation (Fig. 19.3). Hot gases are constantly rising in convection currents to the upper clouds, cooling off, and falling back again. If Jupiter was not rotating or was rotating at a much slower rate, the effect would be a mottled appearance from the rising convection cells. This is in fact what is seen at and near the Jovian poles where the actual motion is not as great as at lower latitudes. At the equator, however, the rapid rotation forces the vertical convection currents into horizontal bands, parallel with the equator.

These atmospheric bands alternate dark and light. The dark ones are reddish brown and the light ones are white or pale yellow. The Pioneer 10 and 11 spacecraft confirmed that the light bands have cooler temperatures and are higher in the atmosphere than the darker bands. The atmosphere circulates as warm gases rise and form cool, white clouds in the upper atmosphere. After cooling, the gas tends to descend on either side of the light-colored bands, forming the darker bands at lower altitudes. The cause of the specific coloration of these light and dark bands is unknown but is believed to result largely from temperature variations.

The surface of the upper clouds, which is the only part of Jupiter that

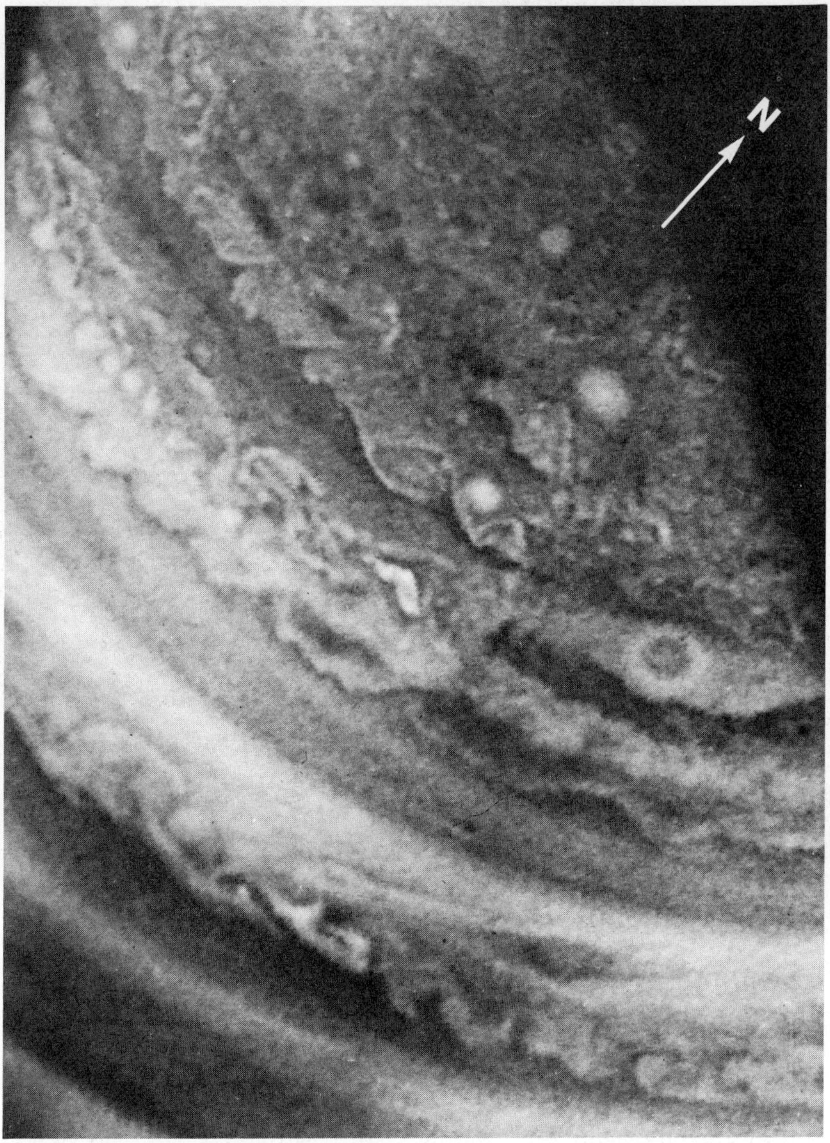

19.3 The Jovian clouds: An enlargement of a Pioneer 11 photograph. Note that as the polar regions are approched, the belts appear to break down into swirling scalloped and oval structures and then disappear completely. The spiral features are believed to be caused by wind shears between adjacent counterflowing jet streams passing each other at close to 500 kilometers per hour (NASA).

19.4 The giant planet Jupiter photographed by Pioneer 10, showing details of the planet's cloud tops never before seen. Note the sharp nucleus of cloud formation near the equator. These are believed to be stretched out by jet streams with velocities of about 380 kilometers per hour (NASA).

we can see directly, features not only colorful striations but also swirls, streaks, spots, loops, and other distortions. Pioneer 11 took close-up photographs of this "surface," revealing a wealth of detail. Many hurricane-like storms, powered by internal heat, are believed to be present in the violent atmosphere. Some are huge and long-lasting, with wind speeds of several hundred kilometers per hour. The most famous feature, the Great Red Spot (Fig. 19.6), is believed to be an immense storm covering about 14,000 kilometers in width and from 30,000 to 40,000 kilometers in length. This oval-shaped disturbance is about three times the size of the earth, and

19.5 Jupiter as photographed by Pioneer 11 two million kilometers from the planet. The spacecraft was above the Jovian north pole on its way to Saturn. Note how the dark belts and bright zones disappear near the polar region. The Red Spot is just visible in the lower right (NASA)

is known to have lasted over 300 years. It towers many kilometers over the surrounding clouds, which are deflected around it as they stream past, leaving a swirling, trailing tail. At such a height, the temperature would be low enough for phosphorus to condense and it is suggested that this may be responsible for the reddish color.

Thus, we have a picture of a very violent and turbulent atmosphere. We can see only the uppermost part of this layer of gases, but as with the sun, we can build a model of Jupiter's structure based upon our knowledge of the Jovian density, mass, gravitational field, and other physical characteristics. Data from the Pioneer spacecraft have been invaluable in this research.

According to the calculated model, the cloud layer is only about 1000

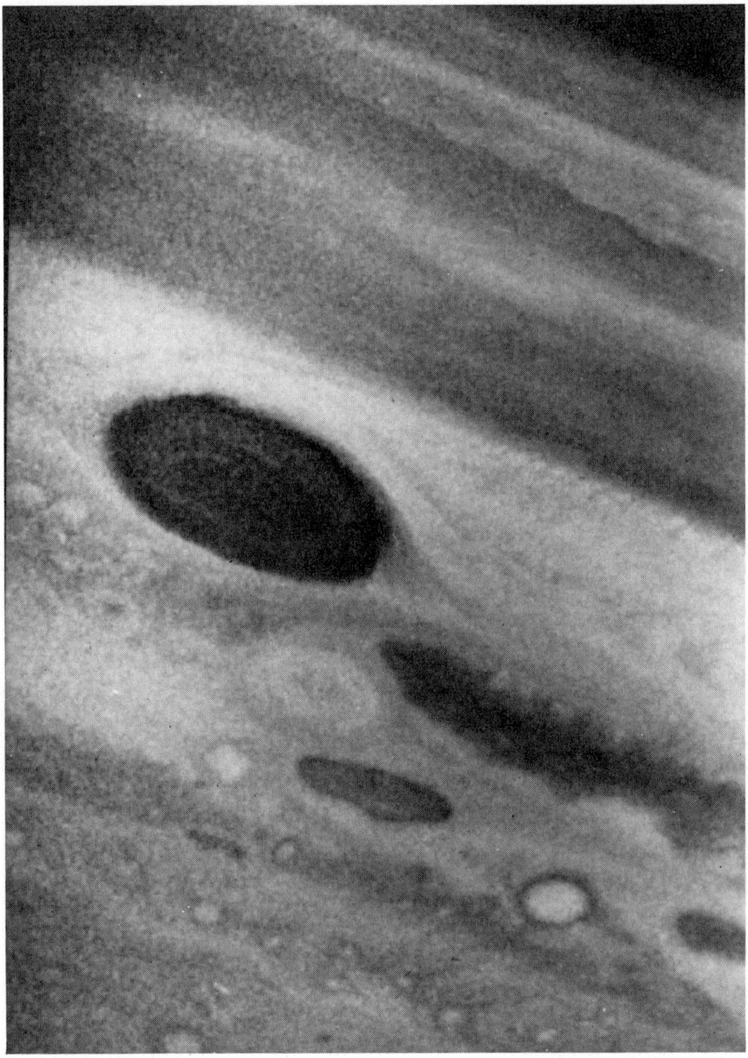

19.6 The Great Red Spot. This is the closest picture yet obtained of this feature. The streaming white clouds north of the Spot move in an opposite direction from those south of it. This is believed to cause the Spot itself to rotate counterclockwise, and details of such motion can be seen within the Spot. Other turbulent areas can also be detected. The photograph was taken by Pioneer 11 (NASA).

kilometers thick. As in the sun and stars, its main components are hydrogen and helium. Jupiter also has small amounts of methane, ammonia, water, and traces of other substances. The elements or compounds that cause the darker coloration in the girdling bands are not yet known although it is believed that the white and light yellow bands are frozen ammonia crystals.

Below the clouds, almost to the very center of the planet, is a very thick layer composed primarily of liquid hydrogen. In the inner part of this stratum is a form of liquid hydrogen that can exist only under extremely high pressures and is consequently unattainable in terrestrial laboratories. This is liquid hydrogen in a metallic state, which is believed to play a part in the creation of Jupiter's strong magnetic field.

Deep within this large mass of liquid hydrogen, a small rocky core of iron and silicates may exist. If Jupiter's chemical composition is similar to the sun's (and it is believed that it is) there should be a small amount of these heavier elements, but like the earth and other planets, they are believed to have collected near the center because they are heavier. This central rocky core must be relatively small because although Jupiter's volume is 1000 times that of the earth, its mass is only 318 times. In other words, its density is far lower than that of the earth or of any of the other terrestrial planets, all of which are composed mainly of the heavier elements.

Jupiter receives only $1/27$th the solar energy received by the earth but emits twice as much as it receives. At the top of its clouds the temperature is several hundred degrees below zero, but it is estimated that the temperature of the tiny solid core is 30,000 K. Thus, Jupiter is far from being a cold body, although it is very slowly cooling off. Its radiation, unlike the sun's, is merely the left-over heat from its original coalescence.

Because of its composition and size, Jupiter has sometimes been described as "the star that failed." If Jupiter had been able to gather more mass during the original formation of the solar system, it would have had the internal pressures and temperatures necessary for nuclear combustion. The sun, however, is 1000 times more massive than Jupiter, so at best Jupiter would probably have evolved into a very dim, small red star.

Jupiter's unexpectedly huge and powerful magnetic field was first detected by Pioneer 10 in December 1973 when it flew within 81,000 miles of the planet. From its radio emission, Jupiter was thought to have such a field, but its complexity and strength were unknown at that time. Near the planet the field strength varies from 10 to 40 times that of the earth's. It is also upside-down, i.e., a compass on Jupiter would point to the south pole. The magnetic field extends out some 10 million miles, making an enormous magnetosphere in which particles from the solar wind are trapped much in the manner as those encountering the earth's magnetosphere. If Jupiter's magnetosphere was visible to the naked eye, it would appear about twice the size of the moon on the celestial sphere.

This magnetosphere's tail extends beyond the orbit of Saturn, some 10,000 times Jupiter's radius. (By contrast, the earth's magnetotail is only 1000 times the earth's radius.) On the sunward side, the Jovian magnetosphere is constantly bombarded by the solar wind pushing it in and out. This "kneading" causes the magnetosphere to emit bursts of electrons, which have been traced as far away as the orbit of Mercury. Thus, Jupiter's influence is felt in many ways throughout the solar system.

The radiation within the magnetosphere is 5000 to 10,000 times that of the terrestrial Van Allen Belts and would be even higher were it not for the four giant moons. These, as well as tiny Amalthea, all lie within the magnetosphere and sweep up many of the high-energy particles. Because they are subjected constantly to this intense radiation (many times greater than is necessary to kill a man), should future missions to Jupiter be manned, they would have to land on one of the tiny outer moons, beyond the deadly radiation of the magnetosphere.

Although no life forms are known to exist on Jupiter, the organic chemicals on which life is based do exist in the Jovian atmosphere. Any life (as we know it) would have to float in this atmosphere, remaining at a temperature level hospitable to its existence. Considering the great turbulence present on Jupiter, such stability is difficult to imagine.

Saturn

Resplendent in the sky as seen through a small telescope, Saturn's ring system is its most obvious feature (Fig. 19.7). Although solid in appearance, the rings can best be described as a concentric series of belts made up of tiny individual satellites, each in its own orbit around Saturn. These chunks of matter, with diameters ranging from about four to thirty centimeters (1½ inches to one foot) are believed to be either ice particles or ice-covered particles. There are many millions of them, all orbiting Saturn in a very thin plane a few kilometers thick, directly above the planet's equator.

Extending almost from the planet itself out to a diameter of 270,000 kilometers (171,000 miles), these belts of satellites are actually four separate rings, all in the same plane. The two brightest ones are clearly visible through a telescope. They are separated by a dark region called the *Cassini Division.* The third ring, lying closer to the planet than the two bright ones, is called the *Crepe Ring,* and the fourth ring, which was discovered only in 1969, reaches almost to the planet itself. These rings are probably either a moon that never formed or one that came too close to Saturn and was destroyed by the crushing tidal forces.

19.7 Saturn and its ring system (photograph from the Hale Observatories).

Observe Jupiter or Saturn (or both) through a small telescope. Make a drawing of your observations.

Although the rings appear substantial when viewed through a telescope, they are extremely thin. If we were to represent the width of the rings by the size of a large sheet of newspaper, their thickness would be proportional to the thickness of the sheet of paper! In fact, the rings are so thin that stars can actually be seen shining through them. A spacecraft passing through them from top to bottom would take no more than a few hundredths of a second.

Due to the orientation of their plane with our line of sight here on earth, the rings, as well as the rest of Saturn, present a changing picture to us over approximately 29½ years, the period of Saturn's orbit around the sun (Fig. 19.8). During the period in which the rings are viewed edge-on from the earth, they entirely disappear from view for several hours, because of their thinness. Their nature, therefore, was an enigma to the early observers. Galileo, who first discovered them, never really knew what they

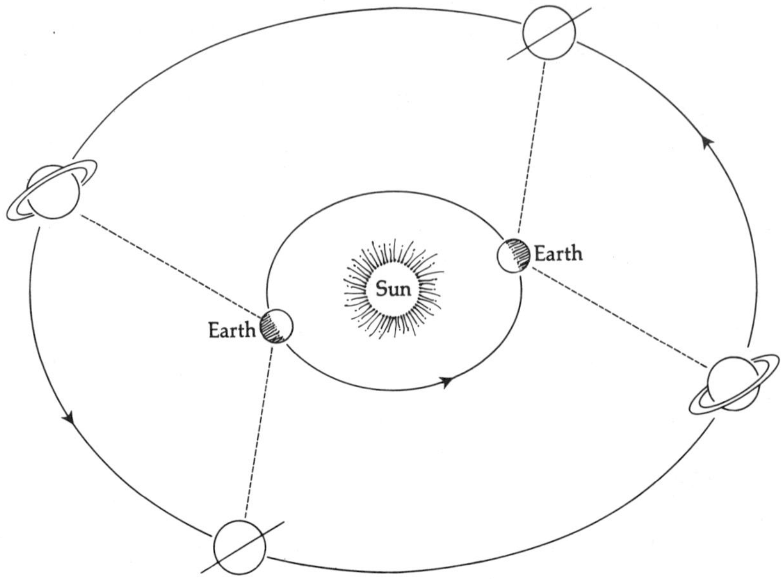

19.8 Orientation of the rings of Saturn as seen from the earth, showing Saturn at different places in its orbit.

were. Through his telescope, the rings first appeared as two small bodies on either side of the main planet and when they disappeared, he wondered if Saturn had "swallowed its own children."

Saturn is the second largest planet in the solar system, almost as large in diameter as Jupiter. However, its mass is less than a third that of Jupiter's and we therefore have an enormously bloated, lightweight giant. Saturn's density is so low that it would actually float on water if a large enough ocean were available! Its mean density is comparable to ordinary

19.9 The rings of Saturn opening from an edgewise view towards a maximum presentation of the southern face. A full cycle of presentations—edgewise, southern face, edgewise, northern face, edgewise—requires about 29½ years, the period of Saturn's orbit (courtesy of New Mexico State University Observatory).

hardwood. Because it has so little mass, it lacks the gravity to "squeeze" itself into as compact a sphere as Jupiter has.

Saturn resembles Jupiter in its composition and general appearance. Both planets are primarily composed of hydrogen and helium, the major elements of the primordial cloud which formed into the solar system. We can only see Saturn's cloud tops but it is believed to have a deep atmosphere. Because Saturn, like Jupiter, radiates about twice the energy it receives from the sun, a hot interior is indicated.

Saturn rotates once every ten hours, almost as fast as Jupiter, and this causes a similar flattening and a banding of the visible clouds. As with Jupiter, the bands run parallel to the planet's equator, but appear less striking, with shades of yellow predominating. Tremendous windstorms and other atmospheric violence are also indicated, caused by the rapid rotational velocity and convection currents from the hot interior.

Radio astronomers have detected strong but infrequent bursts of radio emission from Saturn. They are about one-eighth the intensity of

19.10 The emersion of Saturn from behind the moon as seen from the Earth, November 13, 1967. The time given below each photograph is Universal Time (courtesy New Mexico State University Observatory).

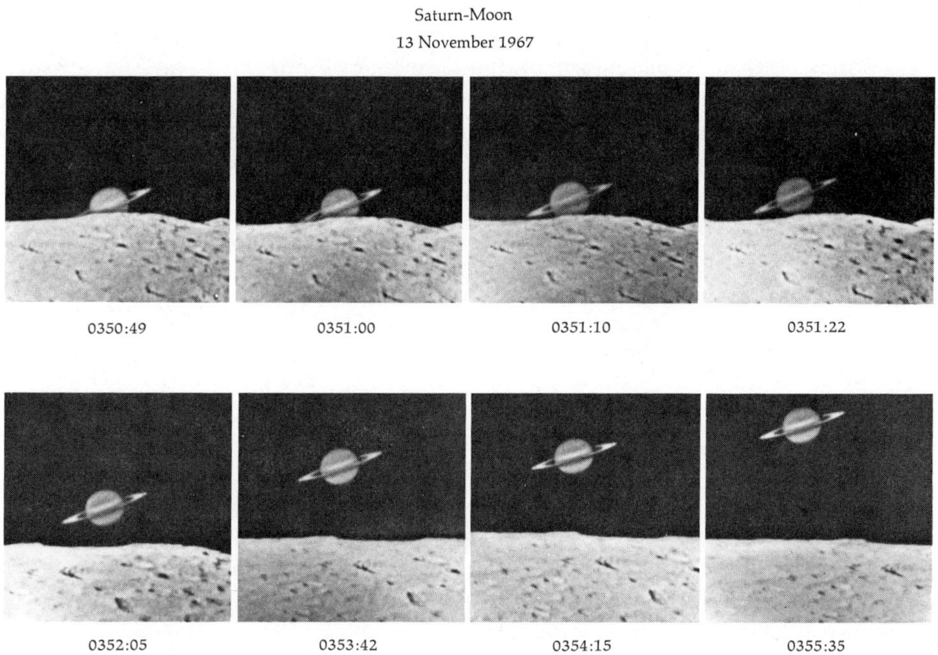

those received from Jupiter. This indicates the presence of a relatively strong magnetic field and encircling radiation belts, probably about one-eighth the strength of Jupiter's.

Well outside the ring system, ten moons are known to revolve around Saturn. Evidence of an eleventh moon seems quite positive although not completely confirmed. Most of Saturn's moons are relatively small bodies and because of this and their great distance from us (Saturn is twice as far away from the sun as Jupiter), we know very little about them. However, one moon stands out as a most unusual object. That is Titan, the sixth moon out from Saturn, and perhaps the largest moon in the solar system. It is three times larger than any other moon of Saturn and is, in fact, nearly 1000 kilometers larger than Mercury. It is estimated to be about 5800 kilometers in diameter. However, it is less massive than Ganymede.

Through the telescope, the surface of Titan is not always completely visible. Reddish-brown clouds obscure the ground. Its atmosphere is composed primarily of methane and hydrogen and has a pressure about four times that of the Martian atmosphere, approaching an atmosphere as thick as the earth's. Although the surface may be ice and the temperature may be way below freezing at all times, some of Titan's conditions may be similar to those on the very young earth. For this reason, there is strong biochemical interest in this moon. Some exobiologists believe that the kind of atmosphere Titan appears to have at present is the kind that the earth had in its very primitive stage when life was just evolving.

Thus far our knowledge of Saturn is based mainly upon observations from the earth or from earth-orbiting spacecraft. No spacecraft has yet flown close to this distant body. However, Pioneer 11, having already flown by Jupiter, is due to arrive there in September 1979. Other spacecraft are also scheduled to visit the planet in future years, including *Voyager*, which was launched in mid-1977. A much greater understanding of this giant planet, its rings, and its moons is in prospect.

Uranus

From the earth, Uranus is visible to the naked eye if the viewing conditions are excellent. It has an apparent magnitude of +5.7. The ancients must have seen it, but because of its extremely slow motion through the sky (only 4½°) along the ecliptic each year), mistook it for a very faint star. In fact, it was plotted as a star on sky charts long before William Herschel identified it as a planet in 1781.

Through the telescope, Uranus appears greenish in color with no distinct markings. Its deep atmosphere is believed to be composed mostly of hydrogen and helium, although 1% of methane probably causes the greenish hue. Because it is more than twice the distance from the sun as Saturn, its upper clouds probably have temperatures below −300° F. Little is known of its interior, although it may well have a solid body of rock and ice below the atmosphere.

Describe a day on Uranus (assuming that an astronaut could be stationed on its outer cloud layer). What sort of calendar can you devise for this peculiar planet? What would be the difference between a sidereal and solar day on Uranus?

Uranus' greatest peculiarity is the orientation of its equator to its orbit. It "rolls along on its side" in its 84-year orbit about the sun (Fig. 19.11). The planet's axis is only 8° from its orbital plane (the earth's is 66½°). If you were at one of Uranus' poles, you would see the sun for 42 years and then live in darkness for the next 42 years.

The years 1976 and 1977 brought many new and unexpected discoveries about Uranus. The length of its day (rotational period), long believed to be about 10.8 hours, is now known to be about twice that figure, making it more like the earth and Mars in this respect. It was also found to be about 4000 kilometers larger in diameter than previous measurements (51,800 kilometers).

19.11 *Uranus revolves around the sun with its axis tilted 98° from its orbital plane instead of in a more nearly vertical position like other planets. This means that its pole is tipped 8°* below *the plane of the ecliptic. Thus each pole has 42 years of darkness followed by 42 years of sunlight.*

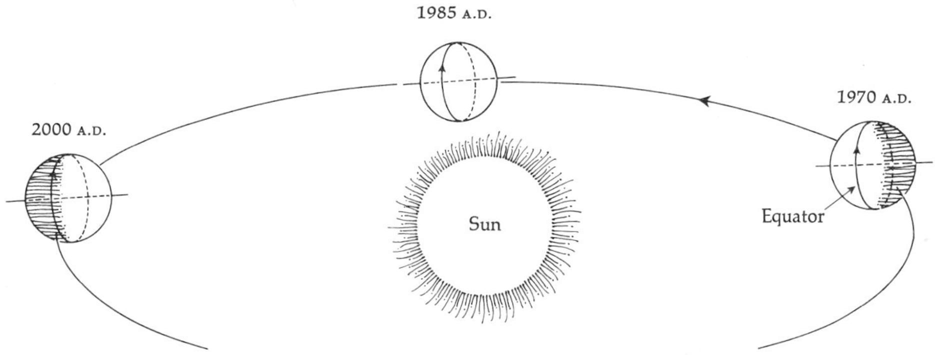

But perhaps most amazing was the discovery of an extensive series of belts or rings of small bits of matter orbiting closer to Uranus than its five previously known moons (Fig. 19.12). Assumed to be similar in composition to Saturn's rings, the rings of Uranus are much narrower, covering a total band width of only about 7000 kilometers (the largest of Saturn's rings is more than 25,000 kilometers wide).

This discovery, made early in 1977, was most unexpected. Scientists attempting to measure the size of Uranus by timing its occultation of a star discovered a whole series of "dips" in the starlight's intensity before and after Uranus itself blocked off the radiation. Since the spacing and duration of the light minimums before and after the main occultation were almost equal, these "mini-occultations" are believed to be Saturn-like rings rather than individual moons.

The newly discovered rings, like the five Uranian moons, are in the plane of Uranus' equator. Whatever caused Uranus to orbit on its side, rather than in a more upright position like the other planets, was probably also responsible for the moons and rings to follow this same strange orientation.

Neptune

The presence of an eighth planet was not even suspected until the nineteenth century. Its prediction and subsequent discovery caused an international scientific uproar. The French and English both claimed the honor of discovery.

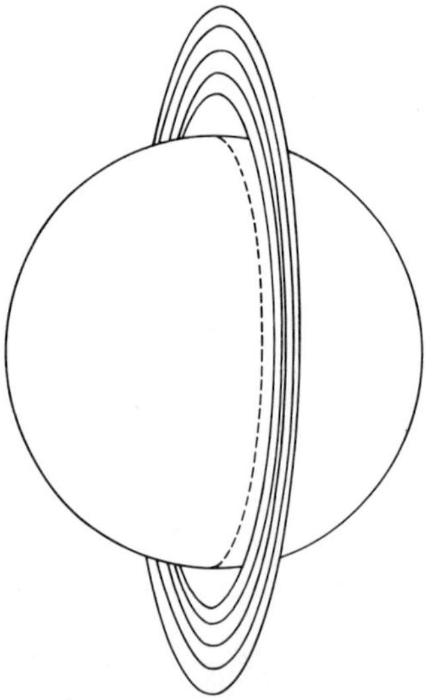

19.12 The rings of Uranus: a schematic drawing. The dotted line is the Uranian equator.

Studies of Uranus' orbit indicated that the gravitational pull of some unknown body was causing Uranus to deviate from its predicted orbit. In the mid-nineteenth century, two young astronomers, John Adams in England and Urbain Leverrier in France, calculated the position of this perturbing body. Adams tried unsuccessfully to get leading astronomers in his own country to search for the unknown object. Not until Leverrier published his predictions did the English astronomers respond. However, in 1846, the German astronomer Johann Galle, responding to a message from Leverrier, found the new planet within an hour after going to his telescope. Today the credit for the discovery is shared by both Adams and Leverrier and it represented a great triumph for celestial mechanics. Using the laws of Kepler and Newton, Neptune's position had been predicted accurately before its actual discovery.

Neptune has been called the twin of Uranus because of their many similarities. They are about the same size and exhibit the same greenish color, due to the small amount of methane in their atmospheres. Both atmospheres are also believed to be composed mainly of hydrogen and helium.

Over a billion miles farther from the sun (and from us) than Uranus, Neptune cannot be seen with the naked eye. A small telescope or a good pair of binoculars can detect it easily, although nothing more than a small greenish disk will be observed. Its motion among the stars is extremely small, remaining in one constellation for about fourteen years. With a period of 165 years, it will not have completed its first revolution since its discovery until the year 2011.

Neptune has two known moons. The larger one, Triton, may possibly be the largest moon in the solar system, with a diameter of 6000 kilometers. It moves in a retrograde orbit which makes it even more of a maverick. The smaller moon, Nereid, orbits outside Triton's orbit and was discovered in 1949. It has a very eccentric orbit, varying from one to six million miles from Neptune.

There is evidence that Neptune's day, like Uranus', may be longer than previously thought. A 22-hour day is very possible. If it is found that Uranus, Neptune, and Mars all have days similar in length to that of the earth, such a coincidence would have to be examined and explained in any solar system theory. As we have seen, it is very easy to theorize when facts are few. As more information continues to be collected, theories must become more detailed and encompassing.

Pluto

Pluto is an anomaly. More like a terrestrial planet than a giant one, it is by far the smallest planet and might well be numbered among the moons of the solar sytem instead. Indeed, there are some indications that Pluto was once a moon of Neptune.

From Pluto, which is almost 5.9 billion kilometers away (over 3½ billion miles), the sun would appear as an extremely bright star, although it sends this most distant planet 250 times more light than the full moon sends to earth. Because of the brilliant glare even at this great distance, the earth and most of the other planets would probably not be visible. The earth would never appear more than 1° away from the sun and even Jupiter would only be 7° away at its greatest elongation (Fig. 19.13).

As might be expected, Pluto, as seen from the earth, is a difficult object to detect. Even with a large telescope, the tiny planet is no more than a very small yellowish spot of light (Fig. 19.14). Its magnitude is +15. This is only 1/4000 as bright as the faintest star you can see with the naked eye.

Pluto's orbit is tilted 17° to the plane of the ecliptic. Other than Mercury, which has a 7° tilt, no other planet's orbit is inclined more than 3.5° to this plane. Pluto also has the most eccentrically shaped planetary orbit, even coming inside Neptune's orbit at one place, and thereby coming closer to the sun than Neptune. At aphelion it is about 7 billion kilometers away, whereas at perihelion it comes as close as 4½ billion kilometers. Even though Pluto can come within the orbit of Neptune, there is no chance of collision between the two planets because of the high inclination of Pluto's orbit to that of Neptune's (Fig. 19.15).

Pluto's magnitude has been found to vary by as much as 20% over a period of about 6⅓ days. This is interpreted as a measure of the planet's rotational period, the assumption being that not all the surface areas reflect sunlight equally, causing light variation as the planet rotates.

Earlier estimates of Pluto's size were largely based upon the amount of light reflected, using as a comparison the reflectivity of the lunar surface.

19.13 An observer on Pluto would see Jupiter only 7° away from the sun on the celestial sphere at its greatest elongation. By comparison, Mercury, as seen from the Earth, can appear as much as 28° away from the sun. But even at that angular distance, Mercury is not easy to detect with the naked eye because of the sun's brilliance.

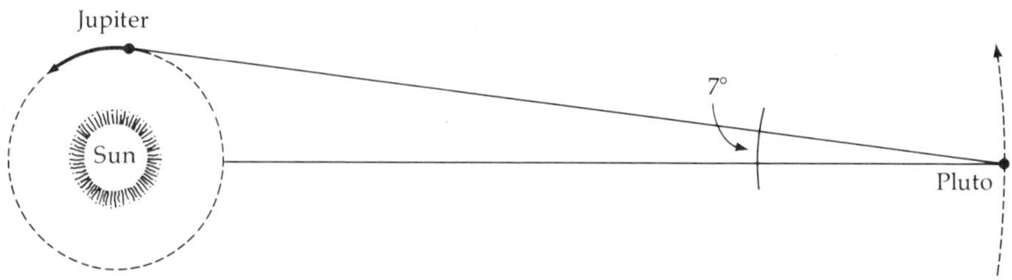

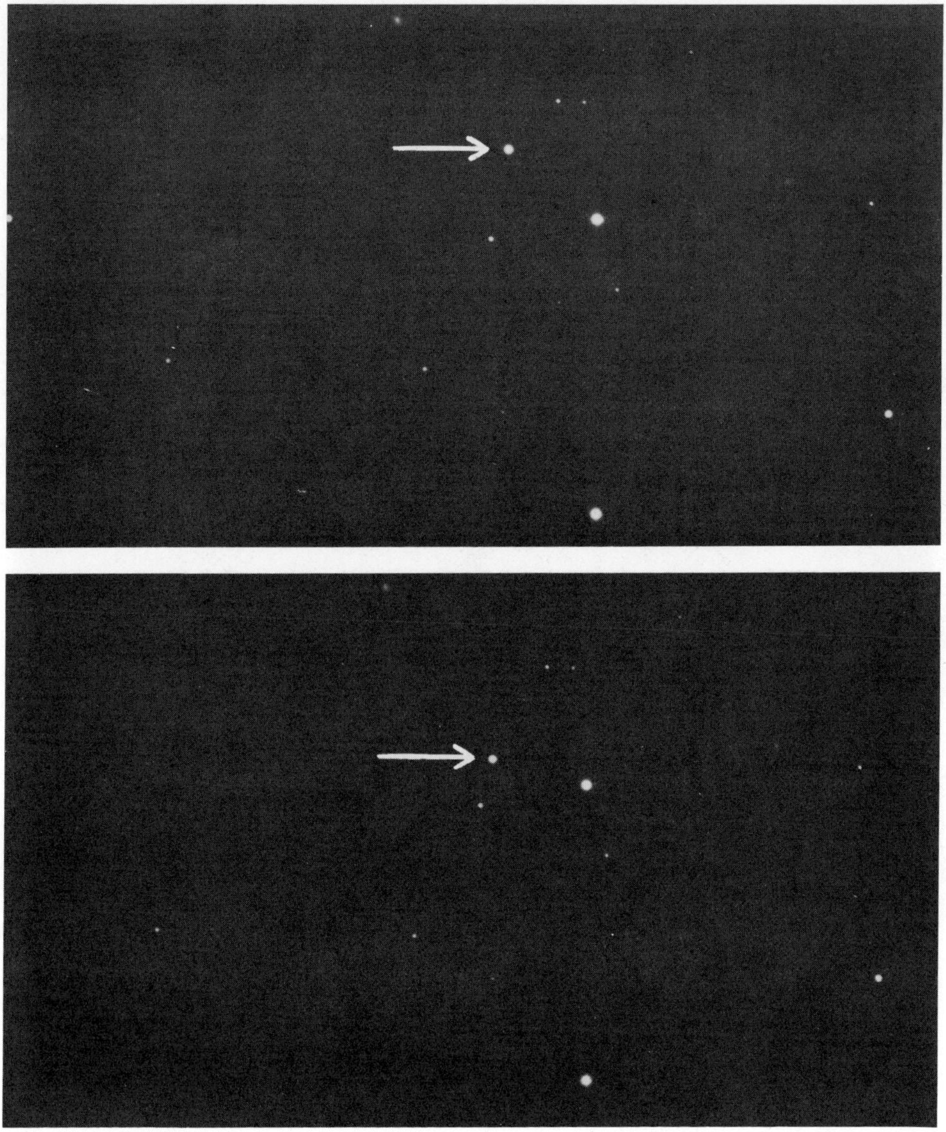

19.14 Pluto moving against background stars. The two photographs were taken 24 hours apart (photographs from the Hale Observatories).

It was believed that Pluto's surface resembled the rocky barren moon. However, in 1976, with the use of spectroscopic analysis, substantial amounts of methane ice were detected on the Plutonian surface. This was

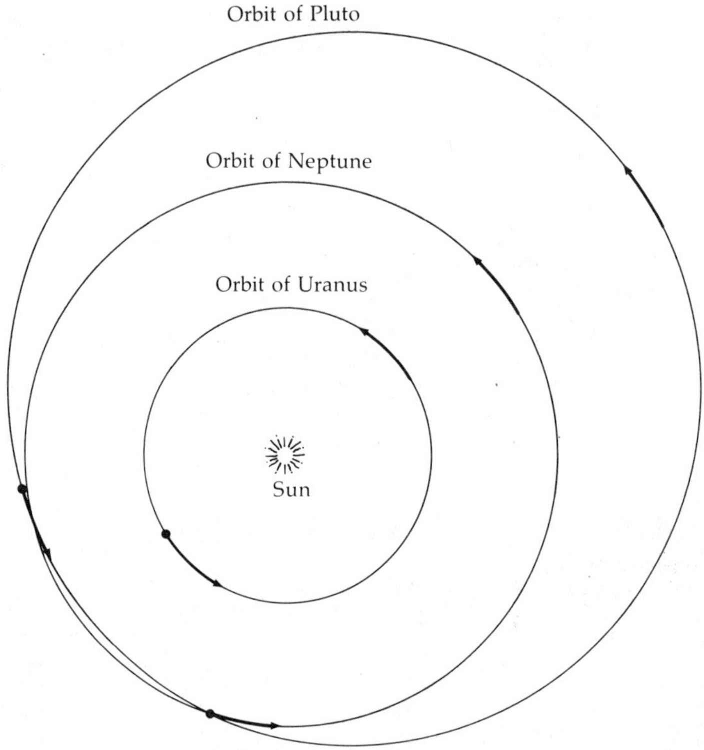

19.15 The orbits of the three outermost planets, Uranus, Neptune, and Pluto. The dots represent the approximate locations of these planets in early 1977. Pluto will move within the orbit of Neptune in 1987, never having been so close to the sun since its discovery in 1930.

the first solid methane detected in the solar system, since methane does not freeze until a temperature of −225° C is reached. This gave further proof of the extremely cold temperature calculated for this planet.

However, such an ice-covered body would be able to reflect light better than the previously assumed rocky moon-like surface. Therefore, calculations of Pluto's size based upon this reflectivity had to be revised downward. The brighter each unit of surface area is, the fewer such units are needed to produce the same total luminosity. Pluto may well be only one-third to one-half the size of Mercury, making it far more moon-like than planetary in size.

The search for Pluto began when astronomers thought that Neptune's gravitational pull did not fully acount for the observed perturbations of Uranus in its orbit. Two American astronomers, Percival Lowell and W. H. Pickering, insisted that yet another planet would be found, and Lowell attempted a mathematical prediction of its position similar to that made by Adams and Leverrier for Neptune. Then came the hunt for "Planet X," which was far more difficult than the quick discovery of Neptune. Clyde Tombaugh, a young astronomer at the Lowell Observatory, after many months of diligently studying and comparing photographs containing millions of stars, finally located the faint planet fairly close to where Lowell had predicted it would be.

Oddly enough, the discovery was a complete fluke. Pluto is too small to be responsible for any measurable perturbation of Uranus in its orbit.

---

*From an ephemeris or other source (*The Observer's Handbook *published yearly by the Royal Astronomical Society of Canada is excellent for this and other such data), find the Right Ascension of each of the planets for a specific date. Using a series of concentric circles to represent the orbits of the planets, draw a straight line from the sun at the center out to Pluto's orbit. This is the Vernal Equinox, or $0^h\ 0^m$ on the chart. Ninety degrees from that line (as measured from the sun), $6^h\ 0^m$ is located. Plot the position of the planets as they appear on this specific date. From the definition of the "ecliptic", how do you determine the Right Ascension of the earth? From your diagram, can you determine which planets were visible from the earth on that date? (See figure below.)*

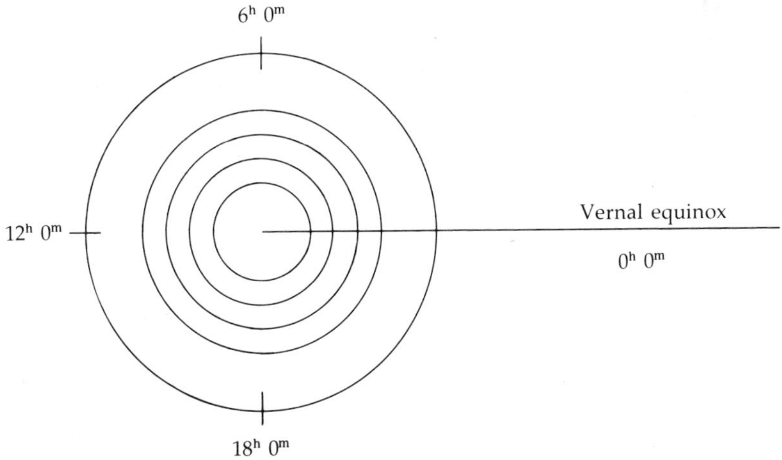

The very small amount of deviation that supposedly could not be accounted for by Neptune's gravitational pull has been found to be within the margin of error of these early measurements and really did not exist. It was truly an amazing coincidence that Pluto was found so near to where Lowell predicted. His mathematics was correct but his data were in error. Pluto's discovery was actually the result of Tombaugh's perseverence and labor, spurred on by the earlier prediction.

Although many people have speculated about still more distant planets, if one exists it must be of even smaller mass and fainter luminosity than Pluto. Otherwise, it would have already been detected, since any major unexplained perturbations in the orbit of Pluto would have been cause for yet another search.

In science fiction stories, one frequently reads of a hypothetical planet, a "counter-earth," which always remains on the opposite side of the sun from the earth, and is therefore never seen by "earthlings." However, the presence of such a planet would easily have been detected long ago by its gravitational effect on the other planets.

Discussion Questions

1. Compare the surface gravity of the earth and Saturn. Even with this similarity, why would Saturn still present problems to future space explorations?

2. Why is Pluto regarded as a terrestrial planet even though it lies beyond the major planets?

3. What are some of the reasons that Pluto is considered by some to be an escaped moon of Neptune?

4. Give at least one unusual fact about each of the nine planets, i.e., something unique to that planet alone.

5. If you could observe the earth from Mars with a telescope, make a diagram of how the earth would appear at different places in its orbit. How would the moon appear?

20 *Minor Members of the Solar System*

Billions of objects which are neither stars nor planets exist in the solar system. They are "minor" in size and mass, but their importance as the most probable representation of the original raw material of the solar system is making them major objects of research. We have already become acquainted with some of them, such as the moons of the various planets and the rings of Saturn and Uranus. However, there are many types of objects in the solar system that are not as closely associated with one of the planets and must therefore be studied separately.

Asteroids

Write out the numerical sequence: 0,3,6,12,24. . . . doubling the number each time. Now add 4 to each number in the series and then divide each answer by 10 (see Fig. 20.1). The resulting numbers correspond very closely with the distances of the first seven planets from the sun (in AUs). Only the distances to Neptune and Pluto fail to follow this quasi-numerical relationship, which is known as *Bode's Law.* Note however, that there is no planet at 2.8 AUs from the sun. Because of Bode's Law, many eighteenth-century astronomers were certain that such a body must exist there and they diligently searched the skies for it. The search was greatly intensified after the discovery of Uranus, whose distance from the sun fit this "law" so well.

On January 1, 1801, the Italian astronomer Giuseppe Piazzi, at his observatory in Sicily, finally discovered a small body which was found to be orbiting the sun almost exactly 2.8 AUs away. Hardly big enough to be called a planet (955 kilometers in diameter), it proved to be the largest of a group of thousands of objects called *asteroids,* which circle the sun in a wide belt between Mars and Jupiter. The first asteroid was named Ceres, after Sicily's patron goddess.

Shortly afterwards several others were found in that same distance range. This spurred the search for more, and, particularly after the advent

	+4	÷10	*Planetary distances from the sun in AUs*	
0	4	0.4	Mercury	0.39
3	7	0.7	Venus	0.72
6	10	1.0	Earth	1.00
12	16	1.6	Mars	1.52
24	28	2.8	Asteroids	2.65
48	52	5.2	Jupiter	5.20
96	100	10.0	Saturn	9.54
192	196	19.6	Uranus	19.2
384	388	38.8	Neptune	30.1

20.1 Bode's law.

of photography, the number of newly discovered asteroids rose rapidly. Today there are about 2000 that have been named and whose orbits are well known. However, the total number is estimated to be about 60,000 to 100,000. The great majority are small chunks much less than a kilometer in diameter ranging down to the size of mere pebbles. However, although there is a large number of asteroids, if all of them were swept together, their mass would be less than one-thousandth that of the earth. Their cumulative size would be less than that of the moon.

The asteroids are very widely scattered throughout an *asteroid belt* between Mars and Jupiter. The known ones rarely pass within a million miles of each other. In 1972 and 1973 neither Pioneer 10 nor Pioneer 11 ever sighted even one asteroid while passing through this region en route to Jupiter and Saturn.

Until recently the origin of the asteroids was believed to have been either the debris of a planet that had exploded long ago or the result of collisions between many sub-planetary bodies. Presently the latter theory is the more favored one, largely because there is no known mechanism by which a planet can explode. The sub-planetary bodies are believed to have originally been formed by the accretion of primeval matter. The disturbing gravitational pull of Jupiter probably prevented the formation of a large planet in this region. Eventually, collisions between these small planetoids led to fragmentation so that practically all the asteroids are nonspherical. The asteroid Eros, for example, measures 13 x 15 x 36 kilometers and is believed to have a rougher surface than any planet or moon studied by radar. However, the very few larger asteroids (those 150 kilometers or more in diameter) are found to be practically spherical, having somehow escaped the fragmentation caused by frequent collisions.

Although most of the asteroids orbit within the asteroid belt, some of them have eccentric orbits that take them well out of this region into other parts of the solar system. It is estimated, for example, that there are some 20,000 asteroids that cross the orbit of Mars and come into the inner part of the solar system. Some of these have approached within several million miles of the earth and have been extremely useful in plotting the exact size of the solar system.[1]

Occasionally an asteroid may come much closer to the earth and may even enter its atmosphere. For example, on August 10, 1972, a 13-foot, 1000-ton asteroid came within 60 kilometers (36 miles) of the earth's surface. The friction generated by its contact with the upper atmosphere caused it to flash across the sky with a luminosity brighter than the full moon. It was seen for hundreds of miles as it "skipped" off the earth's atmosphere and headed out into space again. It has been estimated that had it crashed into the earth's surface, it might have left a crater 100 meters wide.

There is also a group of asteroids that lead and follow Jupiter around in its orbit, keeping 60° away from the giant planet at all times. Known as the Trojan Asteroids because their principle members bear the names of the Trojan heroes, they appear much darker than the regular asteroids and are believed to have had a different evolution than those nearer the sun. At least 1000 Trojan Asteroids are known (see Fig. 20.2).

Comets

The most spectacular minor members of the solar system are the comets (Fig. 20.3). Observed from earliest times, they were most frequently regarded as omens of disaster (Fig. 20.4). Some early philosophers even held

---

[1] From Kepler's laws, the solar system can be precisely plotted in terms of astronomical units. Thus Mars is 1.52 AU from the sun and Jupiter is 5.20. In order to determine the linear distance to these planets, we must know the number of miles or kilometers there are in an astronomical unit. This cannot be discovered by measuring the distance to the sun directly. Not only is the glare and heat too great, but also there are no background stars visible in the daytime to be used for reference points for parallax triangulation.

If the distance in miles of a close object can be determined at night (for example, the asteroid Eros, which made a famous close approach in 1931), then the direct translation from miles to astronomical units can be made. This situation is similar to having a road map of the United States in which the familiar index relating inches to miles is missing. If, however, we can determine the distance between any two relatively close cities, we can tell immediately how many miles are represented by an inch and thus get the distance between any other two points on the map.

Since the advent of radar, it is no longer necessary to make the long painstaking observations and calculations of the linear distances to close objects. Man-made probes to the planets accurately tracked by radar now give us this information.

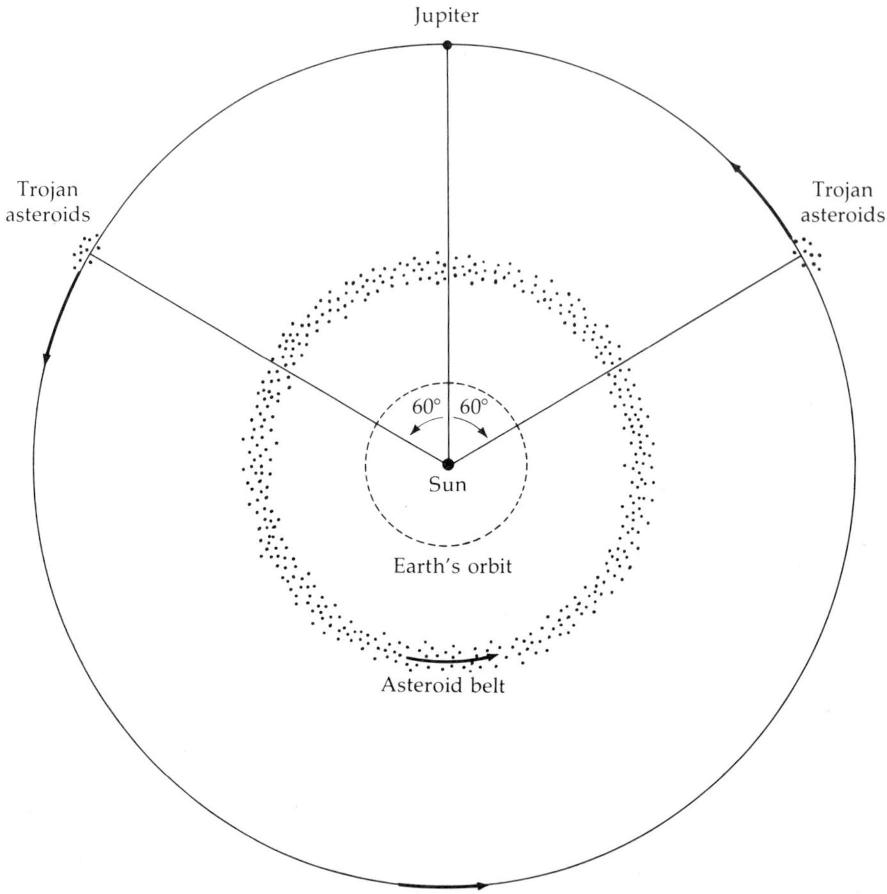

20.2 The Trojan Asteroids lead and follow Jupiter around in its orbit, always maintaining a 60° separation from the planet.

that they were sins of mankind burning as they rose through the atmosphere. It was not until the sixteenth century that astronomers realized that comets were actually distant objects, far outside our atmosphere. Tycho Brahe, unable to find a parallax for a comet, was able to establish that they could not be in the earth's atmosphere but must in fact travel far beyond the moon (Fig. 20.5).

About 1700 Edmund Halley noted the resemblance between several bright comets which had appeared at approximately 75-year intervals. He assumed that these were one and the same comet returning periodically, and applying Newton's newly developed theory of gravitation, predicted that the comet of 1682 would return again in 1759.

20.3 The Ikeya-Seki Comet, named after the two amateur astronomers who discovered it in 1965 (photograph from the Smithsonian Astrophysical Observatory).

20.4 Halley's Comet in 1066 as depicted on the Bayeux Tapestry. Crowds watch the comet in awe while King Harold is told of this terrible omen in the sky.

Halley died in 1742, but when the comet did return as he had predicted it was named in his honor. It has since returned about every 75 years and is expected again in 1986. Halley had proven that comets were normal members of the solar system revolving about the sun, and not some capricious bodies or omens of disaster.

The orbit of Halley's Comet is diagrammed in Figure 20.6. As can be seen, its orbit differs radically from those of the planets. This is true of most other comets as well. Whereas the orbits of planets are nearly circular and lie close to the same plane, comet orbits are inclined at all angles and are often extremely elongated. Some have such highly elongated orbits that they travel far out beyond the planets and take millions of years to make one revolution.

In keeping with Kepler's laws, their velocities increase as they approach the sun. By the time they reach the general vicinity of the earth they are traveling at about 40 kilometers per second. The more elongated the orbit, the more rapidly the comet will whip around the sun (perihelion)

20.5 *Comet Kohoutek. A mosaic of two 3-minute time exposures taken with the 48" telescope of Hale Observatories (photograph from Hale Observatories).*

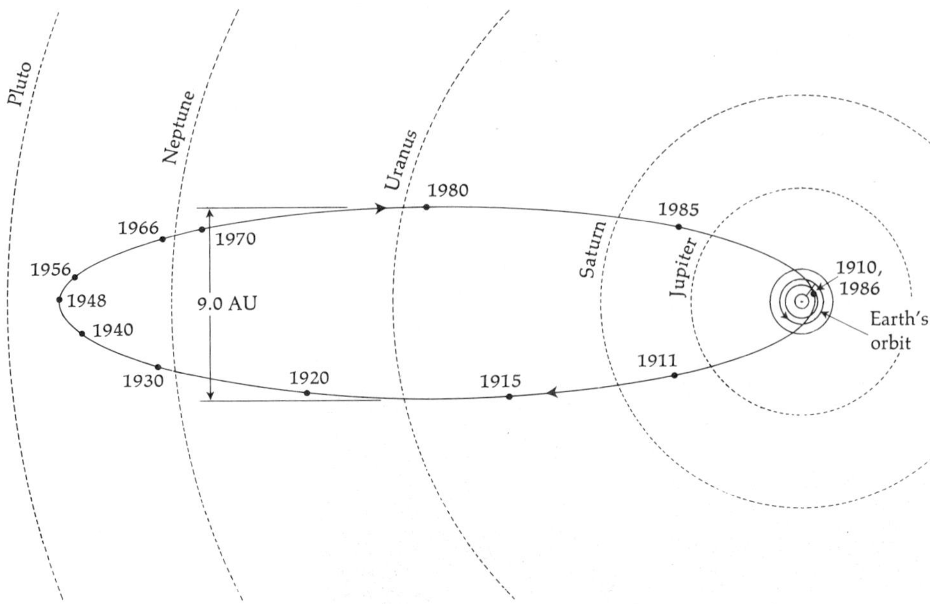

20.6 The orbit of Halley's Comet. Note that the comet is currently on its way back to the sun, near the orbit of Uranus.

before starting the long journey out into space again. There are some comets with periods of revolution of a million or more years that race 180° around the sun in just a few hours, taking the rest of their very long period to complete the other 180°.

When a comet appears in the sky, it may remain visible for many weeks, slowly changing its position among the stars night after night as it travels in its orbit about the sun. The more elongated its orbit, the closer it will come to the sun as it whips around perihelion. This proximity usually produces a relatively bright comet, generally with a long tail, and therefore one that can be seen easily with the naked eye. Such long-period comets appear unexpectedly, since at the time of their last visit to the earth's neighborhood, there was no one here to record their appearance or to calculate their orbits. Comet West, which appeared brilliantly in 1976, will not return again for at least a million years. On the other hand, there are comets such as Encke's Comet, which has such a short period that it appears as a faint telescopic object every few years.

A comet is visible only when it is near the sun. Far away from the sun it is merely a nonluminous mass of frozen gases (particularly carbon monoxide) and water ice intermixed with microscopic bits of rock. It has aptly been described as a "dirty snowball." This mass of matter is the *nucleus* of the comet (Fig. 20.7). As the comet approaches the sun, some of the material in the nucleus begins to diffuse and vaporize, forming a *coma* which, together with the nucleus, constitutes the comet's *head*. The coma first appears as a small, round, diffuse glow which brightens and greatly enlarges as the comet approaches the sun. The coma can in some cases be as much as a hundred thousand miles in diameter, whereas the nucleus itself remains relatively tiny, often no more than a mile or two in diameter.

Some of the matter that diffuses from the nucleus permanently streams away. Propelled largely by the solar wind and by radiation pressure, this streaming material produces the often spectacular *tail.* The tail grows longer as the comet gets closer to the sun (Fig. 20.8) and can reach lengths of more than 100 million miles. In the sky these tails can appear to extend over as much as half the celestial sphere. Since the propelling force for the tail material comes from the sun, the tail always appears pointing essentially away from the sun. This means that when a comet is departing the neighborhood of the sun, the tail goes first (Fig. 20.9).

Many comets exhibit two slightly separated tails. One is composed of

20.7 *Parts of a comet. The tiny nucleus and the coma make up the "head" (Photograph from Hale Observatories).*

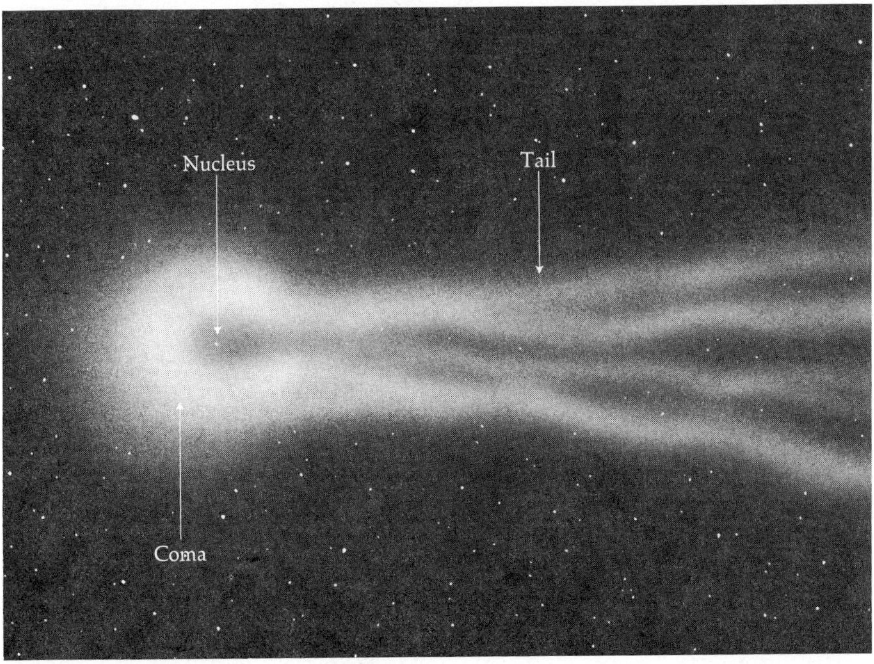

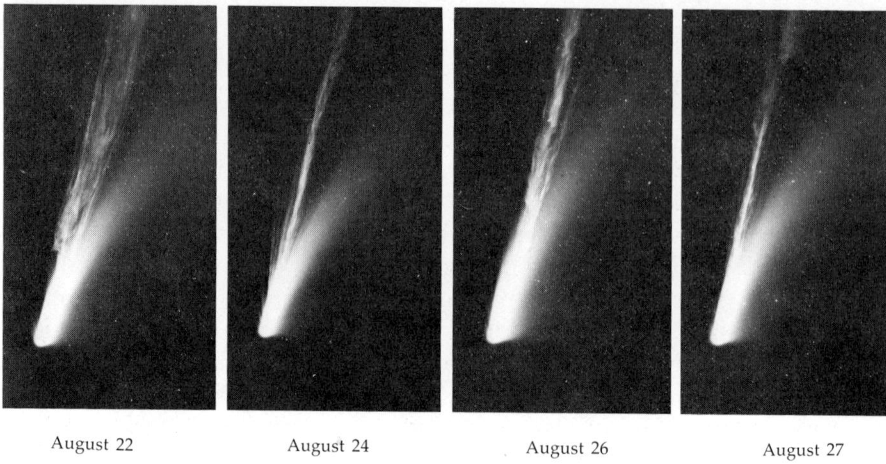

20.8 Comet Mrkos in 1957. Note the two tails streaming from this comet (photograph from Hale Observatories).

dust particles which are pushed away from the comet by the sun's radiation, and is usually yellow, reflecting the sun's color. The other tail is made up of gases from the comet and is pushed away by the solar wind particles. It is usually bluish in color, caused by the interaction of these two materials.

20.9 The comet's tail always points away *from the sun. This is due to the solar radiation and the solar wind.*

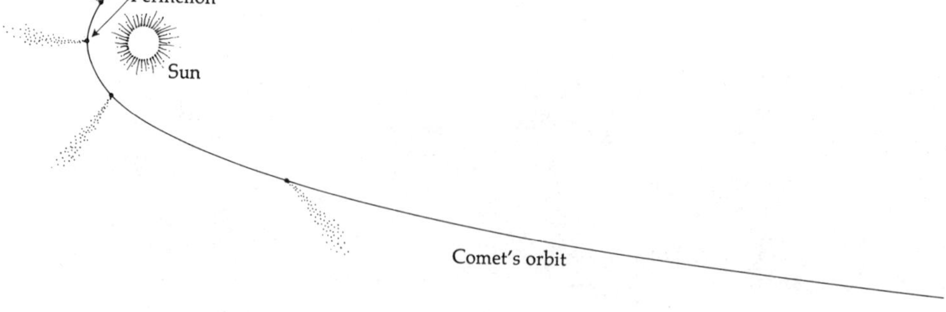

In general, the more elongated the comet's orbit is, the more spectacular its tail will be. As we have seen, such comets come relatively close to the sun, causing them to become unusually bright as the intense heat from the sun diffuses and vaporizes material in the nucleus. The actual matter making up the coma and tail of the comet, although appearing to be substantial because of its fluorescence, is extremely rarified. Only the nucleus itself is dense enough to occult the light of a star, and such blocking of starlight is extremely brief, indicating the very small size of the nucleus. It is not known whether the nucleus is composed entirely of discrete chunks of matter or has a solid core at its center.

We also do not know the mass of any comet, but their masses cannot be more than a billionth that of the earth, since no comet has been known to produce any observable gravitational perturbation on any planet. Nevertheless, a collision of a comet with a planet could cause great devastation in a local area, as well as destroy the comet. Although the probability of a collision between the earth and a comet is very small, such an event may have occurred in the past. In fact, there is some evidence that the area devastated in Siberia in 1908 may have been caused by such an occurrence.

While collisions with comets themselves would be very rare events, the earth does cross the *orbits* of some comets regularly every year (Fig. 20.10). Bits of cometary debris are scattered in broad bands paralleling

20.10 If the earth's orbit intersects the orbit of a comet, the earth will pass through the cometary debris each year. The result is a meteor shower. For example, each year early in May, the earth passes through the orbit of Halley's comet, producing the η Aquarid meteor shower. This shower is named after the constellation from which the meteors appear to radiate.

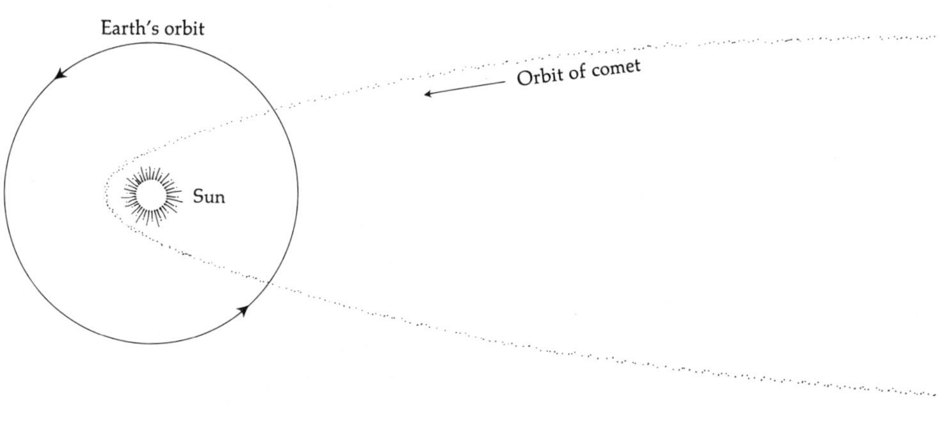

these orbits. When the earth passes through this debris, these tiny bits of matter collide with the molecules of our upper atmosphere and produce brief streaks of light in the sky as they disintegrate. These are called *meteors* but are sometimes nonscientifically referred to as "shooting stars" or "falling stars"—misleading names from much earlier times. As we will see on page 439, not all meteors are caused by the debris of a comet.

Using the list of meteor showers, plan to observe one of them. Plot the path of the meteors through the stars on the celestial sphere during the late evening and early morning hours. How many were you able to detect per hour? Where is the radiant?

However, when the earth does cross the path of such cometary debris, so many particles may be intercepted that a *meteor shower* is produced. The meteors all appear to originate from approximately the same point on the celestial sphere as the minute particles of debris hurtle into the atmosphere. This point is called the *radiant* of the shower. (Fig. 20.11) The shower derives its name from the constellation in which the radiant is located, not from the comet itself since in some cases this is unknown. For example, early in May we may observe the "Aquarid" meteor shower whose radiant is located in Aquarius. This shower is produced by the debris from Halley's Comet. In Figure 20.12, other meteor showers are listed. Note that although the term "shower" is used, most of these events consist of less than 50 meteors per hour—hardly a "shower" in ordinary terms.

Thus, as a comet moves around the sun, it loses some of its mass on each trip (the matter in the tail never returns). After a great many circuits nothing is left but a thin trail of debris. If, therefore, comets only live a relatively short time, why are there any left? Even the long-period comets must have lifetimes very much shorter than that of the earth. Are they then new acquisitions to the solar system? One of the most accepted theories today is that the comets are as old as the solar system but that most have remained in nearly circular orbits lying far out beyond Pluto. From time to time a few of these, because of the gravitational pull of the planets and stars, are slowly drawn from their circular orbits into elliptical ones, bringing them much closer to the sun. Once they are thus brought into the regions of the sun and planets, their days are numbered.

This postulated cloud or reservoir of comets is believed to extend at least 1000 times beyond the limits of the planetary system. Even though it is estimated that there may be some 100 billion comets still existing in this

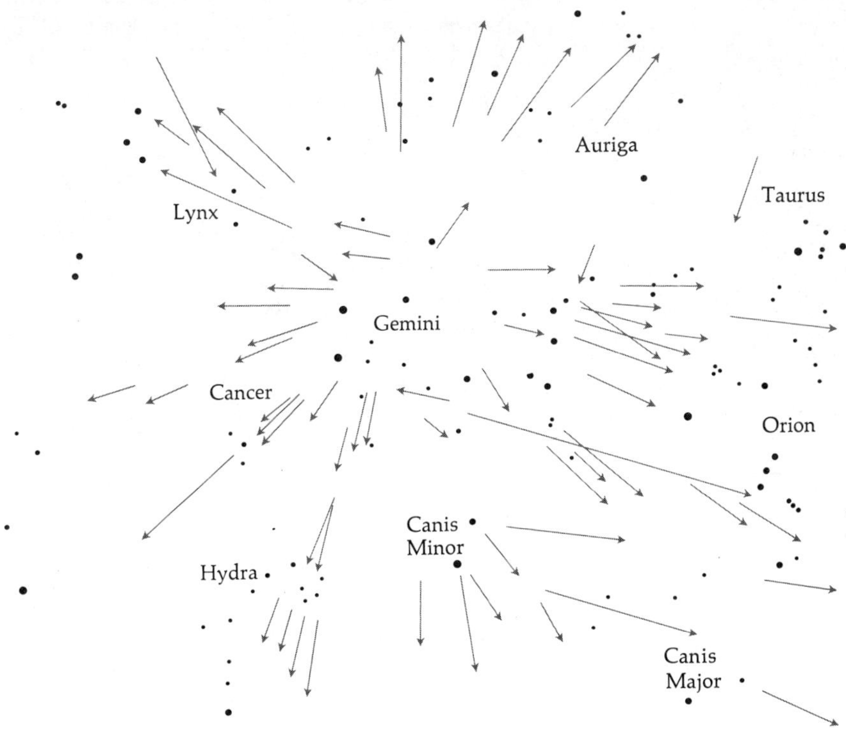

20.11 Apparent paths of meteors of the Geminid shower, which appear to be coming from the "radiant." The effect is similar to the apparent converging of railroad tracks as they are seen disappearing into the distance.

cold, dark region, their total mass is probably no more than $^1/_{100}$ to $^1/_{10}$ that of the earth.

Meteorites

All bodies in the inner solar system show some degree of cratering. Even the earth still has some remaining evidence of an earlier bombardment, although erosion has long since erased most of the clues. The Barringer meteor crater (Fig. 20.13) in Arizona is perhaps the most famous meteoritic crater on our planet because it is one of the best preserved. About 200 other such craters have been identified on the earth.

Current theory of the evolution of our solar system holds that soon after the formation of the planets and moons, about four billion years ago,

Major meteor showers					
Shower	*Date of Maximum Number of Meteors*	*Hourly Rate for a Single Observer*	*Position of Radiant R.A.* *h*	*m*	*Declination Degrees*
Quadrantids	Jan. 3	40	15	28	+50
Lyrids	April 21	15	18	16	+34
η Aquarids	May 4	20	22	24	00
δ Aquarids	July 28	20	22	36	−17
Perseids	Aug. 11	50	03	04	+58
Orionids	Oct. 21	25	06	20	+15
Taurids	Nov. 4	15	03	32	+14
Leonids	Nov. 16	15	10	08	+22
Geminids	Dec. 13	50	07	32	+32
Ursids	Dec. 22	15	14	28	+76

20.12 *Major meteor showers.*

heavy cratering occurred on all bodies with solid surfaces. The chunks of matter that had not already been incorporated into one of the larger bodies were constantly perturbed in their orbits about the sun. The gravitational attraction of the protoplanets, especially Jupiter, caused most of them to either escape the solar system entirely or to crash into one of the larger bodies, creating still more craters.

Such bombardment decreased rather rapidly as most of the debris was thus swept up. In about the next two billion years the cratering decreased to only 1/10,000 as much and has stayed at about that level ever since. Occasional meteoritic bombardment occurs on the earth and no doubt on other bodies in the solar system.

One such meteoritic fall occurred on March 8, 1976 in China's Kurin province. The meteor appeared first as a red fireball the size of the moon, then was seen to explode high in the sky with a brilliant flash of light. Within a minute after entering the earth's atmosphere, it had crashed, scattering meteoritic debris over a 500 square kilometer area (200 square miles). No such previous fall is known to have covered such a large area. The more than 100 fragments that were recovered weighed almost 6000 lbs and the largest fragment weighed close to 3900 lbs. This large single meteorite, the heaviest ever observed to fall, made a crater in the frozen ground 6½ meters deep (about 21 feet). It was traveling at an estimated 500 meters (over 1600 feet) per second when it impacted with the earth.

Such violent and unexpected occurrences must have always been extremely frightening to witnesses (Fig. 20.14), and yet, although stories of

20.13 The Barringer meteorite crater in Arizona. The crater is 1.2 km (¾ mile) in diameter and 180 meters (600 feet) deep. The iron-nickel meteorite that caused this crater has been estimated to have weighed some 250,000 tons and to have been traveling at 16 km. (10 miles) per second when it crashed (photograph from the Yerkes Observatory).

such phenomena were well known, until early in the nineteenth century, scientists refused to believe that stones and metallic chunks could actually fall from the sky. They attributed the weird appearance of these objects to the effects of lightning. Only when a major fall of meteorites occurred in L'Aigle, France were such "old wives' tales" taken seriously and the phenomena studied.

Meteorites are composed either mainly of iron and nickel or of stone. The stony types are far more frequent but, because of their distinctive appearance, the iron-nickel ones are most frequently recovered. The stony meteorites resemble ordinary stones and unless they are seen falling and quickly found, they are not often recovered. Obviously, in terrain free of stones meteorite recovery is easier; for example, more meteorites are found in Kansas than in Maine.

Probably for this reason, the ice-covered mass of Antarctica is proving to be an excellent place to find stony meteorites. In the past few years there

20.14 *An old drawing of the fall of a meteorite. Such an explosion in mid-air is called a* bolide.

have been several large finds, one of which yielded a stony meteorite of almost one-half ton. Its 33 pieces were found exposed on about two acres of barren ice where the snow cover had been removed by strong winds. The ice in which they were embedded is very old, but the age of the meteorite or when it fell to earth is not yet known.

Whereas the stony meteorite resembles ordinary rock so much that only an expert can identify it, an iron-nickel meteorite is more easily recognizable. It is considerably heavier than an ordinary rock of similar size and therefore the heaviest meteorites are of this type. The largest iron-nickel meteorite is estimated to be about 50 tons! When these relatively heavy objects plummet down through the earth's atmosphere, a fusion crust is formed which gives its exterior a brownish pitted appearance. The interior of the meteorite is very cold, having just come from outer space, but the surface will feel very hot for a short time due to the great friction created by its rapid fall. Frost, however, sometimes appears on the surface shortly after the fall—a result of the extreme cold in the interior.

Final proof of an iron-nickel meteorite is the famous cross-hatched *Widmanstaetten figures* which appear if an iron-nickel meteorite is sliced open, polished, and etched with acid (Fig. 20.15). These lines are believed to have formed when the material of the meteorite, once very hot, cooled off very slowly (about one degree every million years!). Since it is hard to explain how a body as relatively small as the average meteorite could first have been hot and then cooled so slowly in space, the Widmanstaetten figures are cited as evidence that meteorites were originally part of a much larger body.

That is one of the principal hypotheses about the origin of meteorites. If they were part of much larger celestial bodies that somehow broke up, they would resemble the interior of a moon or planet. However, if some meteorites are composed of matter that never formed into planets and moons, they represent the primordial material of the solar system.

Because no scientific records (particularly photographic) were made of the initial atmospheric entry of the Kirin meteorite, its previous orbit before encountering the earth is unknown. However, others, although not as big, have been recorded and all were found to come from within the orbit of Jupiter. This lends credence to the theory that meteorites are former asteroids, perturbed in their orbits into paths that cross the earth's orbit. There are 19 known asteroids, called the *Apollo asteroids,* that do have such orbits and have been found to have similar spectral characteristics as many of the recovered meteorites. Undoubtedly there are many other smaller Apollo asteroids, such as the one that became the Kirin meteorite.

In the early 1970's scientists discovered that several meteorites which had landed in Australia and the United States contained amino acids. These findings gave tangible support to the generally accepted belief that primordial elements in space could eventually form complicated organic molecules which could evolve into reproductive living matter. This also appears to be another indication that life may not be confined to the earth alone.

Thus we see many similarities between asteroids and meteorites. Their compositions, original orbits, and the hypotheses regarding their

20.15 A meteorite sliced in half and etched with acid to show its Widmanstaetten figures (photograph from the Yerkes Observatory).

origin are basically the same. With the growing evidence of the elements of primordial life in such bodies, the study of these small objects becomes all the more important.

Meteoroids and Meteors

The tiniest members of the solar system are the small particles of matter such as those that make up the debris from comets. Other minute particles, although not as densely concentrated, are found throughout interplanetary space. In general, all of these particles are called *meteoroids.* As we saw with the cometary debris, when such particles enter the earth's atmosphere, their rapid entry causes friction with the atoms and molecules they encounter in the atmosphere, resulting in a streak of light in the sky called a *meteor.* A meteor is thus not an object but rather an optical phenomenon caused by the ionized gases from the vaporization of the meteoroid as it hurtles through the atmosphere. If a meteroid is large enough to survive the passage through the atmosphere without completely vaporizing and lands on the ground, it is called a *meteorite.* It is unfortunate that all of these terms are so similar since the objects or phenomena that they represent are so dissimilar. To confuse the matter further, none of them is connected with the science of meteorology, which is, of course, the study of weather.

Some meteoroids are so small (micron-sized) that, unlike the larger ones, they are slowed down by the atmosphere and do not heat up as they descend to the surface. These *micrometeorites* instead float gently down. It is estimated that up to 100 tons of such tiny particles enter the atmosphere each day. We are not aware of them since they enter with a minimum of disturbance to our atmosphere. However, they can be detected as microscopic spheres mixed in with the normal dustfall that pervades our atmosphere.

We are also not very aware of most of the more rapidly entering meteroids, the ones that produce meteors. Since a meteor is an atmospheric phenomenon, the curvature of the earth effectively prevents one person from viewing more than a very limited number per night (Fig. 20.16). The heights at which meteors appear can be easily determined by simultaneously observing them from two relatively close locations (Fig. 20.17).

20.16 Because meteors occur in the upper atmosphere of the earth, each is seen by only a relatively small percentage of observers. Observer A *and observer* B *will not be able to see the same meteors although both can see more distant objects at the same time, such as a satellite orbiting the earth, or the stars, planets, or moon.*

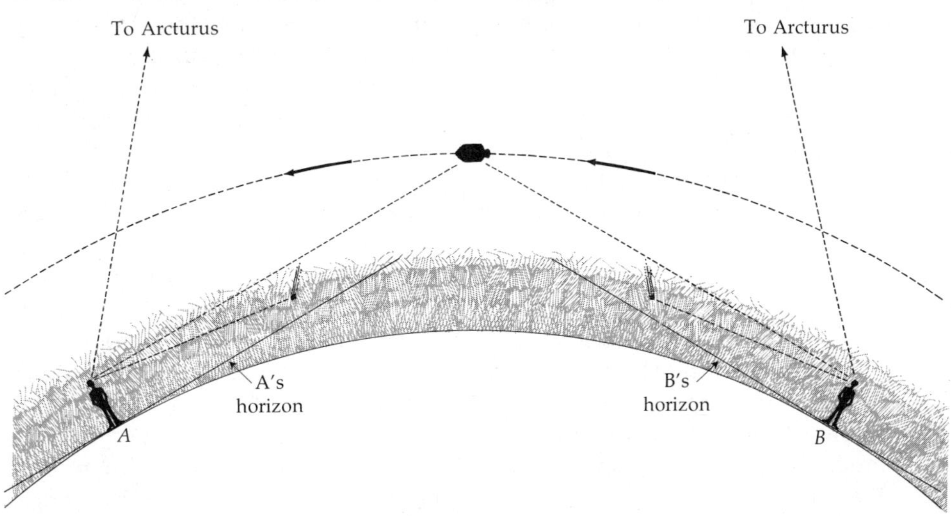

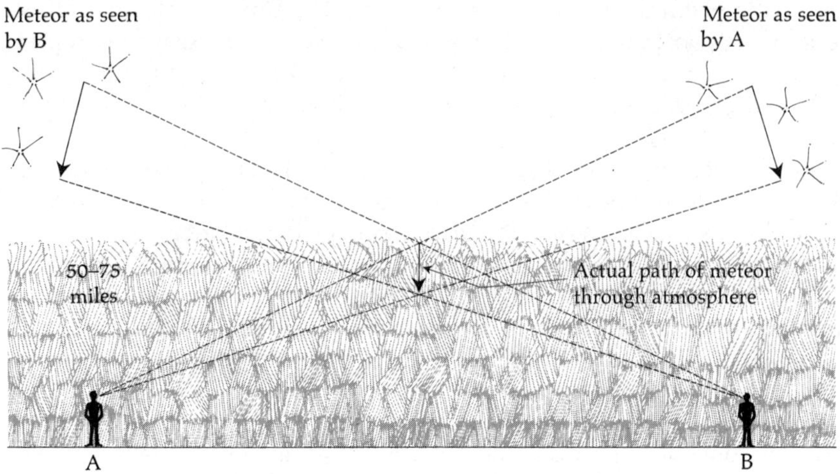

20.17 If a meteor is observed simultaneously from two different locations and its course is plotted against the stellar background, its height can be calculated.

Although they can come from any direction, more meteors are seen after midnight. After that time the observer is on the side of the earth facing the direction the earth is moving in its orbit. Meteoroids on that side approach the earth "head-on" at velocities that combine theirs with that of the earth (Fig. 20.18). Meteoroids on the "pre-midnight" side of the earth must literally catch up with the earth, and therefore enter the atmosphere at a much slower rate. The velocity of a meteoroid upon entering the earth's atmosphere can therefore vary greatly, depending upon the direction of its entry. This applies to both meteor showers from cometary debris and to sporadic meteors as well.

In the case of cometary debris, the particles are traveling as fast as the comet would when in the vicinity of the earth. Because both travel on highly elongated orbits rather than the more nearly circular ones of the planets, their velocities are correspondingly greater than the earth's at this point. Whereas the earth moves at 30 kilometers per second in its orbit, comets move at about 42 kilometers per second when they are about one AU from the sun. The meteor produced by such a particle of debris is a very fast moving streak in the sky whereas a comet is visible for many weeks. Actually they are moving at the same rate. The meteor, however, is much closer to us, only about 50 or 75 miles high; the comet is out among the planets.

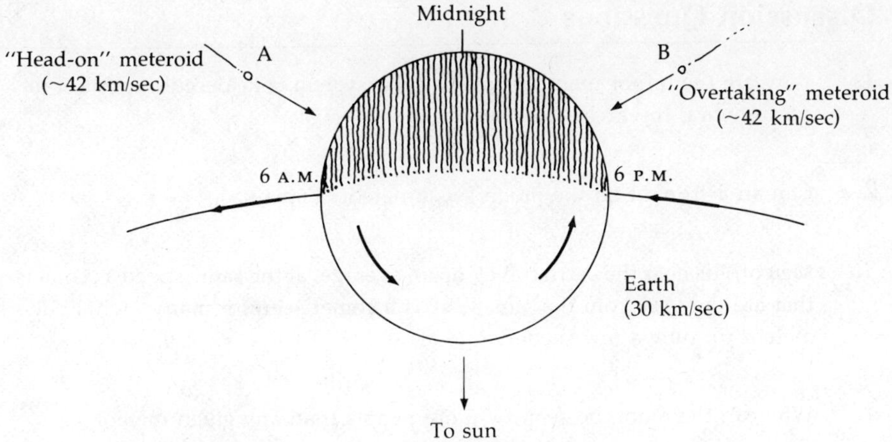

20.18 The meteoroids in the vicinity of the earth have a speed of about 42 kilometers per second. Their impact velocity with the earth's atmosphere depends upon their direction relative to the motion of the earth. Meteroid A *will strike the atmosphere at about 72 kilometers/second (42 + 30), whereas* B*'s impact velocity will be about 12 kilometers/second (42 − 30).*

Occasionally a meteor is so bright that it may momentarily produce the illumination of daylight. These *fireballs* are caused by larger meteoroids that because of their tremendous energies sometimes explode in midair, scattering their pieces over a wide area and making a thunderous noise by an atmospheric shock wave. Such an exploding fireball is called a *bolide.* Fireballs and bolides have been observed since early times and, like comets, have also caused awe and terror (Fig. 20.12). An example of a fireball and subsequent bolide was the Kirin meteoritic shower.

Discussion Questions

1. Why are the minor members of the solar system considered important in astronomical research?

2. Can an astronaut on the moon see a meteor? Explain.

3. Meteoroids near the earth travel, on an average, at the same speed as comets that are one AU from the sun. Why is a comet seen for many nights but a meteor for only a few seconds?

4. Why can the moon be seen by more people than any given meteor?

5. What would the spectrum of an asteroid look like?

6. Why are meteor showers best observed after midnight?

7. How many times bigger in diameter is the earth than the largest asteroid?

8. If the mass of the largest asteroid (Ceres) is one ten-thousandth that of the earth and you weighed 100 pounds on earth, what would you weigh on Ceres?

21 Our Place in Space

The contrast between our present knowledge of the universe—its vastness, its architecture, its complexity—and our knowledge of life in the universe is overwhelming. On one hand, we contemplate a visible universe more than a trillion times greater than that pictured by astronomers half a century ago, and on the other hand, we know of only one form of intelligent life *anywhere,* and that is right here on planet Earth.

Is this contrast only the result of our inability to learn about life elsewhere and our ability to fathom the physical depths of space out to billions of light years? Is perhaps the universe teeming with life, with varied consciousnesses many orders higher and lower than ours? Or do we really represent the highest form of intelligence *anywhere*? Most of us recoil at such a self-centered, egoistic notion. Yet what evidence do we have to the contrary?

So far we have no direct evidence. But very reasonable and even compelling arguments can be put forth in favor of the concept that "we are not alone." One of these arguments takes this form:

Suppose one were to build a model of everything the astronomer can actually observe through the most powerful telescopes, out to the most distant galaxies and quasars. And suppose we made this cross-sectional model as large as the United States itself. Placing the earth and sun somewhere in Kansas in this model would put the most distant objects at either coast. On a model even of this enormous size, the earth would still be totally invisible, even in the world's most powerful electron microscope. The tiniest bit of dust under the front leg of an ant in Peoria, Illinois would be thousands of times larger than the earth on the scale of this model.[1]

Is it reasonable that this tiny, sub-microscopic speck somewhere on the plains of Kansas would be the sole possessor of life, consciousness, and intelligence anywhere between New York and San Francisco? Our reason and our emotions rebel against this idea. It makes us entirely too special—

[1] To make this model represent the expanding universe, it would be sufficient to have New York and San Francisco separate from each other by less than a millimeter each year!

if not indeed a cosmic freak. No, our common sense tell us, there must be other specks—in Nevada, in Utah, in Florida. Perhaps somewhere in the cracks of a sidewalk in New York other sub-microscopic specks exist that mark the presence of life on this model.

Complete the following table:

Our Place in Space	
The furthest observed quasar	
The Virgo cluster of galaxies	
The size of the Local Group of galaxies	
The distance to the Andromeda galaxy	2 million light years
The size of the Milky Way	
The distance to Alpha Centauri	
Distance to Pluto	39 AUs
Distance to the sun	
Distance to the moon	
Diameter of the earth	12756 kms
Width of the United States	
Human height	
Diameter of an auto tire	
Thickness of this sheet of paper	
Diameter of the hydrogen atom	
Diameter of a proton	
Diameter of an electron	

How many times larger than the earth is the visible universe? How many times smaller than the earth is the hydrogen atom?

The Idea Whose Time Has Come

Until quite recently astronomers studiously avoided discussions of life elsewhere in the universe. Statistical speculations about the frequency of planetary systems around other stars and the probabilities of life on such planets were not considered appropriate topics in astronomy. Almost overnight this has changed. Serious discussions and papers on these subjects are now honored in scientific circles. All this despite the fact that there is no direct proof whatsoever of any extraterrestrial life. However, the indirect arguments are so strong that very serious attempts are being mounted to search for such life.

Communication with other intelligences seems to be a far more promising approach than direct visits (unless a very advanced civilization knows how to manipulate space and time in a manner totally unknown to us). The limiting speed of light places drastic but not insuperable limitations on direct communication. However, a civilization on a very distant planet may be deliberately or unknowingly directing radio signals toward us. The latter is the more likely if radio and television communicattion forms a part of their culture. Then, like us, whether they wish to or not, they would be creating a constant outward flow of radiation in that part of the electromagnetic spectrum. Such signals could be detected with large and sensitive radio telescopes, even over distances of hundreds of light years. Such detection would constitute proof that a civilization of at least our equal is located close to the star from which the signals would appear to be coming.

A civilization somewhat more technically advanced than ours, however, might have pinpointed our sun as a target star at which to beam a scheduled repetitive coded message in the hope that someday we would be advanced enough to decode and answer it. A meaningful communication can take place only between civilizations of comparable technological de-

---

Complete the following table:

Our Place in Time

Age of the universe	20×10^9 years
Age of the sun	
Cosmic year	
Precession of earth's axis	
Time since earliest evidence of man on earth	
Time since earliest recorded history	
Time since invention of telescope	368 years
Human life span	
Time for light to come from Alpha Centauri	
Time for earth to revolve around sun	one year
Time for earth to rotate on its axis	
Time for light to come from sun	eight minutes
Time between human heartbeats	
Pulsar rate	
Time for light to cross an atom	
Shortest lived subatomic particles	

For what percentage of man's existence on earth was he unaware of stars fainter than those visible with the naked eye?

---

velopment. It is a sobering thought that a far more advanced civilization (which is certainly possible around stars that are millions of years older than the sun) might long ago have abandoned radio as a means of communication just as we have long since graduated from smoke signals and jungle drums. They may have a means of communication that does not even involve the electromagnetic spectrum and which would be as incomprehensible to us as a television set would have been to a caveman. We have had radio for less than a century—a twinkling of the eye on an astronomical scale. What means of communication might a much older, more advanced civilization have perfected? Obviously we cannot even conceive of it, but neither could an Australian bushman have conceived of an electronic computer.

However, such an advanced civilization might decide to "step down" to our level and use what to them would be extremely primitive techniques. So far, no such signals have been received and verified, but it is conceivable that a vast communications network between advanced civilizations exists, of which we are as unaware as an aborigine is unaware of communication by satellite. Difficult to believe as such concepts may be, the lessons of history and science have been too striking for us to pooh-pooh these ideas. We must never forget that there will be a 21st century wisdom and a 30th century wisdom. Just imagine what a wisdom of A.D 3,000,000 would be like!

The UFO Phenomenon

To many, the idea that actual visits from members of advanced civilizations instead of communication over vast distances might have taken place and are taking place now is a very appealing one. These people base their thinking on two interconnecting hypotheses: one, that "we are not alone," and two, that a truly advanced civilization would have mastered space travel and could roam at will throughout the galaxy. They point to the continuing flow of UFO reports from responsible people from many parts of the world as direct evidence of celestial visitors. While this idea is very appealing and might even turn out to be correct, there is not at present enough evidence of such visits that pass rigorous scientific tests to make them any more than speculation.

This is not to make nonsense of the UFO phenomenon. It does mean, however, that the UFO phenomenon is more complex than we think. Serious students of the subject (and there are many around the world) have made statistical studies of the flow of reports over the past thirty years or more, as well as detailed studies of individual UFO reports from responsible witnesses from many countries. They are convinced that after the great many reports of balloons, aircraft, meteors, twinkling stars, planets, etc., are deleted, there still remains a remarkable remainder of truly baffling and thought-provoking UFO reports. It is these and not the IFOs (Identified Flying Objects) that serious investigators refer to when they use the term UFO. Failure to distinguish between substantiated UFO reports and the very high percentage of reports, which can be explained and hence are not UFOs at all (if one knows what they are they are no longer "unidentified"), has produced a great deal of misinformation and ridicule over the past years, particularly in the popular press.

In science, it is imperative that one define one's terms before proceeding. A proper definition of UFO, generally accepted by students of the subject, is:

> *The reported perception of an object or light seen in the sky or on or near the ground, which is not only mystifying to the original witnesses but which remains unidentified even after close scrutiny of all available evidence by persons who are technically capable of making a positive identification, if one is possible.*

Some people hold that there are no UFO reports that can satisfy this definition, that all reports arise from misidentification of natural phenomena abetted by a vivid imagination. We must remember, controversial issues in science cannot be settled by debate but by continued research. Fortunately, increasing scientific attention is being directed to the UFO phenomenon by social scientists as well as by physical scientists.[2]

Whatever one may personally believe about UFOs, it is an undeniable fact that UFO reports exist, continue to be made, and that the same basic types of reports come from many countries. The UFO phenomenon is global in scope, and the people who report UFOs come from all walks of life. There is a popular misconception that UFO reports are largely made by uneducated persons. Surveys have shown conclusively that this is not so.

Another popular misconception is that UFOs indicate *extraterrestrial* visitation. It should be stressed that in science it is of the greatest importance to distinguish between an *observed* phenomenon and the *theory* to explain the phenomenon. The reality of a phenomenon—such as sunspots or quasars—does not stand or fall on the validity of a specific theory about them. UFO reports are an *observed* phenomenon, distinct from the *theory* that they are extraterrestrial in origin. The reports can be classified into

---

[2] A Stanford University survey published in 1977 indicated that a majority of astronomers responding to the questionnaire believed that the UFO phenomena is worthy of scientific study.

NOCTURNAL LIGHTS: A light seen in the night sky whose appearance and behavior defies conventional explanation.

DAYLIGHT DISC: A daytime UFO sighting, most generally described as oval or disc-like (hence, the name), whose behavior does not correspond to that of aircraft, balloons or meteorological phenomena.

RADAR (RADAR-VISUAL): A radar return which defies normal explanation. This category also contains the especially important subset of those radar cases which reportedly are confirmed visually (e.g., a pilot whose visual observation is confirmed by the tower radar).

CLOSE ENCOUNTERS: A Close Encounter UFO is the reported sighting of a UFO at close range, of the order of not more than a few hundred feet, thus enabling the witness to have a "good look."

CLOSE ENCOUNTERS OF THE FIRST KIND: A close encounter in which there is no interaction with the witness or the environment.

CLOSE ENCOUNTERS OF THE SECOND KIND: A close encounter which interacts with the environment by leaving physical traces of its presence. These may take such forms as burned vegetation, rings or holes in the ground, or physical effects of animate and inanimate matter. Disturbances of vehicles and electronic gear or disturbances of animals and people (burns, eye discomfort, nausea, etc.) are examples of this interaction.

CLOSE ENCOUNTERS OF THE THIRD KIND: A close encounter in which the presence of apparently intelligent beings (or programmed robots) is reported.

21.1 *Classification of UFO reports.*

several categories, but any useful classification system should be independent of any theory of origin. A classification of the great majority of UFO reports (as defined above) has gained acceptance and serves as a useful tool. It is outlined in Figure 21.1.

No one would deny that these UFO reports exist. Catalogues of hundreds of well-witnessed examples of each of the categories are available. It

21.2 *An artist's rendition of a typical UFO as reported by numerous witnesses.*

is not known what physical or other stimulus has given rise to them since we can at present only study UFO reports and not UFOs, a fact which is often overlooked.

It would seem that Close Encounters of the Third Kind would fully substantiate the existence of extraterrestrial life. Has NASA, which has as one of its principal aims the search for extraterrestrial life, overlooked a "good bet" in ignoring this field? Not really, since the case is by no means that clear. We must remember that we still only have *reports,* not authenticated photographs of extraterrestrial beings. Nor have we documented interviews with them (although there are anecdotal accounts of communication with them). Also, even though there are hundreds of catalogued cases, we have no more than a few witnesses to each case.

One might then, well ask, what other explanation than "extraterrestrial life" could we be dealing with, assuming that *all* the people who have reported Close Encounters are not demented. One possibility is that UFOs may represent a phenomenon with which our present day science cannot cope, any more than the science of Galileo's day could have coped with a miniaturized computer or a radio set. How would Galileo have reacted to *reports* of such devices?

There are also some people who think that the UFO phenomenon may represent an interface of our "reality" with that of another dimension, a "parallel universe" so to speak. This is an idea which is not as bizarre as it might seem, in view of our discovery of black holes and the possibility that black hole "singularities" may open into other universes.

The UFO phenomenon, which as a phenomenon is quite real, plus the history of science both should teach us the precious value of an open mind. In the past, the human race has come upon new phenomena which at first were totally puzzling, but which were eventually solved and added to the wealth of total information. As Erwin Schroedinger, the father of modern quantum mechanics, wrote, "The first requirement of a scientist is that he be curious. He should be capable of being astonished and eager to find out." Above all, the scientist should beware the trap of "temporal provincialism"—that state of mind that feels that the science of that particular day holds all the answers, and rejects any phenomenon that does not "fit." The facts and observations that do not "fit" into the present-day scheme of things are the ones that invariably lead to scientific breakthroughs. The UFO phenomenon may well be one of these.

Our Place in Space

We finish this book with the vision of a universe that may well be teeming with life with which we have as yet had no contact outside of ourselves.

21.3 An example of the purported effects of a Close Encounter of the Second Kind. This parched circle of about 40 feet in diameter was found in an otherwise unmolested large soy bean field in Iowa in July 1969. Shortly before this discovery, several witnesses had seen a "traditional" saucer-shaped, lighted craft gliding away from the scene (photograph from Center for UFO Studies).

We continue to be basically confined to our small spaceship, the Earth. Unless significant breakthroughs make interstellar communication and travel possible, we are still essentially "alone" in space. It behooves us to nurture and protect this one inhabitable world that we know.

21.4 An artist's conception of activity at a possible manned, modularized space station in earth orbit. The modules could house the various equipment, functions, and activities of the space station. They would be carried into earth orbit by the Space Shuttle, which is shown in the upper left (NASA).

Although our small planet is some 4½ billion years old, traces of the human race go back hardly more than two million years. The sun, we now know, has enough fuel to maintain our climate for several billion years more. Therefore our two million years of mankind represents about 1/2000 of our potential life on this earth. Whether we will achieve even a small portion of this potential depends upon our ability to solve our social and ecological problems. If we do, then the time will come when our descendants will look back across the vast panorama of time perhaps with greater condecension than that with which we view the ancient caveman.

They will perhaps see us as those primitives present at the dawn of mankind's existence, possessing only an elementary knowledge of the nature of the universe. Perhaps by then they may even know how the limita-

21.5 A space colony: imagine yourself in the 21st century. You are a resident of such a space colony, returning home after a vacation on earth. The 32-kilometer (19 mile) long, 6400-meter (4 mile) diameter cylinder at right and its twin at left are seen as they would appear from an approaching spaceship some 32 kilometers (19 miles) away (NASA).

tions of space/time can be overcome. They will certainly know whether we are or are not alone in the universe, and will undoubtedly have entered into a communications network with other intelligences, if they exist.

It is hard for us today to assimilate all the new ideas that are being suggested in response to the new information we have. We must remember that our picture of the universe is based not only on our scientific knowledge but also on our culture and our philosophy. What new discoveries lie ahead no one can say. There may well be civilizations in other parts of our galaxy or in other galaxies that have already accomplished much of what lies ahead for mankind. Others may just be beginning. The universe clearly presents an unending challenge.

Discussion Questions

1. What effects will the possible discovery of extraterrestrial life have on man?

2. In a science fiction story, a message is sent from one civilization to another in the far reaches of our galaxy. The message was described as "the long since dead talking to the not yet born." What did that statement mean?

3. One of the questions at the end of Chapter 1 was: "What is your concept of astronomy?" Now that you have taken a first course in astronomy, how has your answer to that question changed?

4. When discussing life elsewhere, the phrase "life as we know it" is frequently used. What do we mean by this expression? What other forms of life might be possible? Would we recognize such forms as "life," and if so, how?

Appendix I

Scientific Notation

When using very large or very small numbers, scientists and mathematicians employ *scientific notation.* This avoids the necessity of having to write out a long string of zeros, and is easier to interpret quickly. This notation is used throughout this textbook and in practically all other scientific texts.

You will find, for example, that the speed of light is approximately 3×10^5 kilometers per second and that an Astronomical Unit is about 1.5×10^8 kilometers. The mass of a hydrogen atom is given as 1.67×10^{-24} grams.

To put these numbers into "nonscientific" terms, we follow a very simple rule. The superscript number at the upper right of the 10 is called its *exponent.* It indicates the number of places the decimal point must be moved in order to read the full value of the number. For positive exponents, we move the decimal point to the right; for negative exponents, to the left. Thus: 3×10^5 is the same as

$$3\underrightarrow{00000}.00$$

or "three-hundred thousand." In the same way, 1.5×10^8 is

$$1\underrightarrow{50,000,000}.00$$

or "one-hundred-fifty million."

$$1.67 \times 10^{-24} = 0.\underleftarrow{000\ 000\ 000\ 000\ 000\ 000\ 000\ 001}67.$$

We could also write it:

$$\frac{1.67}{1,000,000,000,000,000,000,000,000}$$

However, since it has far more zeros than we normally have words for (i.e., billion, trillion, quadrillion, etc.), we find it easier to say, "one-point-six-seven times ten to the negative twenty-four grams" when giving the mass of the hydrogen atom.

To multiply when using exponents of 10, we simply add the exponents themselves:

$$10^6 \times 10^4 = 10^{10}$$

$$(4 \times 10^6) \times (3 \times 10^4) = 12 \times 10^{10} \quad \text{or} \quad 1.2 \times 10^{11}$$

(Note that the regular numbers are multiplied, not added)

To divide, we subtract the exponents:

$$10^6 \div 10^4 = 10^2$$

$$(4 \times 10^6) \div (3 \times 10^4) = 4/3 \times 10^2 \quad \text{or} \quad 1.33 \times 10^2$$

(Note that the regular numbers are divided, not subtracted)

To add or subtract exponents of 10 (for example $10^6 - 10^4$), it is necessary to put them into regular notation before performing the mathematical problem:

$$10^6 = 1000000$$

$$10^4 = \underline{10000} = 9.9 \times 10^5$$
$$990000$$

Note also that although the exponents may differ by only a few numbers, this difference represents a factor of many thousands. For example, the mass of the earth is about 6×10^{27} grams, whereas the mass of the sun is about 2×10^{33} grams. Although the exponential numbers differ only by 6, they represent a factor of a million. The sun is approximately 332,000 times more massive than the earth.

Appendix II

Table of Metric and English Units of Measurement

1 mile	1.61 kilometers (km)
1 inch	2.54 centimeters (cm)
1 kilometer (km)	1000 meters = 10^5 centimeters = 0.62 miles
1 meter (m)	100 centimeters = 39.37 inches
1 centimeter (cm)	10 millimeters = 0.39 inches
1 millimeter (mm)	100 microns
1 micron (μ)	10^{-4} centimeters
1 angstrom (Å)	10^{-8} centimeters
1 gram (g)	0.035 ounces (oz)
1 kilogram (kg)	1000 grams = 2.2 pounds (lbs)
1 milligram (mg)	0.001 grams
1 lb	453.6 grams
1 oz	28.35 grams
1 ton (metric)	10^6 grams

Appendix III

Table of Frequently Used Astronomical Constants

Velocity of light	$c = 2.998 \times 10^{10}$ cm/sec = 186,282 miles/sec
Astronomical Unit	AU = 1.496×10^{13} km = 92.9×10^{6} miles
Light year	l.y. = 9.46×10^{12} km. = 6×10^{12} miles = 63240 AU
Parsec	pc = 206265 AU = 3.26 l.y.
Kiloparsec	kpc = 1000 pc
Mass of sun	1.989×10^{33} grams
Luminosity of sun	3.90×10^{33} erg/sec

Appendix IV

The Brightest* Stars

	Star	R. A. 1970 h	m	Decl. 1970	Spectral Class	Magnitude App.	Magnitude Abs.	Distance (l.y.)	Star Name
1	Sun				G2 V	−26.8	+4.8	8 (l.-min)	
2	α Canis Majoris**	06	44	−16°41′	A1 V	− 1.42	+1.4	8.7	Sirius
3	α Carinae	06	23	−52 41	F0 Ib-II	− 0.72	−3.1	98	Canopus
4	α Bootis	14	14	+19 20	K2 III	− 0.06	−0.3	36	Arcturus
5	α Centauri**	14	38	−60 43	G2 V	− 0.02	+4.4	4.3	
6	α Lyrae	18	36	+38 45	A0 V	0.04	+0.5	26.5	Vega
7	α Aurigae	05	16	+45 58	G2 III	0.05	−0.6	45	Capella
8	β Orionis**	05	13	−08 14	B8 Ia	0.14	−7.1	900	Rigel
9	α Canis Minoris	07	38	+05 18	F5 IV-V	0.37	+2.7	11.3	Procyon
10	α Orionis	05	54	+07 24	M2 Iab	0.41	−5.6	520	Betelgeuse
11	α Eridani	01	37	−57 23	B5 IV	0.51	−2.3	118	Achernar
12	β Centauri	14	02	−60 13	B1 II	0.63	−5.2	490	Hadar
13	α Aquilae	19	49	+08 47	A7 IV, V	0.77	+2.2	16.5	Altair
14	α Tauri**	04	34	+16 27	K5 III	0.86	−0.7	68	Aldebaran
15	α Virginis	13	24	−11 00	B1 V	0.91	−3.2	261	Spica
16	α Scorpii**	16	28	−26 22	M1 Ib	0.92	−5.1	520	Antares
17	β Geminorum	07	44	+28 06	K0 III	1.16	+1.0	35	Pollux
18	α Piscis Austrinus	22	56	−29 47	A3 V	1.19	+2.0	22.6	Fomalhaut
19	α Cygni	20	40	+45 10	A2 Ia	1.26	−7.1	1600	Deneb
20	β Crucis	12	46	−59 32	B0 III	1.28	−4.6	490	
21	α Crucis**	12	25	−62 56	B1 IV	1.34	−3.9	370	Acrux
22	α Leonis**	10	07	+12 07	B7 V	1.36	−0.7	84	Regulus
23	ϵ Canis Majoris**	06	57	−28 56	B2 II	1.48	−5.1	680	Adhara
24	λ Scorpii	17	32	−37 05	B1 V	1.60	−3.3	310	Shaula
25	γ Orionis	05	24	+06 19	B2 III	1.64	−4.2	470	Bellatrix

* In order of apparent magnitude.
**Star is a multiple star system. Magnitude and spectral class refer to the major star in the system.

Appendix V

The Nearest* Stars

	Star	R. A. 1970 h	m	Decl. 1970	Distance (l.y.)	Magnitude Abs.	App.	Spectral Class
1	Sun				(8 l.-min)	4.8	−26.8	G2 V
2	α Centauri**	14	37	−60°43′	4.3	4.4	0.1	G2 V
3	Barnard's Star	17	56	+04 36	5.9	13.2	9.5	M5 V
4	Wolfe 359	10	55	+07 13	7.6	16.8	13.5	M6 V
5	Lalande 21185	11	02	+36 10	8.1	10.5	7.5	M2 V
6	Sirius**	06	44	−16 41	8.6	1.4	− 1.5	A1 V
7	Luyten 726-8**	01	37	−18 07	8.9	15.4	12.5	M6 V
8	Ross 154	18	48	−23 51	9.4	13.3	10.6	M5 V
9	Ross 248	23	40	+44 01	10.3	14.7	12.2	M6 V
10	ϵ Eridani	03	32	−09 34	10.7	6.1	3.7	K2 V
11	Luyten 789-6	22	37	−15 31	10.8	14.9	12.2	M6 V
12	Ross 128	11	46	+01 01	10.8	13.5	11.1	M5 V
13	61 Cygni**	21	06	+38 36	11.2	7.5	5.2	K5 V
14	ϵ Indi	22	02	−56 55	11.2	7.0	4.7	K5 V
15	Procyon**	07	38	+05 18	11.4	2.7	0.3	F5 IV-V
16	Σ 2398**	18	42	+59 35	11.5	11.1	8.9	M3.5 V
17	BD +43° 44	00	17	+43 51	11.6	10.3	8.1	M1 V
18	CD −36° 15693	23	04	−36 02	11.7	9.6	7.4	M2 V
19	τ Ceti	01	43	−16 06	11.9	5.7	3.5	G8 V
20	BD +5° 1668	07	26	+05 28	12.2	11.9	9.8	M4 V
21	CD −39° 14192	21	15	−39 00	12.5	8.7	6.7	M1 V
22	Kapteyn's Star	05	11	−45 00	12.7	10.8	8.8	M0 —
23	Kruger 60**	22	27	+57 33	12.8	11.8	9.7	M4 V
24	Ross 614**	06	28	−02 48	13.1	13.1	11.3	M5 V
25	BD −12° 4523	16	29	−12 35	13.1	12.0	10.0	M5 V
26	van Maanen's Star	00	47	+05 16	13.9	14.3	12.4	F wd
27	Wolf 424**	12	32	+09 12	14.2	14.4	12.6	M6 V
28	CD −37° 15492	00	03	−37 30	14.5	10.3	8.6	M3 V
29	BD +50° 1725	10	09	+49 36	15.0	8.3	6.6	M0 V

* In order of distance.
**Star is a multiple star system. Magnitude and spectral class are given only for the major star in the system.

Appendix VI

List of the Constellations

Name (Latin)	*Name or Description (English)*	*Abbreviation*	*Approximate R. A. (hours)*	*Decl. (degrees)*
Andromeda	Princess	And	1	+40
Antlia	Air pump	Ant	10	−35
Apus	Bird of Paradise	Aps	16	−75
Aquarius	Water Bearer	Aqr	23	−15
Aquila	Eagle	Aql	20	+5
Ara	Altar	Ara	17	−55
Aries	Ram	Ari	3	+20
Auriga	Charioteer	Aur	6	+40
Boötes	Herdsman	Boö	15	+30
Caelum	Graving tool	Cae	5	−40
Camelopardus	Giraffe	Cam	6	+70
Cancer	Crab	Cnc	9	+20
Canes Venatici	Hunting dogs	CVn	13	+40
Canis Major	Big dog	CMa	7	−20
Canis Minor	Little dog	CMi	8	+5
Capricornus	Sea Goat	Cap	21	−20
Carina	Keel of ship	Car	9	−60
Cassiopeia	Queen	Cas	1	+60
Centaurus	Centaur	Cen	13	−50
Cepheus	King	Cep	22	+70
Cetus	Whale	Cet	2	−10
Chamaeleon	Chameleon	Cha	11	−80
Circinus	Compasses	Cir	15	−60
Columba	Dove	Col	6	−35
Coma Berenices	Berenice's Hair	Com	13	+20
Corona Australis	Southern Crown	CrA	19	−40
Corona Borealis	Northern Crown	CrB	16	+30
Corvus	Crow	Crv	12	−20
Crater	Cup	Crt	11	−15
Crux	Southern Cross	Cru	12	−60
Cygnus	Swan	Cyg	21	+40
Delphinus	Porpoise	Del	21	+10
Dorado	Swordfish	Dor	5	−65
Draco	Dragon	Dra	17	+65
Equuleus	Little Horse	Equ	21	+10
Eridanus	River	Eri	3	−20

Name (Latin)	*Name or Description (English)*	*Abbreviation*	*Approximate R. A. (hours)*	*Decl. (degrees)*
Fornax	Furnace	For	3	−30
Gemini	Twins	Gem	7	+20
Grus	Crane	Gru	22	−45
Hercules	Son of Zeus	Her	17	+30
Horologium	Clock	Hor	3	−60
Hydra	Sea Serpent	Hya	10	−20
Hydrus	Water Snake	Hyi	2	−75
Indus	Indian	Ind	21	−55
Lacerta	Lizard	Lac	22	+45
Leo	Lion	Leo	11	+15
Leo Minor	Little Lion	LMi	10	+35
Lepus	Hare	Lep	6	−20
Libra	Scales	Lib	15	−15
Lupus	Wolf	Lup	15	−45
Lynx	Lynx	Lyn	8	+45
Lyra	Lyre	Lyr	19	+40
Mensa	Table Mountain	Men	5	−80
Microscopium	Microscope	Mic	21	−35
Monoceros	Unicorn	Mon	7	−5
Musca	Fly	Mus	12	−70
Norma	Carpenter's level	Nor	16	−50
Octans	Octant	Oct	22	−85
Ophiuchus	Serpent Bearer	Oph	17	0
Orion	Hunter	Ori	5	+5
Pavo	Peacock	Pav	20	−65
Pegasus	Winged Horse	Peg	22	+20
Perseus	The hero Perseus	Per	3	+45
Phoenix	Phoenix	Phe	1	−50
Pictor	Easel	Pic	6	−55
Pisces	Fish	Psc	1	+15
Piscis Austrinus	Southern fish	PsA	22	−30
Puppis	Stern of ship	Pup	8	−40
Pyxis (Malus)	Compass on ship	Pyx	9	−30
Reticulum	Net	Ret	4	−60
Sagitta	Arrow	Sge	20	+10
Sagittarius	Archer	Sgr	19	−25
Scorpius	Scorpion	Sco	17	−40
Sculptor	Sculptor's tools	Scl	0	−30
Scutum	Shield	Sct	19	−10
Serpens	Serpent	Ser	17	0
Sextans	Sextant	Sex	10	0
Taurus	Bull	Tau	4	+15
Telescopium	Telescope	Tel	19	−50
Triangulum	Triangle	Tri	2	+30
Triangulum Australe	Southern triangle	TrA	16	−65
Tucana	Toucan	Tuc	0	−65
Ursa Major	Big Bear	UMa	11	+50

Name (Latin)	Name or Description (English)	Abbreviation	Approximate R. A. (hours)	Decl. (degrees)
Ursa Minor	Little Bear	UMi	15	+70
Vela	Sail of ship	Vel	9	−50
Virgo	Virgin	Vir	13	0
Volans	Flying fish	Vol	8	−70
Vulpecula	Fox	Vul	20	+25

Appendix VII

Star Charts

To use the star charts:

1 Select the chart that is nearest to the current date. All of them are for 9:00 p.m. Standard Time. However, each chart may be used at 7:00 p.m. one month later than the specified date or at 11:00 p.m. one month earlier.

2 Hold the chart over your head so that the north point is facing north. The cross (+) at the center of each chart represents the overhead point (zenith) for an observer at about 40° north latitude.

3 The dotted line on each chart is the ecliptic; the dashed line is the celestial equator. The shaded area represents the Milky Way as seen with the naked eye.

4 Each night all the stars appear to revolve around the North Star (Polaris) which is the last star in the handle of the Little Dipper. If you face in the direction of Polaris, you are facing north.

5 The stars are indicated by dots. Some of the ones included in these charts are:

Star	*Apparent magnitude*	*Constellation*	*Star*	*Apparent magnitude*	*Constellation*
Sirius	−1.6	Canis Major	Aldebaran	1.1	Taurus
Vega	0.1	Lyra	Pollux	1.2	Gemini
Capella	0.2	Auriga	Spica	1.2	Virgo
Arcturus	0.2	Boötes	Antares	1.2	Scorpius
Rigel	0.3	Orion	Deneb	1.3	Cygnus
Procyon	0.5	Canis Minor	Regulus	1.3	Leo
Altair	0.9	Aquila	Castor	1.6	Gemini
Betelgeuse	0.9	Orion			

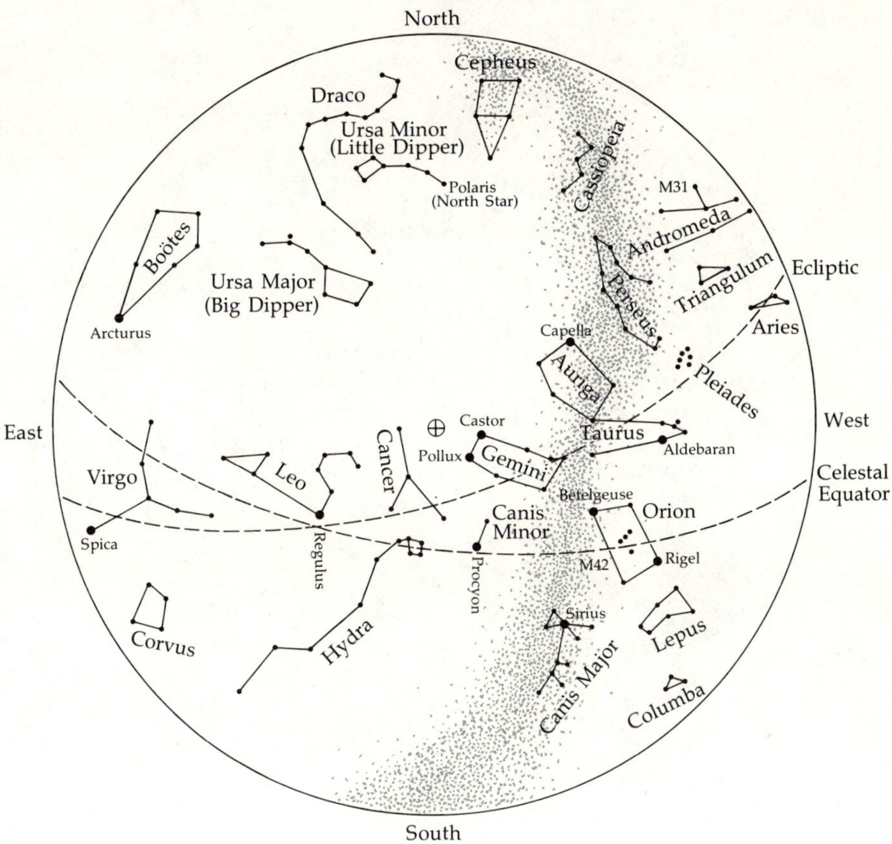

March 21st
9:00 pm

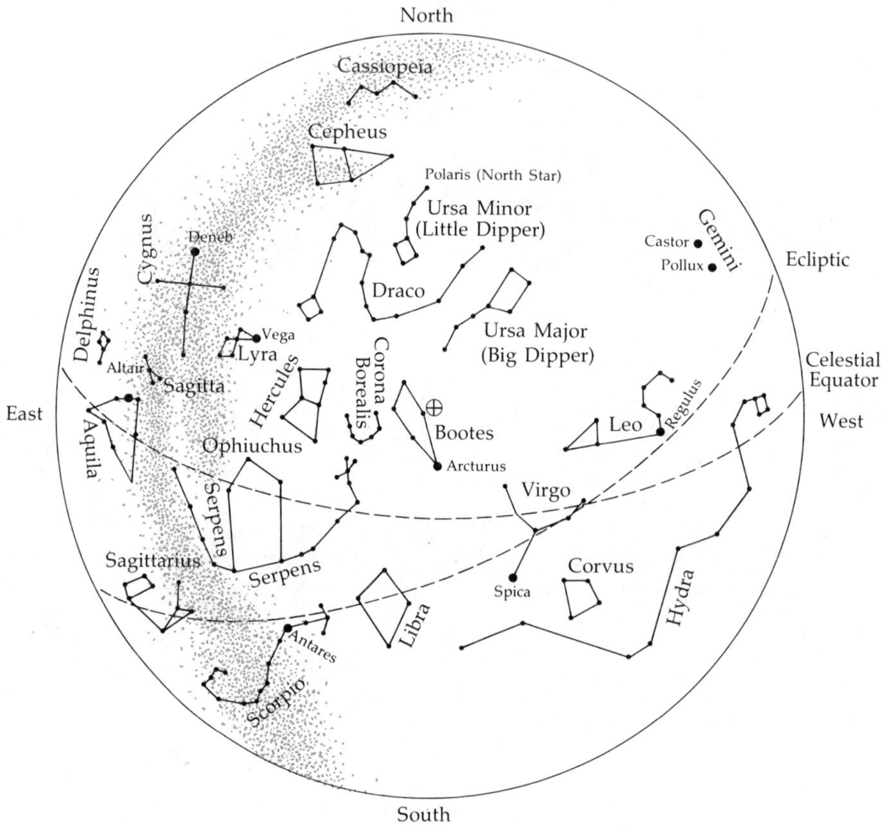

June 21st
9:00 pm

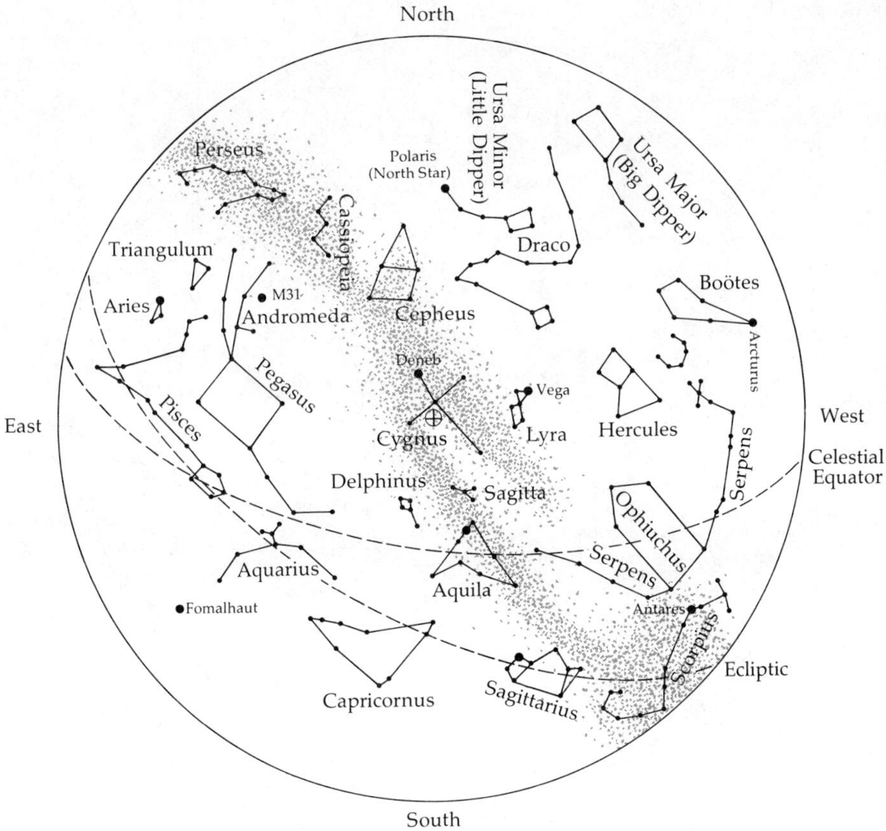

September 21st
9:00 pm

December 21st
9:00 pm

Appendix VIII

Suggested Reading

There are many excellent books on all fields of astronomy and many more are constantly being written as new areas are being explored and developed. Probably, therefore, the best way to keep abreast of the current state of the science is to become familiar with several journals and periodicals. Recommended are:

Astronomy published monthly by Astromedia Corporation, 411 E. Mason St., Milwaukee, Wisc. 53202

Science News published weekly by Science Service, Inc., 1719 N. St., N.W., Washington, D.C. 20036

Scientific American published monthly by Scientific American, Inc., 415 Madison Ave., New York, N.Y. 10017

Sky and Telescope published monthly by the Sky Publishing Corp., 49-50-51 Bay State Road, Cambridge, Mass. 02138

Appendix IX

Answers to Chapter Questions

Chapter 2

5 669.6 million mph; 1078.1 million km/hr.

6 They have equal absolute brightness but *A* is farther away.

Chapter 3

7 312 minutes to Pluto; 4½ years to nearest star.

8 12,000 km/sec recession

9 Dimmer one is $\sqrt{10}$ times farther away.

10 1500 million light years; 66,960,000 mph

11 Velocity of light is always *c*; velocity of spaceship can be determined by measuring Doppler shift of spectral lines of the light. They will be shifted toward the blue end of the spectrum.

12 *B* is twice as far away as *A*.

13 4½ billion light years

14 400,000 times

Chapter 4

3 Positive curvature

Chapter 5

4 Betelgeuse; Rigel

5 At sunset; at noon; at sunrise

6 At the equator; at one of the poles

9 About 90°; at the North Pole

10 At the horizon; Sept. 21 and March 21; northern sky

Chapter 6

3 Star with smaller parallax

4 Row *A*: approaching, receding, receding
Row *B*: neither, sun, neither
Row C: receding, receding, approaching

7 *A*; 100; *A*; 10 times

8 Four magnitudes

9 More than 10,000 times brighter

10 10 parsecs; 2 parsecs

11 *B*; *A* has greater absolute magnitude

14 Two years

Chapter 7

9 Shorter; higher

12 *B*; *A*; ultraviolet

13 $1/3000$ gram; 84,000 homes

17 a) A; b) K; c) K; d) A star is whitish, K star is orange-red e) A

Chapter 8

4 F star

5 Parallax and age

11 Star *B*

12 Energy output is 27 times that of the sun

Chapter 11

5 50 AU/sec; 540,000 mph

6 30 cosmic years

7 $1/(2 \times 10^8)$; less than one inch per year; path of earth would be a seemingly straight line with a very slight wobble in it.

Chapter 12

3 Moth will scatter and block some light; otherwise it is out of focus and cannot be seen; similar to observer sitting at prime focus of 200-inch telescope.

7 100 times farther

Chapter 13

1 *a*, *b*, *e*

4 Pluto—304 minutes or about 5 hours
Mars—about $2\frac{2}{3}$ minutes

18 512 earth years

Chapter 15

4 To the left

10 1:00 A.M. in New York City

11 7 P.M.

Chapter 16

1 Near the meridian

2 First quarter

3 "New" earth

8 a) Waxing crescent; b) Full; c) Half

9 $28\frac{1}{2}°$

Chapter 17

2 In autumn around September 21st

3 Sun would rise and set once a month (earth time); earth would stay in the same place in the sky all the time.

Chapter 18

5 Mercury and Venus

8 No, not in the temperate zones, but at the poles, it could be seen at midnight.

9 Mercury and Venus

Chapter 20

7 Earth is 16.5 times larger

8 2.7 lbs

Glossary

Absolute Magnitude The apparent magnitude a star would have if it were at a distance of ten parsecs.

Absolute zero The coldest temperature possible; the point at which all molecular motion stops; 0 K or −273°C.

Absorption lines Dark lines which appear on a continuous spectrum marking the position of deficiencies in radiation at those discrete wavelengths; caused by the absorption and scattering of energy by atomic processes.

Alpha particle The nucleus of a helium atom.

Altitude The angular distance of a celestial object above the horizon.

Altitude-azimuth system A local system of celestial coordinates measured along and perpendicular to the horizon.

Ångstrom (Å) One-hundred-millionth of a centimeter (10^{-8} cm).

Angular distance The separation of two points (or objects) on the celestial sphere as measured in degrees.

Angular momentum The quantity of rotational motion of a spinning body.

Annular eclipse A solar eclipse in which the apparent size of the moon is less than that of the sun, leaving a ring of light, or *annulus,* visible.

Aperture Diameter of the primary lens or mirror of an optical telescope.

Aphelion The point in an object's orbit about the sun at which it is farthest from the sun.

Apogee The point in an orbit of an earth-circling object which is farthest from the center of the earth.

Apparent magnitude Measurement of how bright a celestial object appears to the observer.

Asteroids Minor planets, most of whose orbits lie between Mars and Jupiter.

Astrology The belief that the relative positions of the planets, sun, and moon on the celestial sphere influence terrestrial affairs.

Astrometry The branch of astronomy that deals with the measurement of positions of objects on the celestial sphere.

Astronomy The mathematical and physical study of the universe.

Astronomical unit (AU) The mean distance between the sun and the earth; approximately 93 million miles or 1.5×10^8 kilometers.

Astrophysics The branch of astronomy that deals with the physics of the universe and the objects in it.

Atom The smallest particle of an element which still keeps the characteristics of that element.

Aurora borealis The sporadic display of bands of light in the northern night sky caused by the interaction of charged particles from the sun, the earth's magnetic field, and the earth's upper atmosphere. In the southern hemisphere, the phenomenon is called the *aurora australis.*

Autumnal equinox That intersection of the ecliptic and the celestial equator at which the sun crosses the equator from north to south; the sun's position on about September 21st.

Azimuth A celestial coordinate measured from the north point, eastward along the horizon, to a line drawn from the zenith to the horizon, and passing through the object.

Bailey's beads During a total solar eclipse, the effect produced by the last rays of sunlight shining between the lunar mountains immediately before or after totality. When one "bead" shines brighter than the others the "diamond ring effect" is produced.

Balmer hydrogen lines Those emission or absorption lines in the hydrogen spectrum which arise from transitions between the second and higher energy levels.

Barred spirals Spiral galaxies in which the arms emanate from the ends of a bar-like structure passing through the center of the galaxy.

Barycenter The common center of mass of two objects revolving about each other as a consequence of their mutual gravitational attraction.

BD Number The catalogue number of a star in the *Bonner Durchmusterung.*

Big bang theory The cosmological theory that states that the universe had its origin in the violent explosion of a primeval mass.

Binary stars A pair of stars revolving around their common center of mass; a double star.

Blackbody A perfect radiator; absorbs and then re-emits all radiation that falls upon it.

Black hole A theoretical model of an extremely compacted ball of matter so dense that its gravitational field curves space-time in its immediate neighborhood to such an extent that radiation cannot escape.

Bolide An extremely bright meteor which explodes in mid-air.

BS Number The catalogue number of a star in the *Yale Catalogue of Bright Stars.*

Carbon cycle The nuclear process by which hydrogen is transformed into helium deep in the interiors of the more massive stars.

Celestial equator An imaginary line on the celestial sphere that is a projection of the terrestrial equator.

Celestial mechanics The branch of astronomy that deals with the study of motion and gravitational interactions of bodies in space.

Celestial pole Points on the celestial sphere which are the projections of the north and south poles of the earth.

Celestial sphere The apparent sphere of the sky.

Center of mass The central point in a body or system of bodies that represents the weighted mean of the masses of the system.

Cepheids A particular class of variable star noted for its period-luminosity relationship.

Chromosphere The atmospheric layer of the sun lying immediately above the photosphere.

Cluster of galaxies A grouping of mutually related galaxies.

Comet A loose aggregate of matter which revolves around the sun, usually in a highly inclined, eccentric orbit.

Conjunction The position of a celestial object, usually applied to a planet, in which the sun and object lie in the same direction from the earth and are both essentially in the same plane with the earth.

Constellation A configuration of stars as seen on the celestial sphere, usually representing some mythological character.

Continuous spectrum A spectrum consisting of a smooth gradation of color from violet to red.

Coriolis effect The apparent deflection of projectiles moving across the surface of the earth; it is caused by the rotation of the earth.

Corona The outermost atmosphere of the sun.

Cosmic rays Atomic nuclei (mostly hydrogen nuclei or protons) moving through space at speeds close to that of light.

Cosmology The study of the origin and evolution of the universe and its development.

Dark nebula A cloud of obscuring matter in interstellar space.

Declination The angular distance of an object north or south of the celestial equator.

Degenerate matter Matter of extremely high density in which the nuclei and electrons still maintain their identity although the matter no longer obeys the ordinary gas laws.

Density The ratio of the mass of an object to its volume.

Diamond ring effect At the time of a total solar eclipse, the effect produced by the last rays of sunlight shining between the lunar mountains immediately before or after totality.

Differentiation (geological) Separation of different kinds of materials into layers in a planet's interior.

Doppler effect (radiation) The change in perceived frequency (wavelength) of the radiation from a source due to its relative motion toward or away from the observer.

Dwarf A main sequence star, especially applied to stars on the lower main sequence.

Eccentricity The ratio of the distance between the foci to the length of the major axis in an ellipse.

Eclipse The blocking of all or part of the light of one body by another passing in front of it.

Eclipsing binary A binary system in which each star periodically passes in front of the other, blocking off all or part of its light in our line of sight.

Ecliptic The intersection of the plane of the earth's orbit and the celestial sphere; hence, the apparent yearly path of the sun on the celestial sphere.

Electromagnetic spectrum The array of electromagnetic waves arranged in order of increasing wavelength.

Electron A negatively charged subatomic particle normally revolving around the nucleus of an atom.

Ellipse A mathematical curve described by a point traveling so that the sum of its distances from two fixed points (foci) is constant.

Ellipticals Galaxies whose contours are elliptical and which contain little, if any, interstellar material.

Elongation The angular distance on the celestial sphere between the sun and a planet as seen from the earth.

Emission lines Discrete, bright spectral lines that correspond to the emission of discrete radiation from atoms.

Emission nebula An interstellar cloud whose radiation arises from fluorescence.

Ephemeris A chronological tabulation of the positions of celestial bodies.

Equation of time The difference between the apparent and mean solar time at any specific time.

Equatorial system The system of celestial coordinates that uses right ascension and declination.

Equinoxes The points of intersection on the celestial sphere of the ecliptic and the celestial equator.

Eruptive variable A star whose changes in light are erratic or explosive.

Escape velocity The velocity an object must have in order to become free from the gravitational field of a body.

Evening star The name sometimes given to any bright planet when it is seen in the early evening sky; the term is most frequently applied to Venus.

Fireball An extremely bright meteor.

Flare (solar) An eruption on the sun accompanied by an intense local brightening.

Flash spectrum The emission spectrum produced by the chromosphere of the sun, momentarily visible at the beginning and end of totality during a solar eclipse.

Fluorescence The process by which radiation is absorbed at one wavelength and re–emitted at another; usually the conversion of ultraviolet radiation into visible light.

Focal length The distance from a telescopic mirror or lens to the point at which the image is produced.

Frequency In astronomy, the number of electromagnetic waves passing a given point in a given period of time.

Galactic cluster An "open cluster" of stars.

Galaxy A vast system of stars ranging from 5000 light years to over several hundred thousand light years in diameter and containing from millions to hundreds of billions of stars and often also containing large quantities of dust and gas.

General relativity That part of Einstein's theory of relativity that deals with nonuniform motion.

Geodesic The shortest distance between two points in any given geometry.

Giant (star) A star of high luminosity and large diameter.

Gibbous moon That phase of the moon which occurs just before and just after the full phase.

Globular cluster A compact, spherical system of many thousands of stars.

Globule A small, relatively dense, dark nebula, possibly the forerunner of a protostar.

Granulation The mottled structure of the photosphere of the sun.

Gravitation The attractive force exhibited between any two material particles.

Greenhouse effect The heating of an enclosure by trapped infrared radiation.

Gregorian calendar The calendar instituted by Pope Gregory incorporating the present system of leap years.

Harmonic law Kepler's third law of planetary motion.

Harvest moon The full moon that occurs shortly before or after the autumnal equinox.

Hayashi line Evolutionary track of a star on the H-R diagram before it reaches the main sequence.

HD Number The catalogue number of a star in the Henry Draper Catalogue.

Heliocentric Sun-centered.

Helium flash The sudden onset of helium burning in the core of a red giant star.

Hertzsprung gap A section in the upper part of the H-R diagram where few stars are found and most of these are unstable ones.

Hertzsprung-Russell diagram The diagram of the spectral classes of stars plotted against their absolute magnitudes.

Hidden side of the moon The side of the moon that is always turned away from the earth.

Hour angle The westward distance along the celestial equator from the local meridian to the hour circle passing through the object being studied.

Hour circle A circle on the celestial sphere that passes through both the celestial poles.

H-R diagram See *Hertzsprung-Russell diagram.*

Hubble constant (or parameter) The rate of increase of recessional velocity of galaxies with observed distance.

Hunter's moon The full moon that follows the harvest moon.

Image orthicon One of the electronic devices whereby an optical image is transformed electronically into a signal which can be amplified and reconverted into a visual image; a TV camera tube.

Inferior conjunction The position of an inferior planet when it is closest to the earth.

Inferior planet A planet closer to the sun than the earth.

Inverse square law The apparent intensity of a light source varies inversely as the square of its distance from the observer.

Ion An electrically charged atom, caused by the addition or loss of one or more electron.

Ionization The process by which an atom loses one or more of its electrons.

Irregular galaxy A galaxy that has neither a spiral nor an elliptical shape and is not symmetrical.

Isotropic The condition in which everything is the same in all directions.

Kepler's law The three laws of planetary motion discovered by Johannes Kepler.

Law of universal gravitation The mathematical statement of the force of attraction between two material particles.

Leap year Those years that contain 366 days; a method for maintaining correspondence between the calendar year and the year of the seasons.

Lens A piece of glass shaped so as to refract light to form an image.

Light Electromagnetic radiation visible to the eye.

Light year The distance that light travels in a vacuum in one year; 1 LY = 9.46×10^{17} cm or about 6×10^{12} miles.

Line of nodes The line connecting the nodes of an orbit.

Local group The cluster of galaxies to which the Milky Way galaxy belongs.

Luminosity A measure of the total amount of electromagnetic radiation emitted by an object per second.

Lunar eclipse An eclipse of the moon; occurs when the moon passes through the earth's shadow.

Lyman lines of hydrogen Emission or absorption lines in the hydrogen spectrum which arise from transitions between the lowest energy level and higher energy levels.

Magellanic clouds Two small galaxies close to and associated with the Milky Way galaxy.

Magnetosphere The sphere of influence of the earth's magnetic field.

Magnitude system A system for expressing the ratios of light intensities from celestial objects.

Main sequence The relatively narrow diagonal band on the H-R diagram that represents the majority of stars.

Major planet One of the larger planets, Jupiter, Saturn, Uranus, or Neptune; sometimes called a *Jovian planet.*

Maria (Latin: seas) The name originally given to the dark areas seen on the moon (Singular: *mare*).

Mascon Mass concentration below the surface of the moon.

Mass The total amount of matter in an object, measured either by its gravitational effect on other objects or by its inertia effects.

Mass-Luminosity law The relationship between the mass and luminosity of main sequence stars.

Mean solar day The time between successive meridian passages of the mean sun; the average length of the apparent solar day.

Mean sun An imaginary sun which moves uniformly eastward on the celestial equator, not the ecliptic; the basis of mean solar time.

Megaparsec (mpc) One million parsecs.

Meridian The circle on the celestial sphere passing through the zenith and the north and south points of the horizon.

Meson A subatomic particle with a mass between that of a proton and an electron.

Meteor The streak of light caused by the burning of a meteoroid when it enters the earth's atmosphere; popularly called a "shooting star" or a "falling star."

Meteorite That portion of a meteoroid which survives the passage through the earth's atmosphere and strikes the surface of the earth.

Meteoroid A meteoritic particle in space; usually refers to such particles before they encounter the earth.

Meteor shower The occurrence of meteors seemingly radiating from a common point on the celestial sphere.

Micrometeorite A meteoroid which is small enough to be slowed down by the earth's atmosphere so that it does not vaporize but instead falls gently to the surface.

Milky way The faint, ragged band of light that cuts across the sky, actually composed of myriads of stars; the name given to our galaxy.

Molecule Two or more atoms bound together; the smallest particle of a chemical compound that still has the characteristics of that substance.

Morning star The name sometimes given to any bright planet seen in the pre-dawn sky; most frequently applied to Venus.

Neap tide The lowest tide of the month.

Nebula A cloud of interstellar dust or gas.

Neutrino A particle with no mass or charge but which carries energy away during certain nuclear transformations.

Neutron star A hypothetical star of extremely high density, composed entirely of neutrons; closely related to pulsars, according to current theory.

Nodes The points of intersection of the orbit of a body and some reference plane such as the ecliptic or the celestial equator.

Northern lights (See *Aurora borealis*).

Nova (Latin: new) A star that greatly increases its luminosity in an explosive manner.

Objective The light-gathering element of a telescope; can be either a mirror or lens.

Obliquity of the ecliptic The 23½° angle between the plane of the ecliptic and that of the celestial equator.

Occultation The obscuring or eclipsing of one astronomical body by another; usually the eclipse of a star or planet by the moon or another planet.

Olbers' paradox If it is assumed that the stars (or galaxies) are uniformly distributed throughout space and are stationary, the night sky should be bright rather than dark.

Open cluster A loose association of stars, usually from a few dozen to several hundred, traveling together through the galaxy.

Opacity A measure of the ability of matter to block all or part of the radiation passing through it.

Opposition The position of a superior planet when it is on the opposite side of the earth from the sun.

Orbit The actual path in space of a celestial object.

Oscillating universe A cosmological theory which states that the universe alternately expands and contracts.

Parallax The apparent displacement of an object against the background when viewed from two different positions.

Parsec The distance of an object having a stellar *par*allax of one *sec*ond of arc; one parsec = 3.26 light years.

Penumbra The region in which the light of a celestial object appears partially blocked.

Perigee That point of the lunar orbit at which the moon is at the minimum distance from the center of the earth.

Perihelion The point in an object's orbit about the sun at which it is nearest to the sun.

Period A time interval between identical aspects of a phenomenon, e.g., the period of a planet or the period of a Cepheid variable.

Period-Luminosity relation The relationship between the period and the average true brightness of a Cepheid variable star.

Perturbation A small disturbance in the motion of a body caused by the gravitational pull of another object.

Photometry The measurement of an object's brightness.

Photon A unit of electromagnetic energy.

Photosphere The visible surface of the sun.

Plages Bright clouds found in the solar chromosphere, often associated with the sunspot regions.

Planck curves The intensity of electromagnetic radiation as a function of wavelength at given temperatures.

Planet A relatively large body (e.g., the Earth, Jupiter) revolving around a star.

Planetary nebula A star with an extended atmosphere of expanding gas; not related to planets in any way.

Planetoid A minor planet.

Precession of the equinoxes The slow westward progression of the equinoxes along the ecliptic.

Prominence A luminous condensation of matter in the lower corona which frequently appears as a tongue-like flame along the edge of the solar disk.

Proper motion The apparent angular motion of a star on the celestial sphere.

Proton A heavy, positively charged subatomic particle; one of the two main particles of the atomic nucleus.

Proton-Proton reaction The nuclear process by which hydrogen is transformed into helium deep in the interiors of those stars on the lower portion of the main sequence.

Pulsar A celestial radio source characterized by radio pulses of extremely precise periodicity; believed to be a rapidly rotating neutron star.

Quadrature The position of a superior planet when its direction is at right angles to that of the sun as seen from the earth.

Quasar A shortened term for *quasi-stellar object,* a celestial object of extremely high energy output—comparable to or very much greater than that of a galaxy—yet much smaller in size than a typical galaxy.

Radar The transmission of a radio pulse and the reception of its echo.

Radial velocity The speed of an object toward or away from the observer.

Radiant (of a meteor shower) The point on the celestial sphere from which the meteors in the shower appear to be coming.

Radiation All forms of electromagnetic energy.

Radiation laws The three related laws discovered respectively by Stefan, Wien, and Planck.

Radio galaxy A galaxy that emits a greater than usual amount of radio radiation.

Radio source A location on the celestial sphere from which radio energy is received; the object responsible for the radio emission.

Radio telescope A specialized radio antenna and associated equipment for the detection and study of radio sources.

Red giants Very large, cool stars.

Red shift The shift of spectral lines toward the red wavelength region of the spectrum, generally interpreted as the Doppler effect in the receding galaxies.

Reflecting telescope A telescope which uses mirrors to form an image of a distant object.

Reflection nebula An interstellar cloud whose luminosity arises from the reflection of the light of nearby stars.

Refracting telescope A telescope which uses lenses to form an image of distant objects.

Regression of nodes Westward movement of the nodes in the orbit of a revolving body caused by perturbations from another object.

Relativity A concept of the natural world arising from the constancy of the speed of light, in which mass, length, and time are considered relative rather than absolute quantities.

Resolving power A measure of the ability of an astronomical instrument to separate two extremely close objects in the field.

Retrograde motion The apparent backward motion of a planet on the celestial sphere caused by the relative motion of the earth and the planet.

Revolution The motion of one celestial body around another or around their common center of mass.

Right ascension The celestial coordinate measured eastward along the celestial equator in the equatorial system.

Right ascension-Declination system (See *Equatorial system.*)

Rill (or Rille) A trench-like, often irregular, feature of the lunar surface.

Rotation The turning of a body about its own axis.

RR Lyrae A variable star found mainly in globular clusters and surrounding the nucleus of the galaxy; thought to be among the oldest of stars.

Saros A series of similar eclipses which recur after a period of about 18 years.

Satellite Any relatively small celestial object revolving around a larger object.

Scintillation The apparent twinkling of stars caused by the distortion of the light passing through the earth's atmosphere.

Semi-major axis One-half of the longest dimension of an ellipse.

Seeing A measure of the quality of an astronomical image, as affected by the earth's atmosphere.

Shooting star A meteor.

Sidereal period The period of revolution of one body about another with respect to the stars.

Sidereal time The system of time based upon the stars rather than the sun; *Zero hour* is when the vernal equinox is on the meridian.

Skyglow A very faint residual atmospheric luminescence.

Solar constant The amount of solar energy received per square centimeter at the earth's distance.

Solar eclipse The total or partial obscuration of the sun by the moon.

Solar system The effective sphere of influence of the sun; the sun and all the objects revolving around it.

Solar time The system of time based upon the mean solar day; *Zero hour* is 12 hours after the mean sun crosses the meridian.

Solar wind The flow of corpuscular radiation from the solar corona.

Solstices The points on the ecliptic at which the sun is farthest north or south from the celestial equator; the arrival of the sun at these points marks the beginning of summer and winter.

Space-time continuum A coordinate system involving three spatial coordinates and one time coordinate; the time and place of any event can thus be specified.

Space velocity The actual velocity of a star with respect to the sun.

Special relativity That part of the theory of relativity which deals with uniform motion.

Spectral class (or type) Stellar classification according to characteristics of the stellar spectrum.

Spectral lines Discrete lines in a spectrum which are caused by the emission or absorption of energy by atoms of specific elements.

Spectrogram The photograph of a spectrum.

Spectrograph The instrument used to photograph a spectrum; usually attached to a telescope.

Spectroscope The instrument used to visually observe the spectrum of a celestial object or other light source.

Spectroscopic binary A double star in which the components are so close that they can only be distinguished by radial velocity shifts of their spectral lines.

Spectroscopic parallax The parallax deduced from the study of the stellar spectrum.

Spectrum The array of wavelengths or color when radiant energy (usually white light) is dispersed.

Spicules Small spike-like eruptions in the solar chromosphere.

Spirals Galaxies which are pinwheel-shaped, with "arms" spiraling away from their nuclei.

Spring tide The highest tide of the month.

Standard time The time adopted for convenience within a longitudinal zone of 15° average width; the local mean solar time of a standard meridian located in that zone.

Star A self-luminous sphere of gas.

Steady state The cosmological theory which holds that the large-scale picture of the universe is constant no matter when or where it is viewed and that matter is constantly being created to replace the receding galaxies.

Stellar evolution The progressive changes a star goes through during its existence.

Stellar parallax The angle subtended by the radius of the earth's orbit at the distance of a star.

Sunspots Cooler regions on the photosphere which appear dark by contrast.

Supergiant (star) An enormous star, which in some cases may be larger in diameter than the orbit of the planet Mars.

Superior conjunction The position of an inferior planet when it is directly on the opposite side of the sun from the earth.

Superior planet A planet whose distance from the sun is greater than the earth's.

Supernova A cataclysmic stellar explosion in which the luminosity of the star temporarily increases millions of times.

Surface gravity The value of the force of gravity at the surface of a body; determines the weight of a given mass.

Synchrotron radiation Radiation emitted by charged particles being accelerated in magnetic fields to speeds approaching the speed of light.

Synodic period The interval between successive identical appearances in the sky of a planet in relation to the sun and the earth.

Tangential velocity The velocity of a star across the line of sight.

Terminator The line of sunrise or sunset on the moon or planets.

Terrestrial planet Any of the smaller planets, including Pluto.

Tides The ebb and flow of terrestrial waters due to the gravitational forces of the moon and sun; any deforming of one body by the gravitational forces exerted upon it by other bodies.

Time dilation The relativistic increase of an interval of time as a function of relative velocity.

Trigonometric parallax The direct measurement of a star's annual parallactic shift on the celestial sphere.

Triple-Alpha process A series of nuclear reactions by which three helium nuclei (alpha particles) are transformed into one carbon nucleus.

Tropical year The interval of time between successive passages of the sun across the vernal equinox.

Umbra The shadow of a celestial object; the region in which the light is completely blocked.

Universal time (U.T.) Mean solar time at 0° longitude (Greenwich, England).

Van Allen Belt The regions surrounding the earth in which rapidly moving charged particles from the sun are captured by the earth's magnetic field.

Variable stars Stars whose intrinsic luminosity varies.

Velocity curve A plot of the variation of the radial velocity of a star with respect to time.

Velocity-Distance Relationship The relationship between the observed red-shift of distant galaxies (interpreted as a Doppler shift) and their distances from us.

Velocity of escape (See *Escape velocity*.)

Vernal equinox The point on the celestial sphere at which the ecliptic crosses the celestial equator from south to north; spring begins in the northern hemisphere when the sun reaches this point.

Wavelength The distance between successive crests in a series of waves.

White dwarfs A star in its last stages of evolution in which matter has collapsed to the degenerate state.

Widmanstätten figures Fine-lined structure observed on the interior surfaces of iron meteorites after they have been cut, etched, and polished.

Zenith The point on the celestial sphere directly above the observer.

Zodiac The band of 12 constellations lying along the ecliptic.

Zodiacal light A faint luminosity along the zodiac due to the scattering and reflection of sunlight by interplanetary dust; best observed near the horizon after evening twilight and before dawn.

index